Sven Maaløe

Principles of Igneous Petrology

With 291 Figures

Springer-Verlag
Berlin Heidelberg New York Tokyo

Professor Dr. SVEN MAALØE
Department of Geology
University of Bergen
Allegaten 41
N-5014 Bergen

ISBN 978-3-642-49356-0   ISBN 978-3-642-49354-6 (eBook)
DOI 10.1007/978-3-642-49354-6

Library of Congress Cataloging in Publication Data. Maaløe, S. (Sven), 1943– . Principles of igneous petrology. Includes index. 1. Rocks, Igneous. I. Title. QE461.M216 1985   552'.1   84-26829

© Springer-Verlag Berlin Heidelberg 1985
Softcover reprint of the hardcover 1st edition 1985

8700 Würzburg.

2132/3130-543210

Dedicated to
Peter J. Wyllie for his inspiration

Preface

Igneous petrology was to some extent essentially a descriptive science until about 1960. The results were mainly obtained from field work, major element analyses, and microscopical studies. During the 1960's two simultaneous developments took place, plate tectonics became generally accepted, and the generation of magmas could now be related to the geodynamic features like convection cells and subduction zones. The other new feature was the development of new analytical apparatus which allowed high accuracy analyses of trace elements and isotopes. In addition it became possible to do experimental studies at pressures up to 100 kbar.

During the 1970's a large amount of analytical data was obtained and it became evident that the igneous processes that control the compositions of magmas are not that simple to determine. The composition of a magma is controlled by the compositions of its source, the degree of partial melting, and the degree of fractionation. In order to understand the significance of these various processes the relationship between the physical processes and their geochemical consequences should be known. Presently there are several theories that attempt to explain the origin of the various magma types, and these theories can only be evaluated by turning the different ideas into quantitative models. We will so to speak have to do some book keeping for the various theories in order to see which ones are valid. the present book is intended as an introduction to the more fundamental aspects of quantitative igneous petrology.

Apart from using quantitative analysis as an evaluation of ideas, the application of physicochemical equations has a much more significant potential. Many if not most igneous processes can not be observed, and will never be observed. Processes like convection in the mantle and magma chambers, dyke propagation, and magma accumulation in the mantle can not be observed directly. These and other processes can only be understood by developing quantitative models. Hence, the application of physical and chemical principles, by using equations and various constants, is mandatory for the understanding of the generation of magmas.

The attempt to convert an idea into equations is frequently a difficult task, and this represents a real challenge for the present generation of the students of igneous petrology. It is hoped that the pres-

ent book will assist the graduate students to a quantitative treatment
of petrology, although the present book only represents a beginning.

Clearly in order to bring further development to igneous petrology it is nescessary to apply physicochemical principles. However,
the new ideas, the creative insight, is hardly attained by reading text
books on physics. There is still a lot left for the individual petrologist
to discover without the direct aid of the exact sciences, but some
training in these will no doubt prove most useful.

The present book is dedicated to Peter J. Wyllie, because he in
many ways has added to the author's understanding of petrology as
well as experimental work. I would also like to thank Dr. Jim V. P.
Long, Cambridge, whose technical skill in constructing complicated
apparatus like ion and microprobes has been a great experience.
Further I would like to thank some of my Danish colleagues for
their kindness and inspiration, my thanks goes to Dr. Tom S. Petersen, Dr. Ib Sørensen, and Dr. Ella Hoch.

Bergen, April 1985   S. Maaløe

Contents

I Monary Systems

1 Introduction

By comparing the compositions of minerals and rocks with the compositions obtained from experimental work it becomes possible to estimate the PT conditions that control the igneous processes. The estimation of the phase relations of magmas and the magmatic sources, like the mantle and the subducted oceanic crust, therefore, form the basis for an understanding of the generation of igneous rocks. After the PT conditions have been estimated the next step is an analysis of the dynamic processes that control the temperature and pressure conditions. Such an analysis will involve the calculation of the temperature distribution in mantle plumes and ascending convection currents, an estimation of the velocity of dyke propagation, and calculations of the cooling rates of intrusions.

Since the phase relations are fundamental to igneous petrology, we will first consider the phase relations of solids and liquids, and the subsequent chapters will deal with geochemistry and the dynamic and kinetic processes related to magma genesis.

The phase relations of a system are dependent on the thermodynamic properties of the phases that form part of the system. The phase relations could, therefore, not be understood before the thermodynamic theory was developed. It was J. W. Gibbs (1873, 1878) who first developed the thermodynamic theory. Several of the fundamental ideas was already developed before Gibbs did his work, but it was Gibbs that combined the thermodynamic ideas and made the theoretical fundament for thermodynamics. Gibbs papers were rather difficult to understand, and it was not before Lewis and Randall (1923) published their classic book on thermodynamics that the full potential of the theory was generally appreciated. However, the thermodynamic implications for phase relations was worked out in the period 1890–1920 by H. W. B. Roozeboom, R. van der Alkemade, and F. A. H. Screinemakers. The application of thermodynamics to petrology appears attractive, however, the determination of even a binary system requires comprehensive and highly accurate calorimetric work, and the experimental approach is the most accurate and convenient one.

The experimental work was begun already by Hall (1805) and Daubrée (1860) (cf. Loewinson-Lessing 1954), but reliable measurements of high temperatures could not be made before the platinum thermocouple became available in 1886. It was N. L. Bowen who did the pioneering experimental work within petrology, his first paper in 1912 was on the system nephelinite–anorthite. Until about 1960 experimental work was done almost exclusively at the Carnegie Institution in

Washington where Bowen worked. Many fundamental results have been worked out by other members of this institution; we may mention J. F. Schairer, H. S. Yoder, F. R. Boyd, B. T. C. Davis, P. M. Bell, and H. K. Mao. Since 1960 other laboratories was set up which have provided important results like those of C. W. Burnham and P. J. Wyllie. Another laboratory that has provided important results, especially within high-pressure technology is that of G. C. Kennedy. The high-pressure technology was founded by P. W. Bridgman, who worked with the physical properties of materials, while the high-pressure apparatus required for petrological investigations, the piston–cylinder apparatus, was developed by Boyd and England (1960). In Europe improvements in design has been made by W. Johannes (Hannover), and in recent years new apparatus for experimentation above 100 kbar has been developed and used by the Japanese workers S. Akimoto, N. Kawai, S. Endo, and M. Kumuzawa. The development of the diamond cell has provided a dramatic increase in the pressure range and it is now possible to perform experiments at pressures up to 1700 kbar at elevated temperatures (Ming and Basset 1974; Mao and Bell 1978).

The theory of phase relations was fully developed around 1930. The more complicated aspects of phase theory has been applied to petrology by P. J. Wyllie in a series of papers dealing with gas bearing systems, which have some of the most intricate phase relations.

For further reading on phase relations the excellent book of Rhines (1956) is strongly recommended, whereafter the comprehensive, but difficult book by Ricci (1951) may be studied. Textbooks on phase relations within petrology have been made by Bowen (1928), Ehlers (1972), and Morse (1980). An excellent account of experimental techniques has been given by Edgar (1973).

2 Monary Phase Relations

The one component or monary system shows the phase relations and the stability ranges of minerals that form from one component. The diagrams can, therefore, be used to estimate the possible mineral assemblages that are stable within given pressure and temperature ranges.

The minerals may undergo two types of phase changes or phase transitions. They may retain their chemical composition while their structure is changed. This phase transition is called a polymorphic phase change. Polymorphic phase transitions are mostly related to pressure variations, examples are the quartz–coesite and graphite–diamond transitions, while the tridymite–cristobalite transition is caused by a change in temperature (Figs. I-1 and I-2). The minerals may also decompose and form two new minerals, thus, by increasing pressure, albite decomposes into jadeite and quartz (Fig. 1-3):

albite = jadeite + quartz

Another decomposing reaction is the incongruent melting reaction; enstatite melts to form forsterite and liquid:

enstatite = forsterite + liquid

Fig. I-1. The P-T diagram for carbon at pressures up to 600 kbar. Both graphite and diamond display inversions on their liquidus curves (Dickinson 1970)

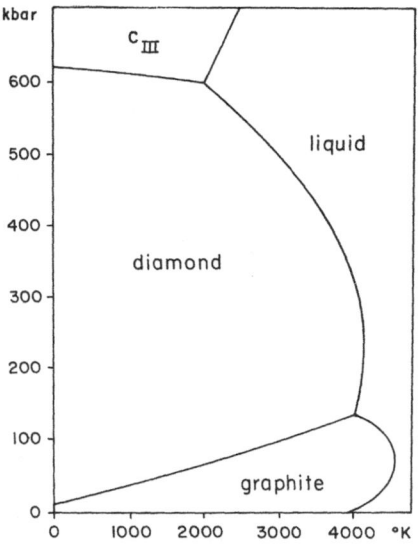

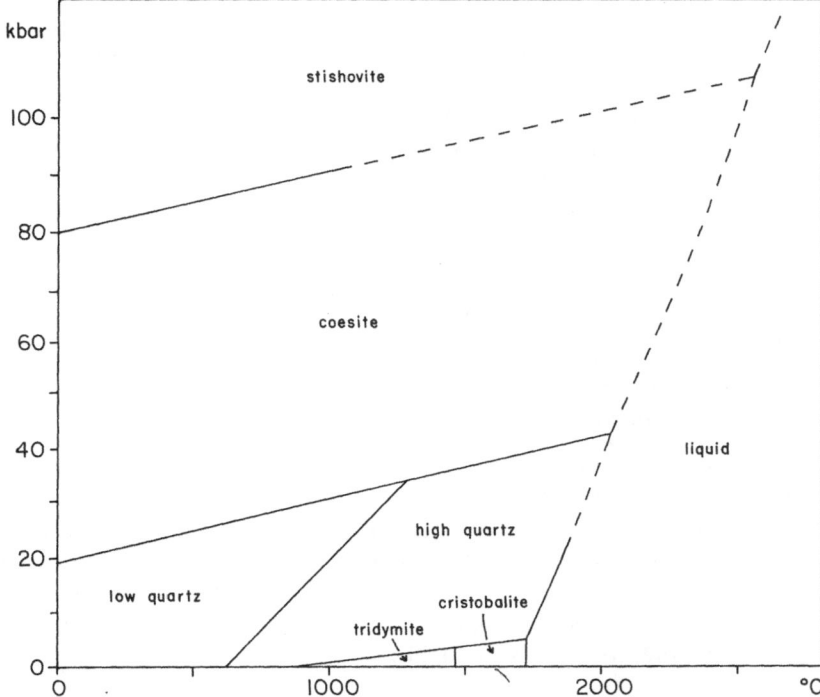

Fig. I-2. The P-T diagram for SiO₂ (silica). The liquidi for coesite and stishovite have not yet been determined, and their liquidi are tentative (Boyd and England 1959)

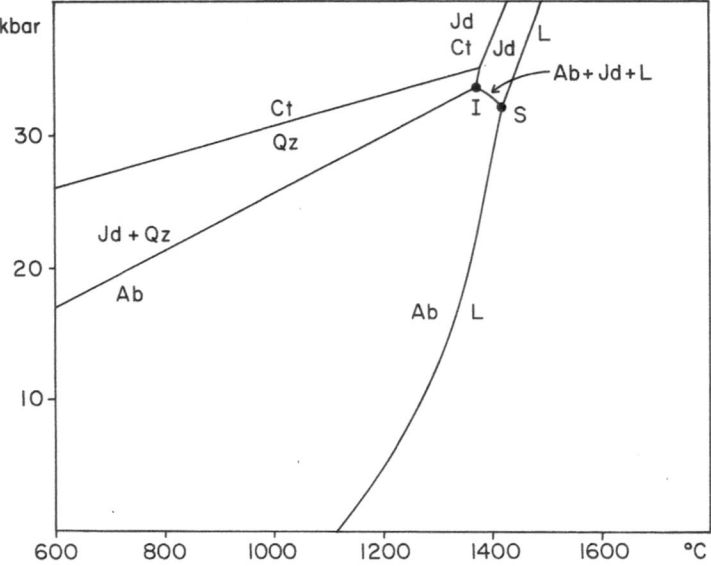

Fig. I-3. The P-T diagram for albite. By increasing pressure albite decomposes into jadeite and quartz, and the diagram is not a strictly monary one, as four phases are in equilibrium at *point I*: (L, S(Ab, Jd, Qz). The *point S* is a invariant point with the equilibrium L, S(Ab, Jd) (Huang and Wyllie 1975)

The general phase relations for a monary system is shown in Fig. I-4. There are four different phases, two solids (S_1 and S_2), one liquid, and one gas phase. Two different phases are in equilibrium along the curves, and three phases are in equilibrium at the points where the three curves intersect. The stability relationships of phases are related to the thermodynamic properties of the minerals, and it can be shown that the number of phases and components are related by a simple rule called the Gibbs phase rule:

$$P + N = C + 2 \tag{1}$$

where:

P: number of phases present
N: degrees of freedom or variance
C: number of components

In order to describe the different phase assemblages it is convenient to introduce a symbolic notation. Let the three phases be represented by G, L, and S. Further let the single phases present be represented by example, g_i, l_i, and s_i. An equilibrium assemblage may, thus, be described by $G(g_i)$, $L(l_i)$, $S(s_i)$. Albite in equilibrium with its melt is, thus, given by L, S(ab), and if a water vapor is present, in addition, the assemblage is described by G, L, S(ab). If quartz and plagioclase is added to this assemblage we have G, L, S(ab, q, pl).

Consider a point like A in Fig. I-4, which is situated within the stability field of S_1. We can both vary pressure and temperature within certain limits, while still retaining the same phase assemblage, $S(S_1)$. This is not the case for the assemblage

Fig. I-4. The basic phase relations of a monary system. There are five univariant curves and two invariant points in this diagram. The univariant curve L, G ends in the critical point. The liquids or gases with temperatures and pressures higher than that of the critical point are sometimes called fluids

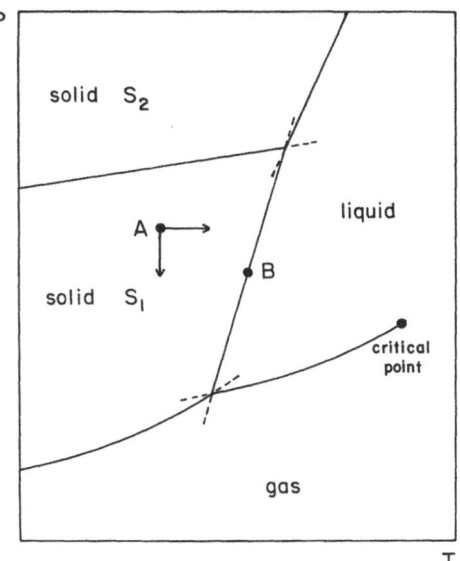

at point B, where S_1 is in equilibrium with its melt, so that we have the phase assemblage L, $S(S_1)$. If the pressure is increased the liquid will disappear, and if the temperature is increased the solid will melt. Thus, when the temperature is changed the pressure must be changed simultaneously if the assemblage L, $S(S_1)$ should remain stable. The temperature must, therefore, be a function of the pressure or vice versa. There is consequently only one degree of freedom, while the phase assemblage $S(S_1)$ has two degrees of freedom. At point A we have:

$$P + N = 3$$
$$N = 2$$

At point B we get:

$$N = 1$$

At the triple points where the three phases are in equilibrium we obtain then $N = 0$. The curves where $N = 1$ are called univariant curves, and the points where $N = 0$ are called invariant points. The field between the univariant curves are sometimes called bivariant sectors.

The slopes of the univariant curves are estimated from Clausius–Clapeyron's equation:

$$\frac{dP}{dT} = \frac{T \Delta H}{\Delta V} \tag{2}$$

where:

ΔH: the latent heat of melting or vaporization
ΔV: the change in volume by melting or vaporization
T: temperature in degrees Kelvin.

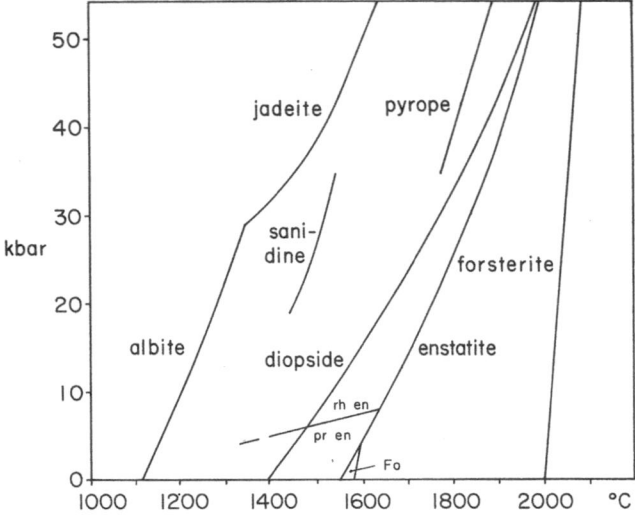

Fig. I-5. Some melting curves for silicates. Their slopes are fairly similar and about 10 °C/ kbar. The boundary between the two polymorphs of enstatite is shown by (*rh en*) for orthorhombic enstatite, and (*pr en*) for protoenstatite. Enstatite melt incongruently at low pressures to forsterite (*Fo*) and liquid

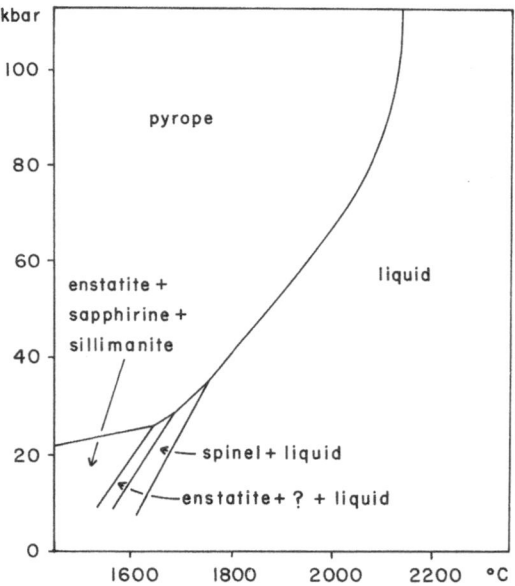

Fig. I-6. The melting curve for pyrope at pressures up to 100 kbar. Apparently there is an inversion in the liquidus at about 100 kbar (Ohtani et al. 1982; Boyd and England 1962)

Table I-1. SiO$_2$ polymorphs and their densities at 1 bar

	g/cm^3	
Quartz	2.65	Trigonal
Tridymite	2.26	Orthorhombic
Cristobalite	2.33	Pseudocubic
Coesite	2.92	Monoclinic
Stishovite	4.30	

By most melting reactions there is an increase in volume, and the melting temperature increases with increasing pressure. For silicate minerals the gradient of the liquidus curve is about 10 °C/kbar (Fig. I-5). Hydrate minerals like the amphiboles, serpentine, and micas decompose by increasing temperature into anhydrous minerals and a H_2O gas phase. At low pressure the dehydration results in an increase in volume, and the reaction volume is positive. At higher pressures the reaction volume becomes negative, because the volume of water vapor decreases with increasing pressure. The slope of the melting curve, therefore, changes from positive to negative. Amphibole is not stable in the mantle at pressures above 20–30 kbar, while phlogopite is stable above this pressure range (Wyllie 1979; Eggler and Baker 1982).

The compressibility of liquids are larger than that of solids and the volume of reaction decreases by increasing pressure. An inversion in the melting curve is, therefore, also possible by the melting of anhydrous minerals, and the melting curves of graphite and pyrope garnet display a change from positive to negative slopes (Figs. I-2 and I-6).

The P-T phase relations for quartz and its polymorphs is shown in Fig. I-2. At low pressures there are three polymorphic transitions due to increased temperature: low quartz–high quartz, and high quartz–tridymite, and tridymite–cristobalite. At high pressures above 20 kbar coesite becomes stable up to pressures of about 80 kbar where stishovite becomes stable. The density of the high temperature polymorphs tend to decrease, while that of the high-pressure polymorphs tend to increase (Table I-1). Coesite has been observed in kimberlite nodules, and in the rocks from the meteor crater of Arizona. Stishovite has not yet been observed in mantle derived nodules. This might indicate that the nodules stem from depths less than 300 km.

3 Phase Changes Within the Upper Mantle

The monary phase changes play a major role for the upper mantle, and we will, therefore, consider the phase transitions of the mantle. The mantle-derived nodules as well as the presence of harzburgite in ophiolite complexes suggest that olivine and its polymorphs may constitute the major part of the mantle. The seismic evidence and the results obtained from high-pressure experiments appear to confirm a high percentage of olivine in the mantle.

The P and S wave velocities of the mantle minerals like olivine, enstatite, diopside, and garnet are so similar that one cannot estimate the mineralogical composition of the mantle from the absolute values of the P and S wave velocities. The mineralogical constitution can, however, be estimated from the seismic discontinuities that occur at depths of about 420 and 670 km (Fig. I-7). These phase changes are caused by solid state phase changes. The origin of these discontinuities can be estimated by a comparison between the pressures of the phase transitions with the depth of the discontinuities. If some of the minerals undergo transformations at the same pressure this comparison will not yield a definitive result, but such coincidences do not appear to be present. It should be noted that a phase transition only will give origin to a discontinuity if the mineral occurs in a substantial amount.

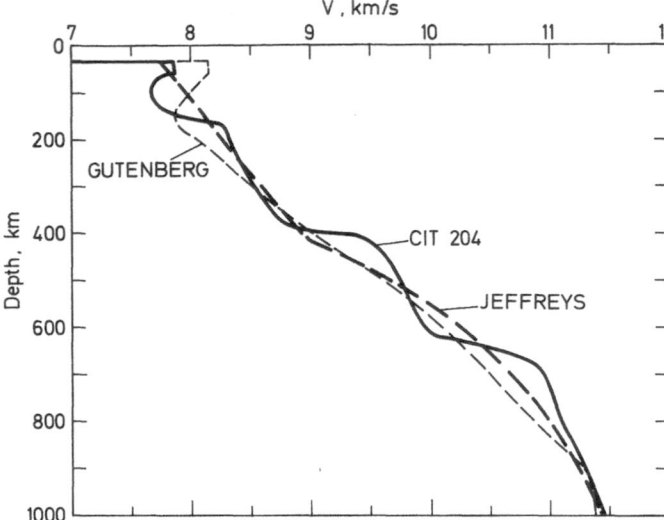

Fig. I-7. Different estimates of the P wave velocities within the mantle to a depth of 1000 km. The CIT 204 model shows the variation considered most relevant using modern results, while the two other curves are early estimates

The solid state phase changes in the mantle are temperature dependent, and the temperature gradient of the mantle should be known with some approximation before the experimental results can be applied. Presently the gradient is only known with some approximation, but the seismic discontinuities are rather well defined, so that a high accuracy is not mandatory. On the other hand, it will be possible to estimate the geothermal gradient with some accuracy when the required experimental data become available.

The estimated depths of the two discontinuities display some variation, but the variations are so small that it is relevant to calculate the seismic average properties of the mantle. Using seismic data the variation in density and pressure was calculated by Dziewonsky et al. (1975), and the temperature distribution was calculated by Dziewonsky and Anderson (1981). The pressure and temperature curves shown in Fig. I-8 are based on their results. The model obtained in this manner is called the parametric Earth model. So far the seismic data has not revealed any major transitions within the lower mantle.

The pressure at the core–mantle boundary is about 1400 kbar, and the temperature is considered at about 6000 °C. At the upper–lower mantle boundary at a depth of 670 km, the pressure is 240 kbar, and the temperature is estimated at about 2000 °C. The experimental conditions required to duplicate these conditions are extreme, but the development of the tetrahedral and octahedral anvil apparatus and the diamond cell has allowed investigations up to 1700 kbar and 1500 °C simultaneously. In the pressure range up to 200 kbar, temperatures up to 2500 °C have been attained (Othani et al. 1982). So far most experimental work up to 250 kbar has been carried out at temperatures lower than those prevailing in the upper mantle, but extrapolations may in some cases be done at higher temperatures.

The presence of high-pressure polymorphs of olivine with a spinel-like structure was proposed already by Bernal (1936). Pioneering experimental work by Ringwood (1958) showed that olivine does change into a new polymorph with a modified

Fig. I-8. The parametric
Earth model at pressures up
to 400 kbar. The *upper
curve* shows the estimated
temperature variation, and
the pressure variation is
shown just below. The *lower
curve* shows the density
variation. There are abrupt
changes in the density at
420 km and 670 km, and
these are related to phase
changes

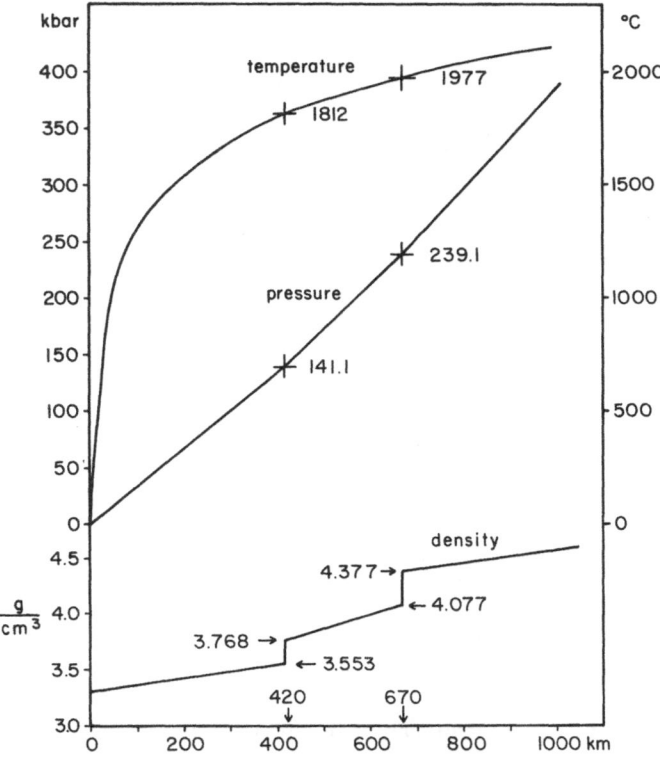

spinel structure at pressure above 100 kbar (Fig. I-9). Subsequent experimental
work has shown that olivine has four different polymorphs at pressures up to
500 kbar. The low pressure polymorph of forsterite is called α-forsterite and has an
orthorhombic structure and a density of 3.22 g/cm³. At 1200°C and 141 kbar α-.
forsterite changes into β-forsterite which has a modified spinel structure and a
density at 1 bar of 3.47 g/cm³ (Akimoto et al. 1976; Kawada 1977). The β-forsterite
polymorph changes into γ-forsterite or sp-forsterite at 1200°C and 220 kbar. The γ-
forsterite has a true spinel structure and a density at 1 bar of 3.55 g/cm³. These
transitions are dependent on both temperature and composition as evident from
Fig. I-10. The β-polymorph is only present in the Fo-rich part of the system, and
the transformation pressures decrease with increasing fayalite content. The pressure
of the α, β, and γ transitions increases with increasing temperature, and the α–β
and β–γ transitions are, thus, exothermic according to the Clausius–Clapeyron
equation. This implies that the ascending convection currents will undergo a temperature decrease of the order of 100°C by the β–α transition.

The γ-polymorph of forsterite with a spinel structure remains stable at 1000°C
up to 250 kbar where it then decomposes into $MgSiO_3$ with perovskite structure
and periclase:

$$\gamma Mg_2SiO_4 = MgSiO_3 (p) + MgO$$

If the olivine contains fayalite we obtain magnesio–wüstite (mw) instead of
periclase. The combined density of $MgSiO_3$ (p) and MgO is 3.55 g/cm³ at 1 bar. The

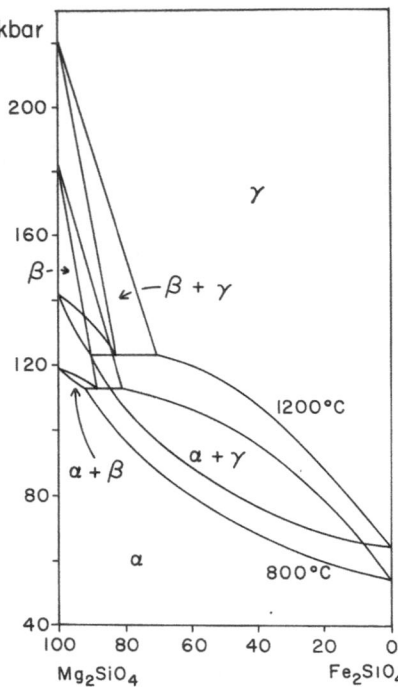

Fig. I-9. The polymorphs of olivine at pressures up to 200 kbar at 800 °C and 1200 °C (Akimoto et al. 1976; Kawada 1977)

temperature dependence of this phase transformation differs from the former ones in that the transformation pressure decreases with increasing pressure. According to Ito and Yamada (1982), the transformation pressure is given by:

$$P \text{ kbar} = 273 - 0.02 \text{ T°C}$$

Mantle material that ascends from the lower mantle into the upper one, like plumes are believed to do, will thus undergo an exothermic reaction and a temperature increase.

In order to relate these phase transformations to the seismic structure of the mantle we must extrapolate the estimated transformations to higher temperatures. The experimental data of Kawada (1977) for the α, β, and γ transitions show that olivine with 90% Fo change from α to β olivine at 1800 °C and 151.5 kbar. This temperature is similar to that of the 420 km discontinuity (Fig. I-8) of the parametric Earth model, and the pressure is only slightly lower. The parametric Earth model suggests an increase in density of 0.22 g/cm³ at a pressure of 141.1 kbar, and can thus be related to the α–β transition of olivine, as the difference in density between the α and β forsterite is 0.25 g/cm³. There is no known phase transition for enstatite, diopside, and garnet at about 150 kbar, and the discontinuity can be related exclusively to olivine. Olivine must consequently be a major constituent of the mantle. The accuracy of the applied data do not allow a meaningful estimate of the actual amount of olivine, but if the data is taken at face value it can be calculated that olivine constitutes about 70% by volume of the upper mantle, a content which is similar to that of lherzolites (Maaløe and Aoki 1977).

Fig. I-10a, b. The phase relations of olivine (**a**) and enstatite polymorphs and reaction products at pressures up to 500 kbar at 1000 °C. The temperature within the mantle is higher than 1000 °C at pressures above 30 kbar, and the phase relations shown here can not be applied directly to the mantle. The ilmenite field shown in (**b**) is tentative and based on an estimate at 1100 °C (Ito and Yamada 1982)

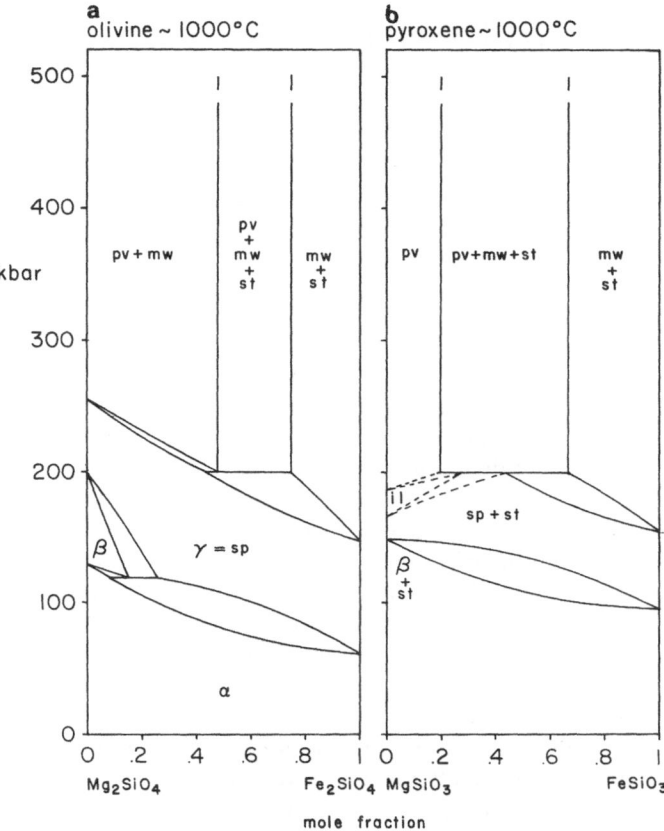

According to the parametric Earth model, the temperature at the 670 km discontinuity is about 2000 °C. Extrapolation of the experimental results of Kawada (1977) suggests that the β–γ transition occurs at a pressure of 266 kbar at 2000°C. On the other hand, the γ-ol = pw + w transition occurs at about 233 kbar at 2000 °C and, thus, at a lower pressure than the β–γ transition. This suggests that the discontinuity at 670 km depth is caused by the decomposition of β-olivine into (Mg, Fe) SiO_3 (pw) and (Mg, Fe) O (w). The density increase for this transition is 0.30 g/cm³, while the β–γ transition only results in a density increase of 0.08 g/cm³. As the parametric Earth model suggests a density increase of 0.30 g/cm³, it is most likely that the 670 km discontinuity is caused by the decomposition reaction. It is, thus, possible that γ-olivine is not present in the mantle.

The various estimates given above are based on extensive extrapolations and are no doubt subject to revision. However, it appears fairly certain that olivine must be the major constituent of the mantle.

The phase relations of enstatite, one of the other constituents of lherzolite, is shown in Fig. I-10b. This diagram is only tentative as the stability fields of the ilmenite polymorph have not yet been estimated at 1000 °C. The phase relations of the ilmenite polymorph have been estimated at 1100 °C by Ito and Yamada (1982) and their results have been used in Fig. I-10b.

II Binary Systems

1 Introduction

The binary system is the fundamental unit within the phase theory for multicomponent systems; the higher ordered systems are built up from binary joins, and the binary phase diagrams display the basic phase relations in a convenient manner as both compositions and temperatures are read directly from the diagrams.

The addition of another component to a liquid of a pure solid results in a change in the melting temperature of the solid. This effect was first observed experimentally by Raoult (1888), and the thermodynamic equations for freezing point depression were worked out by van't Hoff (1887). One of the first binary systems with petrological applications, the albite–anorthite system, was determined by Bowen (1913) and the pioneering work within experimental petrology was carried out by him and his co-workers during subsequent decades at the Carnegie Institution in Washington. Bowen applied the experimental results as a petrogenetic tool; the opposite procedure was attempted by Vogt (1904), who deduced the binary phase relations that prevail during the crystallization of igneous rocks from the mineralogy of these rocks. His results are perhaps mostly of cursory interest today, but it is interesting to notice that his deductions generally turned out to be correct.

2 The Phase Rule

The phase rule states that: $F + N = C + 2$. So for a binary system with $C = 2$, one gets at isobaric conditions: $F + N = 3$. Thus, at isobaric invariant conditions there will be three in equilibrium at a given temperature. A univariant equilibrium implies that two phases are stable along a curve in the T, x coordinate system. If the solid phase displays solid solution in the entire composition range there will only be two phases in equilibrium within this range. The curve that shows the variation in composition of the liquid as function of temperature is called the liquidus curve, and the similar curve for the solid phase is called the solidus curve.

3 The Lever Rule

The amount of liquid and solid in equilibrium might be calculated when the compositions of the two phases are known. Let a binary system be a mixture of the two components A and B, and let the weight percentages of B in the liquid and solid phase be x_b^l and x_b^s, respectively. Using the equivalent definition for x_a^l and x_a^s one gets:

$$x_a^l + x_b^l = 100\% \tag{1}$$

$$x_a^s + x_b^s = 100\% \tag{2}$$

If the total amount of the binary mixture is 100 g, and S and L denote the amounts of solid and liquid, respectively, then:

$$S + L = 100 \text{ g} \tag{3}$$

The values of S and L may now be calculated when x_b^l and x_b^s are known. The total amount of component A and B are n_a and n_b, respectively. As the total amount of these components must remain constant during the crystallization one gets:

$$(1/100)\,(x_a^l\,L + x_a^s\,S) = n_a \tag{4}$$

$$(1/100)\,(x_b^l\,L + x_b^s\,S) = n_b \tag{5}$$

The amount of solid, S, might be expressed explicitly by elimination of L from Eq. (5) using Eq. (3):

$$S = \frac{100\,n_b - 100\,x_b^l}{x_b^s - x_b^l} \tag{6}$$

By elimination of S from Eq. (5) one gets:

$$L = \frac{100\,n_b - 100\,x_b^s}{x_b^l - x_b^s} \tag{7}$$

By division of Eq. (6) by Eq. (7) the result is:

$$\frac{S}{L} = \frac{n_b - x_b^l}{x_b^s - n_b} \tag{8}$$

The significance of the right side of this equation is evident from Fig. II-1. The numerator is equal to p and the denominator is equal to q, so the ratio S/L is given by:

$$\frac{S}{L} = \frac{p}{q} \tag{9}$$

As $L + S = 100$, then S and L are also expressed by:

$$S = \frac{100\,p}{p + q} \tag{10}$$

$$L = \frac{100\,q}{p + q} \tag{11}$$

Equation (9) is called the lever rule, and it defines the value of the ratio S/L. However, it is more convenient to obtain the values of S and L directly using Eqs. (10) and (11).

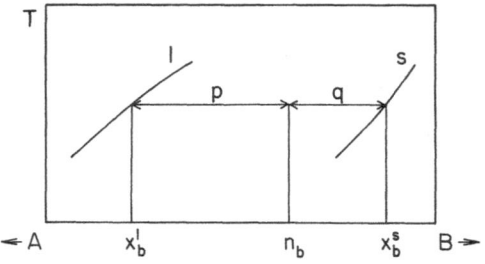

Fig. II-1. The lever rule. The ratio between p and q is defined by the compositions x_b^l, n_b, and x_b^s. The *curves l* and *s* are the liquidus and solidus curves, respectively, for some binary system

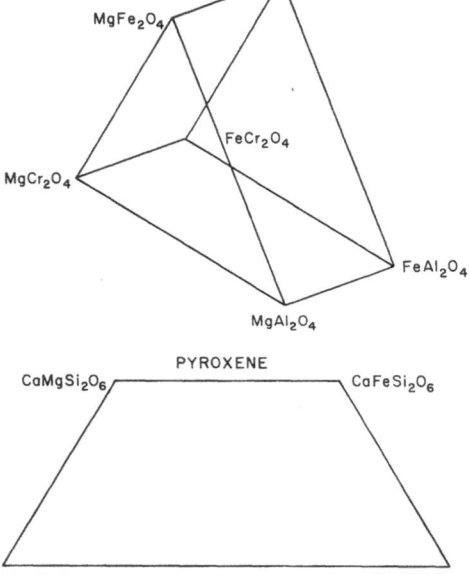

Fig. II-2. The compositional triangle for pyroxenes and the Johnstone spinel prism

Example 1

Using Fig. II-1 the following values are obtained, $x_b^l = 30\%$, $x_b^s = 80$, and $n_b = 60$. Thus, p equals 30% and q equals 20%, and $S/L = 30/20 = 1.50$. The amount of S is calculated from Eq. (10), $S = 100 \, (30/50) = 60$. The amount of L is given by: $L = 100 \, (20/50) = 40$. There are, thus, 60 g solid and 40 g liquid.

4 Binary Isomorphous Systems

By far the majority of rock-forming minerals are solid solutions, their compositions might be mixtures of up to four major components (Fig. II-2). The study of the petrogenesis of rocks, therefore, requires a determination of the phase relationships of solid solutions involving several components. That minerals may be solid solutions of several components complicates the experimental determination of the phase relations of these minerals, but is also an advantage. The composition of a mineral, like quartz, that does not display solid solution with respect to major elements, is constant, and its composition will not point at any specific condition of formation. The compositions of minerals which display solid solution with respect to several components, like pyroxenes and garnet, will vary with the conditions of formation, and their compositions might therefore indicate pressures and temperatures of formation. Presently, the binary systems involving the components CaO, Al_2O_3, MgO, SiO_2, Na_2O, K_2O, H_2O, and CO_2 are fairly well known at 1 bar and at higher pressures. The phase relations of systems with FeO or Fe_2O_3 have been investigated in less detail, probably because of the problems related to control of the oxygen fugacity. Experimental techniques that either monitor or control the oxygen fugacity have been developed in recent years, and detailed investigations of systems with iron are probably forthcoming. Systems with three components have not been investigated in the same detail, the liquidus relations are known for many ternary systems, but the solid solution behavior is not known in the same detail.

Minerals that display solid solution generally have the same crystal structure throughout the composition range, while their lattice dimensions vary with the composition. The neso-silicate olivine is mainly a solid solution of fayalite (Fe_2SiO_4) and forsterite (Mg_2SiO_4), which both have orthorhombic structure and belong to crystal class mmm. The silicate tetrahedrons maintain their relative positions around the Fe^{++} and Mg^{++} cations, while the distances between the tetrahedrons vary with the Fe^{++}/Mg^{++} ratio (Fig. II-3). In other minerals, such as

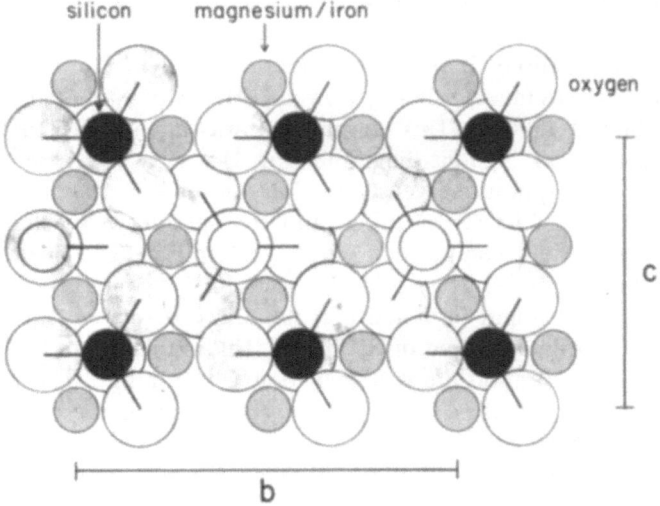

Fig. II-3. The structure of olivine. The relative positions of all the atoms and the silicon tetrahedra remain the same by the substitution of Mg^{++} with Fe^{++}, only the distances between the atoms will vary. The unit cell dimensions are as follows:

	a	b	c
Forsterite	4.756 Å	10.195 Å	5.981 Å
Fayalite	4.817 Å	10.477 Å	6.105 Å

The distances between atoms are thus slightly larger for fayalite, but its density is much larger as iron is heavier than magnesium

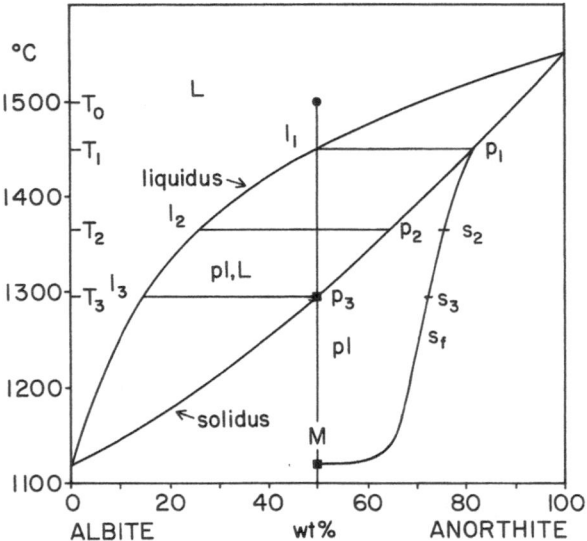

Fig. II-4. The albite-anorthite system. The liquidus and solidus curves are shown as estimated by Bowen (1913) and Greig and Barth (1938). The *curve sf* shows the average composition of the accumulated fractionate for composition M

plagioclase, the situation is more complicated as there is a double substitution, $(Na^+ + Si^{+4})$ which is substituted with $(Ca^{++} + Al^{+3})$.

The phase relations for a binary isomorphous system will be demonstrated from the albite–anorthite system and the fayalite–forsterite system. Both these systems are of the monotonous type, i.e., the melting temperature varies monotonously with composition, other systems may display a temperature maximum or minimum. The phase relations for the albite–anorthite system at 1 bar is shown in Fig. II-4. The system was investigated by Bowen (1913) who estimated the melting temperature of albite at 1100 °C. This temperature was later corrected to 1118 °C by Greig and Barth (1938), and the diagram in Fig. II-4 is based on their estimate.

The liquidus curve shows the variation in the temperature of the liquid as function of its composition, and the solidus curve shows the variation in temperature and composition of the solid phase. The horizontal lines like l_1–p_1 are called tie lines. They join the equilibrium compositions of the solid and liquid phases. It might be noted that the anorthite content of plagioclase is higher than the anorthite content of the liquid. This is related to the fact that the melting temperature of anorthite is higher than that of albite. The relationship between compositions of phases and their melting temperature was stated as a rule by Konowalow (1881), and the rule may be expressed as follows: "The crystalline solutions first separating out always contain a higher content of those components whose addition to the melt raises its freezing point."

5 Equilibrium Crystallization

The mixture M with 50% anorthite will be in a complete liquid state at T_0 (Fig. II-4). By decreasing temperature the liquid will reach the liquidus curve at T_1 where the first nuclei of plagioclase with composition p_1 will be formed. Two widely different types of crystallization may now take place, that is, equilibrium crystallization and fractional crystallization.

By the equilibrium crystallization there is a fast diffusion rate in the solid and liquid phase, so that there is a free exchange of ions between the two phases. The composition of the solid phase is exactly that indicated by the solidus curve. By decreasing temperature the plagioclase becomes more albitic, and the already formed plagioclase crystals are, therefore, assumed to give off $(Ca^{++} + Al^{+3})$, while they receive $(Na^{+} + Si^{+4})$ from the liquid. This model assumes that there is a fast diffusion of these constituents in plagioclase, or that the crystallization rate is very slow compared to the diffusion rate. This situation rarely occurs, neither in experiments nor in nature, as the aluminum ion is fixed in the silicate lattice. The result is that plagioclase crystals nearly always are zoned, even in deep seated rocks like gabbros and granites. On the other hand, the diffusion rate in olivine is apparently relatively large, as olivine rarely displays zoning in gabbroic rocks.

The equilibrium crystallization of liquid of composition M will now be considered. After the liquid has reached the liquidus curve at T_1 and the first plagioclase crystals have formed, the crystallization will proceed while its composition is moving along the solidus curve, and the composition of the liquid moves along the liquidus curve. At T_2 the liquid will have the composition l_2 and the composition of plagioclase is p_2. The crystallization will proceed in this manner until the last amount of liquid has solidified. This will happen at T_3 where the plagioclase crystals have obtained the composition of the mixture. The plagioclase crystals have then varied in composition from p_1 to p_3, and the liquid has moved from l_1 to l_3. The amounts of liquid in the system during crystallization may be estimated at any temperature using the lever rule. At T_2 the liquid contains 26.5% An, and the plagioclase 65.0% An. Thus, according to the lever rule $L/S = 15/23.5$, or $L = 100 \, p/(p+q) = 39.0$. The variation in amount of liquid as a function of its composition is shown in Fig. II-5. For liquids with an initially high content of

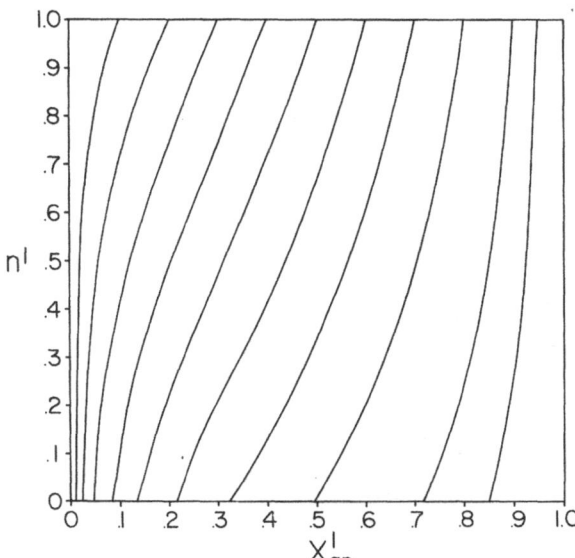

Fig. II-5. The variation in anorthite content of the liquid by equilibrium crystallization, n^l is the amount of liquid, and x_{an}^l the anorthite content of the liquid

anorthite there will be a restricted variation in composition, the variation is largest for liquids with an intermediate anorthite content, and small for liquids with an initially low anorthite content.

6 Fractional Crystallization

By fractional crystallization there is no exchange of ions between crystals and liquid. This might happen in two different manners. The diffusion rate in the solid phase might be far too low for an effective exchange of ions, or the crystals might accumulate on the bottom of a magma chamber, in which case the diffusion rate is too low for exchange of material, because the distances involved are of the order of meters or hundreds of meters. The crystallization may, thus, be of fractional type, even the crystals remain fluentively in the magma. This type of crystallization will be called fluentive fractional crystallization, while the other type generally is called gravitative fractional crystallization. It should be noted that the fractional crystallization in both cases might occur at thermodynamic equilibrium conditions. The lack of equilibrium by fractional crystallization is the diffusive one, not the thermodynamic one.

By fractional crystallization the liquid with composition M will move along the liquidus all the way down to pure albite, and the last plagioclase crystallized will be pure albite. A quantitative account of the crystallization course requires the application of Rayleigh's (1902) fractionation equation, the lever rule cannot be applied as the total composition of the fractionate is unknown. The details will be given in Chap. X, where the obtained results will be considered. The variation in the amount of liquid as a function of its composition is shown in Fig. II-6 for perfect fractional crystallization. For liquids with high anorthite content to begin with there will at first be a very small variation in its anorthite content. However, during the final stage of crystallization the variation is very large. For liquids with a

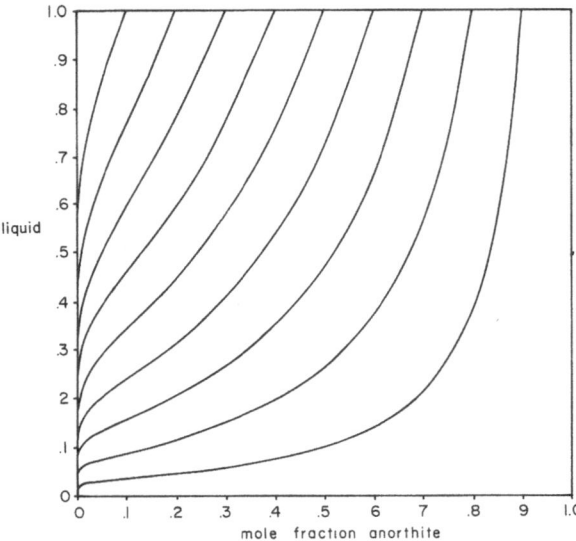

Fig. II-6. The variation in the anorthite content of plagioclase liquids by fractional crystallization of plagioclase

Fig. II-7. The zoning developed by fractional crystallization of three different plagioclase liquids. The zoning has been calculated assuming a spherical shape of the crystals. The initial composition of the liquid is shown by x_i^o

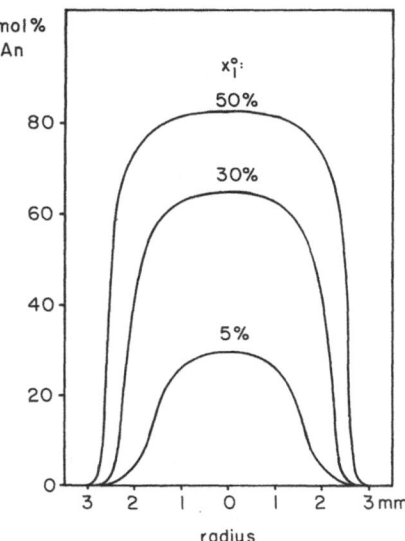

low anorthite content the variation in composition will be small. The anorthite and forsterite contents of plagioclase and olivine, respectively, in basaltic phenocrysts, is about 80% in both cases, and the normative ratios An/(An + Ab) and Fo/(Fo + Fa) for the basalts will initially display a small variation during the fractional crystallization.

The variations in composition for a liquid with initially 50% An is shown in Fig. II-4. The liquid will change composition from 50% An to 0% An, and the composition of plagioclase will vary from 80% An to 0% An. The composition of the plagioclase that crystallizes at any given moment will be on the solidus curve. However, the average composition of plagioclase that has crystallized will be situated on the curve s_f. Thus, at T_2 the composition of plagioclase just forming is p_2, but the composition of the total amount of plagioclase that has crystallized is s_2. By further crystallization the composition of the liquid will move towards pure albite. When the liquid reaches this composition there will still be some liquid left. In the present case it will be 13% of the initial amount. The liquid will remain at the melting temperature of albite, i.e., 1118 °C until the last albite liquid has crystallized. The anorthite content of the plagioclase crystals of a cumulate will vary from p_1 at the bottom to albite at the top. By fluentive fractional crystallization the formed crystals will display the zoning shown in Fig. II-7.

The rocks formed from magmas controlled by fractional crystallization show a continuous variation in the composition of their minerals, this applies, for example, for the andesitic magma series, and this continuous variation was termed the continuous reaction series by Bowen (1928). It will be evident from the phase diagrams for systems containing CaO and NaO that the minerals that are stable at high temperature contain a high CaO content. Andesitic magmas will, therefore, become depleted in CaO by fractional crystallization and the granitic magmas which have a low CaO content may have been formed from an andesitic magma with a high CaO content.

7 Influence of Additional Components and Pressure

The shape of the liquidus and solidus curves will mainly depend on the latent heats of melting and the melting temperatures. The temperature difference between liquidus and solidus for a given composition will increase with increasing latent heat of melting. The variations in shape for the two curves have been worked out in detail by Reismann (1970). It might be anticipated that pressure and additional components would influence the phase relations and change the position of tie lines, but experimental evidence suggests that a liquid with a given an/ab + ab ratio will be in equilibrium with plagioclase with a rather constant an/an + ab ratio. The liquidus and solidus curves for the plagioclase system along the univariant join in the diopside–anorthite–albite system were estimated by Wyllie (1963) from Bowen's (1915) experimental results, and are shown in Fig. II-8. The shapes and temperatures for the two curves are different from those of the albite–anorthite system, but the effect of the diopside component for the composition of plagioclase is relatively small as evident from the Roozeboom diagram shown in Fig. II-9. Also it is evident that the pressure appears to have a moderate effect on the equilibrium composition.

The influence of additional components on the fayalite–forsterite system also appears to be minor. The MgO/FeO ratios for olivines and basaltic liquids being in equilibrium were throughly worked out by Roeder and Emslie (1970), and their results are shown in Fig. II-10, where their data may be compared with the experimental results of the fayalite–forsterite system (Bowen and Schairer 1935). A basaltic liquid of composition "a" will crystallize olivine with a forsterite content of 70% Fo according to Roeder and Emslie's (1970) results. Composition "a" contains 20.8 mol% FeO and 14.2 mol% MgO, and the MgO/FeO ratio is 0.68. A liquid with this ratio in the fayalite–forsterite system will initially crystallize olivine with a forsterite content of 73% Fo. The difference in equilibrium composition between the two systems is small and of the order of ±5% Fo. However, the crystallization temperatures for the basaltic compositions will be substantially lower than those for the fayalite–forsterite system due to additional components. The fayalite–forsterite

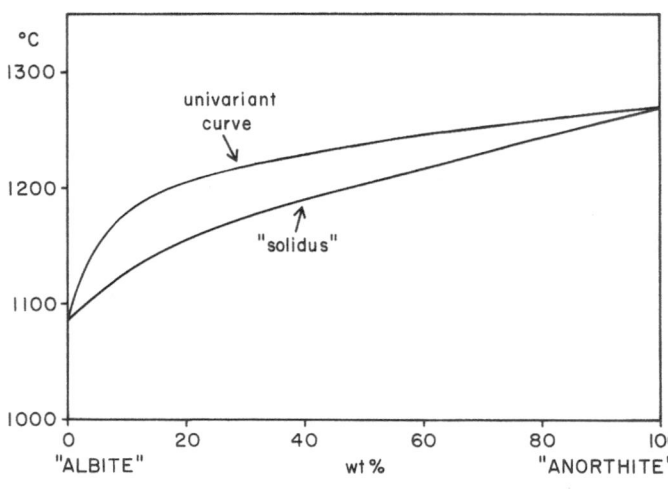

Fig. II-8. The equilibrium relationships for the plagioclase system along the univariant curve L, S(di, pl) in the system albite–anorthite–diopside (Wyllie 1963)

Fig. II-9. The Roozeboom diagram for the albite-anorthite system at 1 bar, 5 kbar and 10 kbar water pressure. The di-ab-an curve shows the variation for the univariant curve in the albite-anorthite-diopside system (cf. Fig. II-8). The anorthite content of plagioclase (an_s) is shown as a function of the anorthite content of the liquid (an_l)

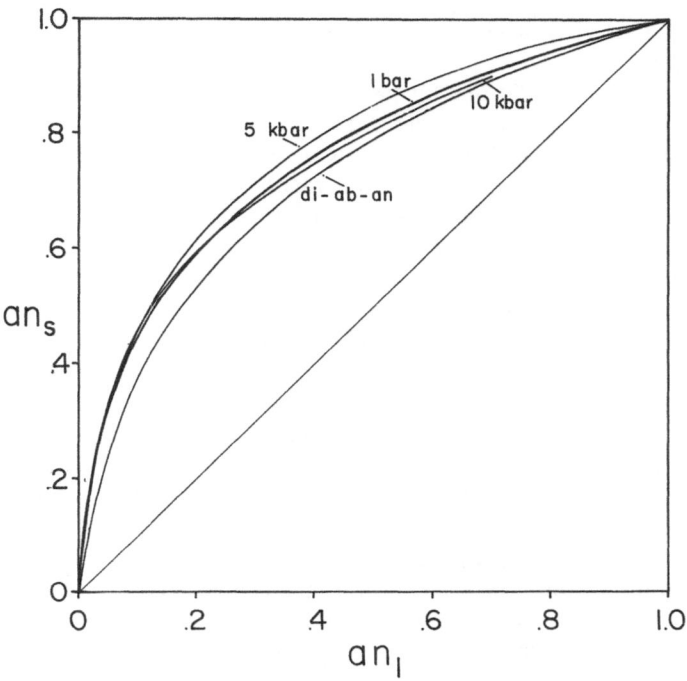

Fig. II-10. The geothermometer of Roeder and Emslie (1970) is compared with the equilibrium relationships of the fayalite-forsterite system. The geothermometer is only considered of some accuracy up to 1350 °C, but it compares favorably with the binary system at higher temperatures

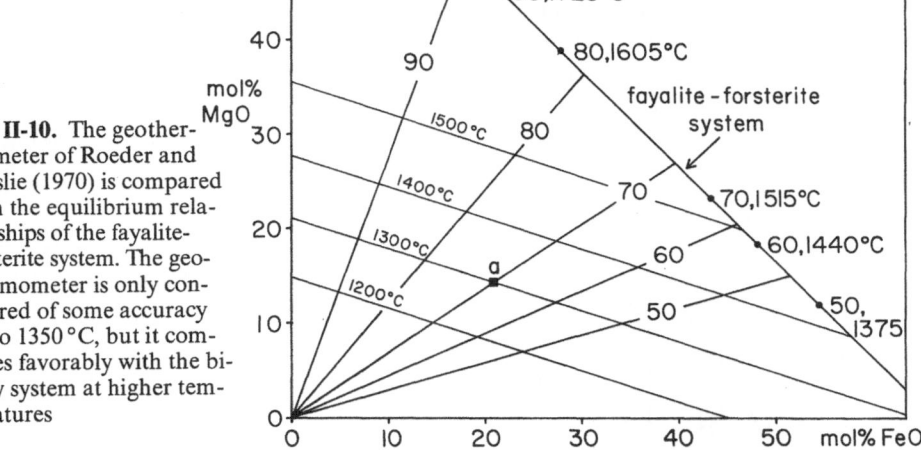

system displays nearly ideal solutions (Larimer 1968; Saxena 1973), and it is evident from Fig. II-10 that additional components only have a small influence on the equilibrium compositions.

The tie lines for the two systems with solid solution considered here show a limited variation in equilibrium compositions by the addition of new components, but this behavior is not necessarily valid for other systems with solid solutions.

8 Partial Melting

The phase relations by partial melting at equilibrium conditions will be demonstrated using the fayalite–forsterite system shown in Fig. II-11. This system is very
similar to the plagioclase system, but as basaltic magmas are generated by partial
melting of the olivine-bearing mantle, partial melting within the fayalite–forsterite
system has important petrogenetic applications.

Partial melting of the mantle at nonequilibrium conditions has been proposed
by some petrologists, but is highly improbable. The velocity of the convection
currents in the mantle may be of the order of 5 cm/yr, assuming a small
geothermal gradient of 10 °C/km for the lower asthenosphere. The mantle material
of an ascending convection current or plume will cross its solidus temperature with
the rate of $5 \cdot 10^{-4}$ °C per year. Equilibrium in basaltic or peridotitic melts at high
pressures and is obtained within hours, or at least days, so the rate of partial
melting in the mantle is several orders of magnitude lower than the time required
for equilibrium. Partial melting at equilibrium conditions is, therefore, considered
most appropriate here and is considered next.

A mixture with the composition m (50% Fo), will at first form a melt with the
composition 18% Fo at l_1. By increasing temperature the liquid will move from l_1
towards l_2, simultaneously the composition of the olivine will move from o_1 to o_2.
The equilibrium partial melting process is the exact opposite of the equilibrium
crystallization. A system that has obtained equilibrium will always be in one given
state quite independent of its prehistory. The variations in amount of liquid may be
estimated using the lever rule, and are shown for three different initial compositions
in Fig. II-12. The increase in amount of liquid with increasing temperature is much
the same, but the shape of the curves vary with the composition of the initial
olivine.

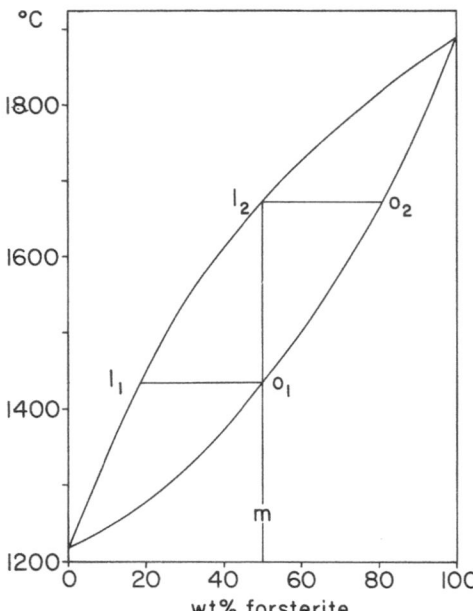

Fig. II-11. The fayalite-forsterite system

Fig. II-12. The variation in liquid composition for three compositions in the olivine system at varying degrees of partial melting

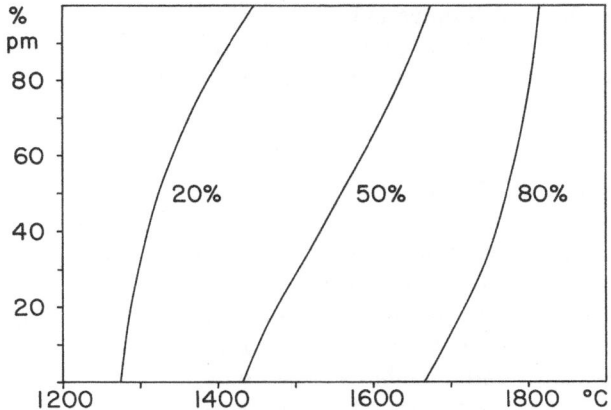

9 Binary Isomorphous System with Minimum

Some systems like the albite–orthoclase system and the akermanite–gehlenite system shown in Fig. II-13 display a temperature minimum. If the solid and liquid solutions were ideal, the system would be of monotonous type. The minimum is due to the heat of solution being positive, while that of an ideal solution is zero. The crystallization within the two lobes is exactly the same as for the albite–anorthite system. The major difference is the presence of the temperature minimum within the system rather than at one of the components. It may be noted that this temperature minimum is different from the minimum of an eutectic system. As in an eutectic system the liquids will fractionate towards the minimum, but the course of crystallization by equilibrium conditions is different. A liquid of composition l_1 will begin to crystallize gehlenite of composition g_1, and by equilibrium crystallization

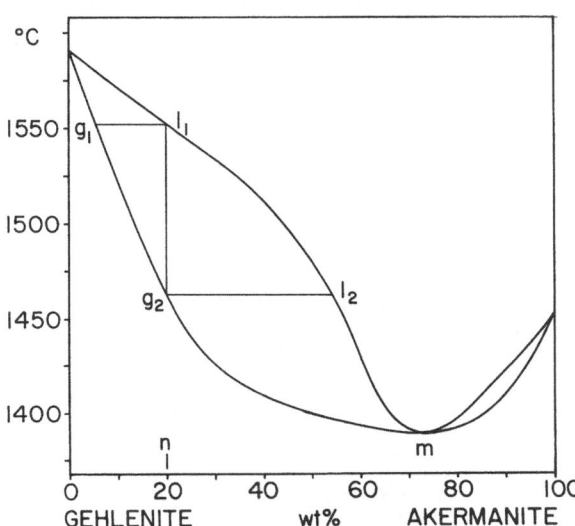

Fig. II-13. The gehlenite-akermanite system at 1 bar (Osborn and Schairer 1941)

the liquid will move to l_2 where the last liquid disappears. By fractional crystallization the liquid will move to m where the last liquid crystallizes melilite of composition m. By partial melting of melilite composition g_2 the first liquid will be formed at l_2 and not at m.

10 Eutectic Systems

There are two types of binary eutectic systems. By the simple one, the solid phases have the composition of the pure end members, and by the other type the solid phases display limited solid solution. Strictly speaking there are no crystals that are formed absolutely pure, but in some cases the deviation in composition from that of the pure end member is less than the accuracy of chemical analyses, and the composition may, therefore, be considered that of the pure end member.

The simple type will be considered first, the anorthite–silica system shown in Fig. II-14 is this type. By the simple type there is no difference in equilibrium and fractional crystallization because the two solid phases have a constant composition. A liquid of composition m (20% SiO_2) will begin to crystallize anorthite at T_1 (Fig. II-14). At a lower temperature like T_2 the liquid will have the equilibrium composition l_2 and a certain amount of crystals have formed. This amount may be estimated from the lever rule: $S = p/p + p = 12.5/32.5 = 0.38$, thus, 38% of the initial liquid has crystallized at T_2. By decreasing temperature the liquid will move further along the liquidus curve for anorthite and move towards the eutectic point e. When the liquid reaches e, 73% of the initial liquid has crystallized. At e tridymite begins to crystallize, and the melt will remain at the eutectic temperature until the last liquid has solidified. This follows from the lever rule:

$$P + N = 2 + 1 \quad \text{(P constant)}$$

$$3 = 3$$

The degree of freedom is zero at isobaric conditions when three phases are present. The composition of the liquid will remain at the eutectic composition by equilibrium

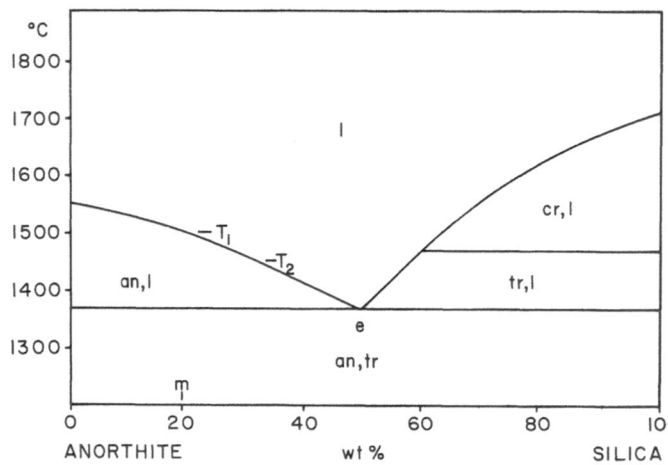

Fig. II-14. The anorthite-silica system at 1 bar (Bowen and Schairer 1947)

crystallization, and their crystallization rates will be given by the composition of the eutectic point. Say, if the composition of the eutectic point is e% silica, and the crystallization rates of anorthite and silica are r_a and r_s, respectively, then the ratio between r_a and r_s will be given by:

$$\frac{r_a}{r_s} = \frac{100 - e}{e} \tag{12}$$

If the crystallization rates have a different ratio, the liquid will move off the eutectic point which is impossible by equilibrium crystallization.

11 Partial Melting

By partial melting the phase relations for a eutectic system with no solid solution will be exactly the opposite of equilibrium crystallization. When the mixture of composition m is raised in temperature it will begin to melt at the eutectic temperature 1368 °C, and the composition of the melt will be that of the invariant point, i.e., the eutectic point. The heat delivered to the system will be used entirely for the melting process until one of the solid phases has disappeared, that is, tridymite when the composition of the mixture is m. After all the tridymite has entered, the liquid will move along the liquidus curve towards T_1 where the last anorthite goes into solution in the liquid. The specific heat of minerals is about 0.25 cal/g °C, and the latent heats about 100 cal/g. A mixture of minerals can, therefore, absorb a substantial amount of heat before it leaves the invariant point, how much will depend on the difference in composition between the invariant point and the composition of the mixture. Considering that the latent heat is about 400 times larger than the specific one, it is likely that partial melting in the mantle and crust will tend to occur at low degrees of partial melting.

12 Eutectic System with Solid Solution

A system with the two solid phases displaying solid solution consists of two solidus–liquidus lobes interrupted by a solvus curve (Fig. II-15). The crystallization relationships for the two lobes are the same as for a system of monotoneous type.

By equilibrium crystallization a liquid of composition m_1 will first crystallize B of composition b_1, and by progressive cooling the crystals will react with the liquid and change their composition along the solidus curve from b_1 towards b_e. The crystallization will stop when the B crystals have obtained the composition of mixture, that is, at b_3. All mixtures with compositions to the right of b_e will crystallize completely at equilibrium conditions before they reach e, and will only crystallize the B phase. The mixture of composition m_1 will only consist of B crystals with composition b_3 during cooling from b_3 to b_4. At b_4 the B crystals begin to exsolve A with composition a_4. During subsequent cooling B will become enriched in the B component and A enriched in the A component. The effect of exsolution is often observed in minerals; augite might exsolve pigeonite, diopside in lherzolite might contain exsolution of spinel, and alkali feldspar frequently contains lamellae of

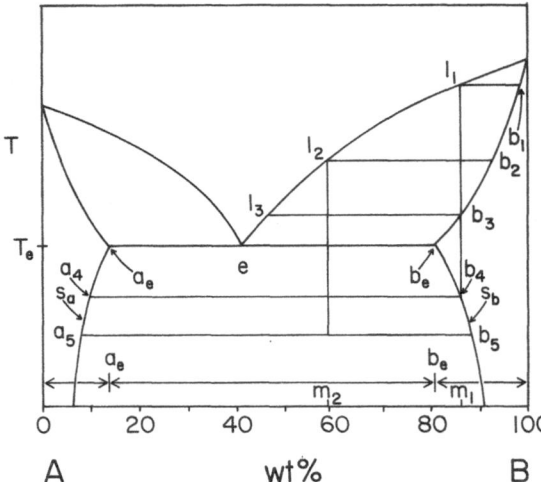

Fig. II-15. A binary system with two solid phases that display limited solid solution

albite. The exsolved phase generally has a lamellar habit and orientation related to the structure of the host. As evident from Fig. II-15 the composition of the two phases A and B will depend on temperature, and the composition of such exsolving phases has applications in geothermometry. Especially the diopside–enstatite solvus has been studied in order to estimate the geothermal gradient of the mantle (O'Hara and Yoder 1967; Nehru and Wyllie 1974).

Liquids with initial compositions between e and b_e, like m_2 will first crystallize B and thereafter reach the eutectic point e where A begins to crystallize with composition a_e. The temperature will remain constant at T_e until all the liquid has crystallized. The crystallization within the A-rich part of the system will be similar to that of the B-rich part. All liquids between A and a_e, and b_e and B will crystallize before the liquid reaches e, while all liquids between a_e and b_e will reach the eutectic point e.

By fractional crystallization all liquids will move towards e independently of their original compositions.

13 Partial Melting

The behavior by partial melting will depend on the initial composition of the mixture. A mixture like m_2 will generate the first liquid at T_e with composition e, and the mixture will remain at this temperature until all A has been melted. Thereafter the liquid will move along the liquids curve to l_2 where the last B is melted. All the mixtures with compositions between a_e and b_e will first generate liquid with composition e before liquid moves along one of the liquidus curves. Mixtures with compositions between A and a_e and B and b_e will generate liquids at temperatures higher than T_e, and will change their compositions continuously by heating.

14 Peritectic Systems

The eutectic crystallization behavior implies that the solid phases remain stable within the whole melting interval of the binary system. Some silicates have a more restricted temperature range of stability, because another silicate becomes more stable within the melting interval. As an example, enstatite is in equilibrium with its melt between 1543 °C and 1557 °C, however, at higher temperatures forsterite becomes the stable phase and enstatite reacts out of the liquid (Fig. II-16). The point where enstatite reacts with liquid forming forsterite is called a peritectic point, and the point is isobaric invariant as there are three phases involved in the reaction: fo+l=en. In ternary and multicomponent systems the peritectic point is generally called reaction point. Enstatite is said to melt incongruently, because the composition of the melt in equilibrium with enstatite at the peritectic point is different from that of enstatite.

Such reaction points occur in some silicate systems and they play an important role in the crystallization of the andesitic series. Studies of granites and andesites suggested that olivine is stable at high temperatures and that pyroxene becomes stable at lower ones. Then at still lower temperatures amphibole appears to form by the reaction of pyroxene with liquid. It was, therefore, suggested by Bowen (1928) that there is a discontinuous reaction series consisting of the following members:

olivine
pyroxene
biotite
muscovite

This sequence is due to the variation of stability of the minerals at decreasing temperatures, and the phase relations of the series are probably related to the existence of a sequence of reaction points.

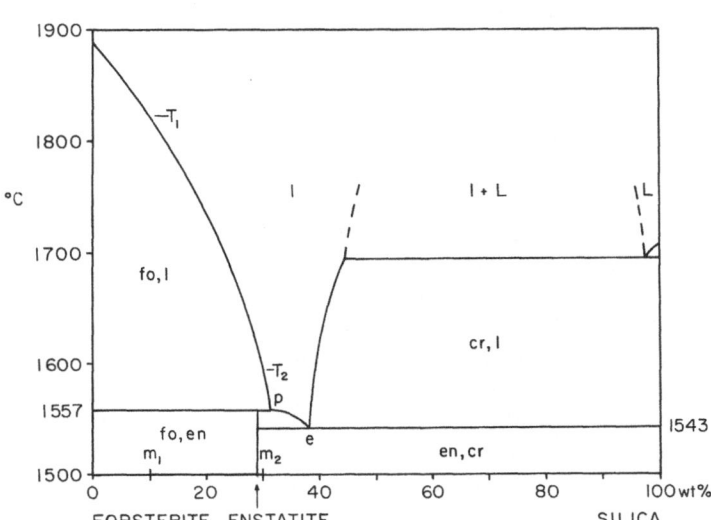

Fig. II-16. The binary peritectic system forsterite–silica at 1 bar (Bowen and Schairer 1935)

The phase relations for a peritectic system where the solid phase has a constant composition will first be considered, whereafter a system with solid solution is dealt with.

15 Equilibrium Crystallization

A binary peritectic system where the solid has a constant composition is shown in Fig. II-16. Forsterite and silica constitute the two end members, and enstatite forms an incongruent compound with the composition 29.92% SiO_2. The peritectic and eutectic points p and e have the compositions 31.67% and 38.68% SiO_2, respectively.

The crystallization courses will be different within the composition ranges fo-en, en-p, p-e, and e-si, and will be considered in this sequence.

A liquid with composition m_1 will first form forsterite at T_1, and will thereafter move along the liquidus curve towards p while the amount of forsterite increases. At p enstatite will begin to crystallize simultaneously with forsterite. Some of the forsterite will react with the liquid forming enstatite, but most of it will remain stable. Using the lever rule it might be estimated that there is 68.42 g forsterite in the liquid when it reaches p, if the initial amount of liquid is 100 g. After the reaction has been completed there will be 66.58 g forsterite and 33.42 g enstatite. The reaction between forsterite and liquid and the crystallization of enstatite will proceed until the last amount of liquid has disappeared. The temperature will thereafter decrease by further extraction of heat from the system, while the amounts of forsterite and enstatite remain constant.

The liquid of composition m_2, between enstatite and p will first form forsterite when it is cooled. Shortly afterwards it will reach p where forsterite begins to react with the liquid forming enstatite. All the forsterite will react out of the liquid, which thereafter moves from p towards e. When the liquid reaches e cristobalite will begin to form, but the amount of cristobalite that will crystallize will be small.

Liquids between p and e will first crystallize enstatite and then cristobalite. The crystallization behavior of the liquids with compositions between e and silica will also be eutectic below 1700 °C.

16 Fractional Crystallization

By equilibrium crystallization the end products varied with the composition of the initial liquid. By fractional crystallization the liquids will always end up in e, quite independent of their original composition.

The liquid of composition m_1 will fractionate towards p crystallizing forsterite. When the liquid reaches p there will only be a few crystals of forsterite left in the liquid as nearly all the forsterite has been fractionated out of the liquid. There will, therefore, be no forsterite left which can react with the liquid, and enstatite will begin to crystallize while the liquid moves towards e. The temperature will, therefore, not remain constant at p by fractional crystallization, instead the liquid will pass the peritectic point p without any retardation. When the liquid reaches e enstatite will be joined by cristobalite. Here the temperature will remain constant as long as there is some liquid left because there are three phases in equilibrium. The crystallization of the liquid with composition m_2 will proceed in a similar manner, except that a small amount of forsterite will fractionate out before the liquid reaches p.

17 Partial Melting

The composition of the liquid formed in a peritectic system by partial melting will depend on the composition of the mixture. All mixtures between fo and en will first form a liquid of composition p, and all mixtures with compositions en and si will form liquids of composition e. A small change in the composition of enstatite-rich mixtures will result in melts of widely different compositions.

By heating mixtures of composition m_1 the first liquid is formed at p, and the temperature will remain constant at 1557 °C until all the enstatite has been melted. At this stage forsterite will form the only crystals suspended in the liquid. By subsequent heating the remaining forsterite crystals will melt, and the last forsterite crystals disappear at T_1.

By partial melting of mixture m_2 a liquid of composition e will form at first, and then the liquid will move towards p, where olivine begins to form. At p some forsterite will crystallize while enstatite is melted. When the last amount of enstatite has disappeared the liquid moves along the liquidus curve for forsterite until the last forsterite has entered the melt.

Partial melting within the composition interval between p and 45% silica is similar to partial melting in an eutectic system. All liquids generated will initially have the composition e.

18 Peritectic System with Solid Solution

A peritectic system where the solids display solid solution is shown in Fig. II-17. Most of the phase relations are essentially the same as for a peritectic system where the solid phases have a constant composition, the new features are an exsolution gap shown by the solvus curves v_a and v_b, and a continuous variation in the composition of the solid phase B being stable at low temperatures.

19 Equilibrium Crystallization

A melt of composition l_1 will, by cooling first, form solid A of composition a_1, and by subsequent cooling the composition of the liquid moves from l_1 to l_3, and simultaneously A changes composition from a_1 to a_3. The last amount of liquid disappears when its composition becomes that of l_3.

The liquid of composition l_2 will crystallize A of composition a_2. With decreasing temperature the liquid will move towards the peritectic point p, and the composition of A moves towards a_p. As the composition of l_2 is between those of a_p and b_p the liquid will start to crystallize B at p while A is resorbed. The mixture of

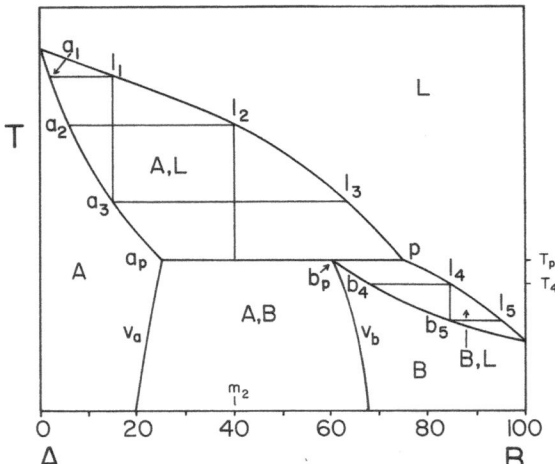

Fig. II-17. A binary, peritectic system with limited solid solution

the two solids with composition a_p and b_b and liquid will remain at constant temperature, i.e., that of the peritectic point T_p until all the liquid has crystallized. At temperatures lower than T_p the compositions of A and B will be defined by the two solvus curves v_a and v_b.

The liquid of initial composition l_3 will first crystallize A of composition a_3. As the liquid reaches p, B will begin to crystallize while A is resorbed. In this case all A will be resorbed, as l_3 has a composition between those of b_p and p. When the resorption of A is completed the liquid will move towards l_4, while the composition of B moves from b_p towards b_4. The liquid will not reach l_4, but will be completely crystallized at a temperature slightly below T_p.

Finally, a liquid of composition l_4 will crystallize in the same manner as l_1, a solid of composition b_4 will first be formed, and the crystallization of the liquid is complete when the liquid reaches the composition of l_5.

20 Fractional Crystallization

A liquid of composition l_1 will fractionate towards p. At this point there will be no A left in the liquid as A is fractionated out of the liquid, and the liquid will, therefore, move towards the composition of pure B. At B there will be a small amount of liquid left, the actual amount will depend on the shape of the liquidus and solidus curves for solid B. By fractionation all liquids will change their compositions towards pure B, irrespective of their initial composition.

21 Partial Melting

The partial melting behavior will be different within the compositional ranges $A-a_p$, a_p-b_b, b_p-p, and $p-B$.

The melting behavior for mixtures with compositions between A and a_p will be identical to that of the albite–anorthite system described above.

By partial melting mixtures with compositions between a_p and b_p, say m_2, the first melt will form at T_p, and the composition of the liquid will be p. The melting mixture will remain at T_p until all B has been melted. The liquid will by increasing temperatures thereafter move from p to l_2, where the last amount of solid A disappears.

A mixture of composition b_4 within the interval b_p–p will begin to melt at a temperature slightly below T_p, and the first liquid will have composition l_4. A difference in the melting behavior between systems with solid phases of constant composition and solid solutions might be noted here. A mixture of silica and enstatite with a composition between the composition of the eutectic point and the peritectic point will begin to melt at the eutectic point L, S (en, cr). When the solid phases display solid solution the initial melting will not occur at the minimum temperature of the system T_B, but at a higher temperature. For example, as just mentioned, the melting is initiated at T_4. When the mixture of composition b_4 is heated above T_4 the liquid will move towards p. At this stage solid A will begin to from while B becomes resorbed in the liquid. If more heat is supplied to the melting mixture, B will become completely resorbed, and the liquid will move from p towards l_3, being in equilibrium with A. The amount of A in the liquid will be small and the last amount of A disappears at some stage between p and l_3.

22 Congruent System

A binary congruent solid phase has a composition situated between those of the two end members of the binary system, and melt without forming a new solid phase. The system nephelinite–tridymite has albite as a congruent phase, the compositional relationship is given by the following reaction:

$$NaAlSiO_4 + 2\,SiO_2 = NaAlSi_3O_8$$

The phase diagram for the binary system at 1 bar is shown in Fig. II-18. Nepheline melts incongruently at high temperatures forming carnegeite. Albite melts congruently at 1118 °C, and the liquidus curve for albite displays a maximum just above the composition of albite. Such a thermal maximum is an inherent property for congruent compounds. It may be shown from thermodynamic considerations that the melting temperature of a congruent phase always represents a thermal maximum. The two points e_1 and e_2 are eutectic points representing the equilibria L, S (ne, ab) and L, S (ab, tr), respectively.

By either fractional or equilibrium crystallization the liquid of composition M_1 will move towards e_1, because the albite component is subtracted from the liquid. All liquids between e_1 and ab will end up in e_1 irrespective of the type of crystallization. Similarly all liquids between ab and tr will end up in e_2. Liquids with the composition of pure albite will crystallize nothing but albite.

It is, thus, evident that the thermal maximum divides the binary system into two isolated sectors. All liquids between ne and ab will fractionate towards e_1, and all liquids between ab and tr will end up in e_2. The thermal maximum, thus, divides the binary system, as it is impossible for a liquid to move from one sector to the other, and the maximum is, therefore, sometimes called a thermal barrier. The im-

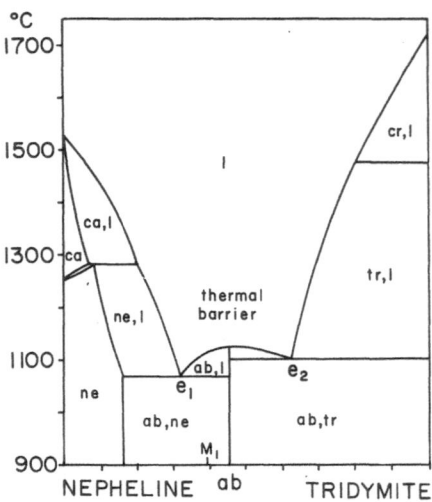

Fig. II-18. The system nepheline–silica at 1 bar. Albite forms a congruent compound in the system and its melting temperature defines a thermal maximum in the system (Morey and Boven 1924)

portance of thermal barriers for petrogenesis was realized by Bowen (1928) and Yoder and Tilley (1962). Before the work of Yoder and Tilley (1960) it was considered that alkalic basalts might be derived from tholeiites by low pressure fractionation. However, the phase relations of the system olivine–tridymite–nepheline shows that the join diopside–albite forms a thermal barrier in the ternary system, separating nepheline normative compositions from enstatite normative ones. It, thus, appears that it is impossible to derive an alkalic basalt from a tholeiitic one at low pressures. The thermal barrier disappears at pressures above ca. 8 kbar (O'Hara 1968), and the transition, thus, becomes possible at higher pressures.

23 Binary Exsolution

Some solids remain stable within large temperature intervals, while other solids either display polymorphic transitions or change their composition with temperature by exsolving a new phase. Such exsolution behavior is known to occur for a variety of rock-forming minerals like plagioclase, alkali feldspar, ortho- and clinopyroxenes, and Fe–Ti oxides. The exsolved new phase occurs in the matrix of the old one, and its crystallographic orientation is generally related to that of the host. Some examples are shown in Fig. II-19.

The chemical composition of the new phase and the host will vary both temperature and pressure. An estimate of the chemical compositions of two solids displaying an exsolution relationship will, therefore, not suffice for a temperature estimate. Both a geobarometer and a geothermometer is required for an estimate of the temperature of formation, and three compositional parameters should consequently be estimated.

Several geothermometers have been worked out both for the continental crust and the mantle. Especially the pyroxene geothermometer has been considered in much detail in order to estimate the geothermal gradient of the mantle. The basic

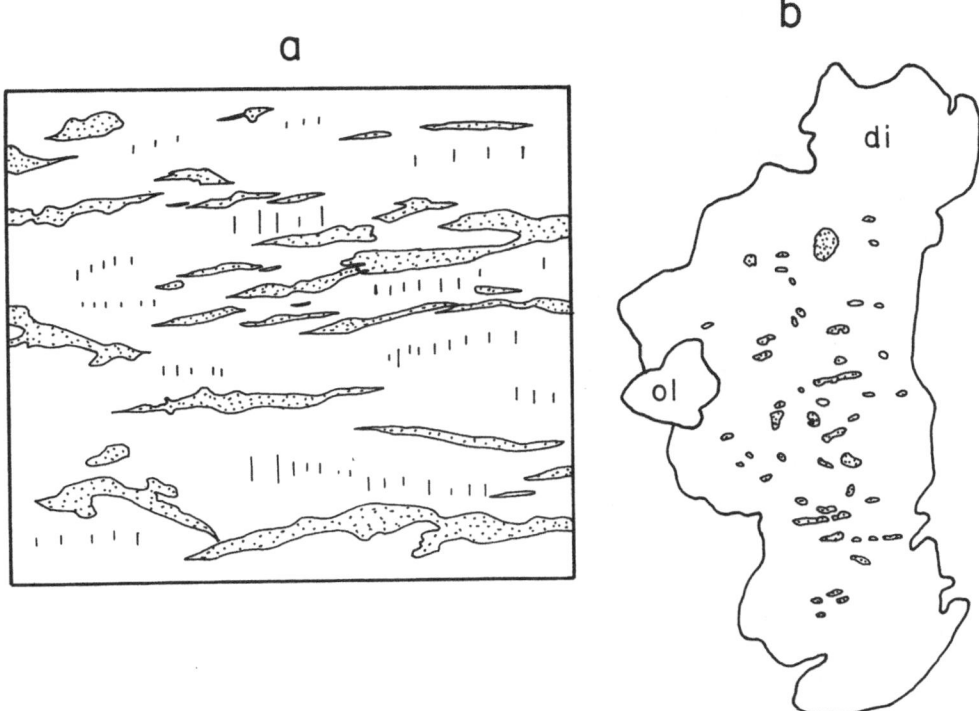

Fig. II-19a, b. Two different exsolution textures developed during decreasing temperature. **a** Microcline perthite with coarse and fine albite lamellae developed perpendicular to each other; **b** Exsolved enstatite in diopside. The diopside grain stem from an olivine eclogite from Oahu, Hawaii. The grain is 4 mm long

phase relations for exsolution will be considered here, while the application has been considered by Boyd and Nixon (1973).

A binary system consisting of the two components A and B displaying exsolution is shown in Fig. II-20. The variation in chemical composition of the two solid phases with respect to temperature is defined by the solvus curve, which has a temperature maximum, called the critical solution temperature. A critical solution pressure is also present in some systems, both types of critical points are termed consolute points. The critical solution temperature might be situated anywhere within the binary system, but is generally near the middle of the binary systems. The solidus curve might intersect the solvus curve, as shown in Fig. II-15, and there is no critical solution point in that case.

The exsolution behavior might be demonstrated by considering the binary mixture of composition n. Let this mixture be completely liquid initially, the following changes will then occur by decreasing temperature. At T_1 solid b begins to form, and the mixture becomes completely solid at T_2. There is only one solid phase in the system in the temperature interval from T_2 to T_3. Beneath the solidus minimum m and the solvus maximum M there is complete miscibility between the two solid phases. When the solid phase B of composition n reaches T_3, it will begin to exsolve solid phase a. By decreasing temperatures the amount of a will increase

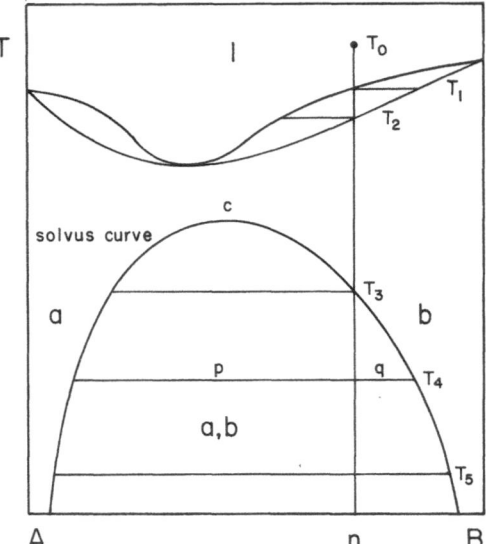

Fig. II-20. A binary system with solid exsolution. The equilibrium compositions of the two solid phases are shown by the solvus curve

while the composition of b will become more B-rich. The relative amount of a and b may be estimated from the lever rule, thus, the ratio of the weight percentages of a and b at T_4 are given by the ratio q/p. The two solvus limbs will be asymptotic towards pure A and B, so that a and b approximates the compositions A and b by decreasing temperature. However, minerals rarely obtain their pure end member composition because the diffusion rate decreases with decreasing temperature. Geothermometers based on the compositions of exsolved minerals should, therefore, be used with some care as the exsolved minerals might have equilibrated at a different temperature from that of the other minerals. The exsolved phase need not remain as inclusions in the host. If the diffusion rate is high then the exsolved phase might form separate grains in the rock, or be situated along the grains' boundaries (Fig. II-21).

Variations in temperature and pressure are frequently observed in metamorphic rocks. If a temperature decrease is followed by a temperature increase, then the exsolution process is reversed. At T_5 the binary mixture n will consist of both a and b. As the temperature is increased solid phase a becomes more rich in B, and solid phase b becomes more rich in A. With increasing temperature the amount of a decreases and the last amount of a disappears at T_3.

The type of exsolution described above assumes that thermodynamic equilibrium is prevalent at all temperatures. At nonequilibrium conditions another type of exsolution occurs called spinoidal exsolution. An excellent account of spinoidal exsolution may be found in Gordon (1968, p. 90).

24 Liquid Immiscibility

The silicate melts consist of monomeric and polymerized silicate anions and metal ions. The silicate tetrahedrons are linked together by bridging oxygens, while the

Fig. II-21. An olivine crystal from the Rhum intrusion. The crystal contains small dendritic and flat exsolutions of chrome spinel. They are concentrated in the center of the crystal, while some of the exsolved material has diffused to the margin of the crystal. The amount of material at the margin is slightly exaggerated for clarity

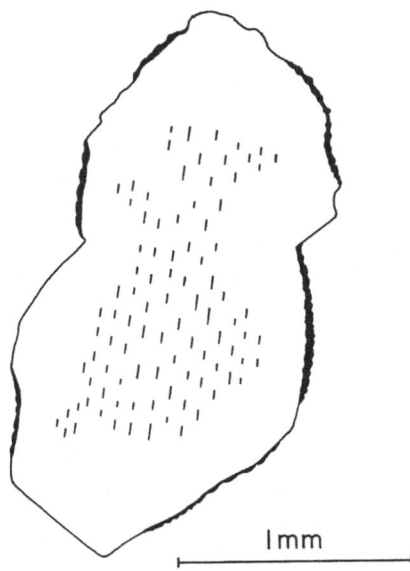

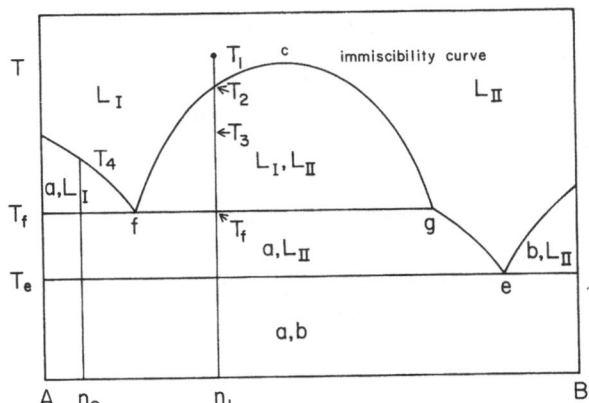

Fig. II-22. A binary monotectic system

cations are situated near the free oxygens so that electrostatic neutrality is maintained (Richardson 1974). If a new type of anion is added to the silicate melt it might either be dissolved in the silicate melt, or it might form a separate liquid phase in the melt, in which case it is said to be immiscible in the silicate melt. Thus, if carbonate, sulfur, or some oxides are added in sufficient amounts two liquid phases might be formed. The solubility of say carbonate in a silicate melt will be dependent on both temperature and pressure, a magma might form one homogeneous liquid at one pressure and two liquids at another.

The phase relations for a binary system with liquid immiscibility will be demonstrated here considering a monotectic system. Another type of binary system with liquid immiscibility is the syntectic system which is shown in Fig. II-24.

The phase relations of the monotectic system are shown in Fig. II-22. The curve separating the field with two liquids from the field with one liquid is called the immiscibility curve, and its maximum point c is termed the maximum point or the consolute point. The binary mixture of composition n_1 will form a homogeneous liquid at T_1. The liquids to the left of the consolute point M are called L_1 and the liquids to the right of this point are called L_{11}. The liquid of composition n_1 will begin to form two liquids when it is cooled to the temperature T_2. At this temperature the liquid L_1 will begin to separate liquid L_{11}, and by decreasing temperature the amount of L_1 will decrease while the amount of L_{11} increases. The two liquids L_1 and L_{11} will coexist down to T_f. At T_f solid a will begin to crystallize from L_1, as the following reaction occurs:

$$L_1 = a + L_{11}$$

The temperature will remain constant at T_f until all L_1 has been consumed. The crystallization of the liquid L_{11} will thereafter proceed in an eutectic-like manner, a will crystallize as the liquid moves from g towards e. Two solid phases will crystallize at e, that is, a and b, and the temperature will remain constant until the last amount of L_{11} has disappeared.

Consider now the crystallization of a liquid with composition n_2. The liquid will begin to crystallize a at T_4 and L_1 and a coexists down to T_f. At this temperature L_{11} will begin to form drops in L_1. As heat is withdrawn from the system more and more L_1 is formed, and ultimately L_1 will disappear whereafter solid a will crystallize from L_{11} until L_{11} has moved to the eutectic e where b begins to crystallize.

It is, thus, evident from the two compositions considered that the point g is a reaction point, and that all liquids will end up in e quite independent of the type of crystallization and the initial compositions of the liquids.

25 Principles of Construction

The phase relations of binary systems are determined by a series of experimental runs at various temperatures at different compositions. These runs define the

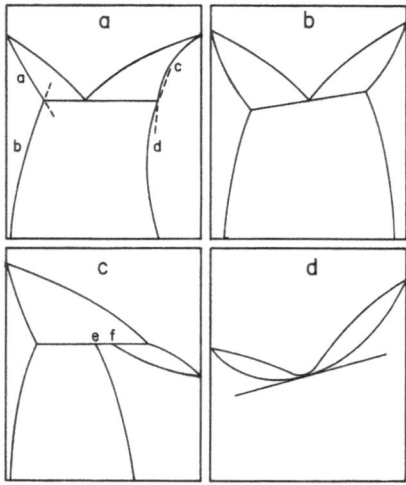

Fig. II-23 a–d. Errors of construction in binary phase diagrams as explained in text

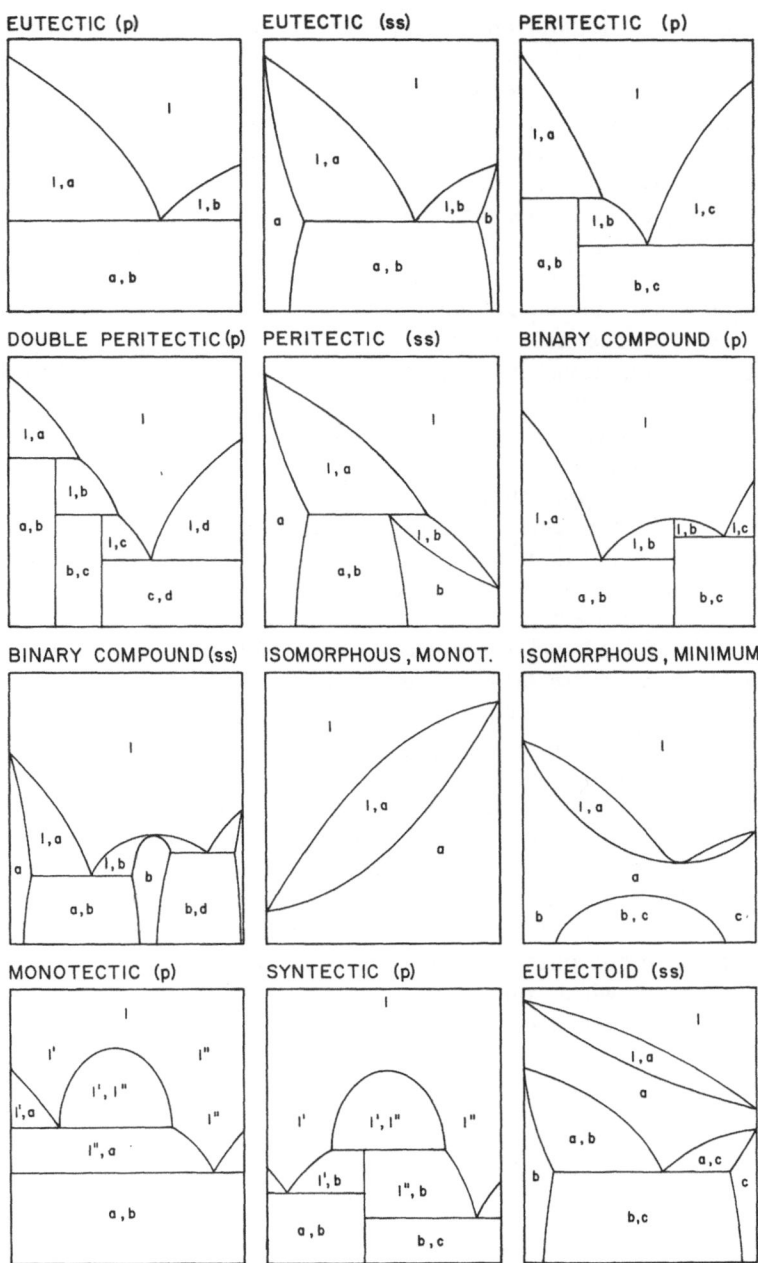

Fig. II-24. A selection of the more common binary phase diagrams. The solid phases may retain a constant composition (*p*) or display solid solution (*ss*)

positions of liquidus and solidus curves, and the phase diagrams are partly constructed on the basis of the obtained run results. The additional information for the construction of phase diagrams is obtained from phase rules, which specifies particular features of the diagrams. These rules are derived from the thermodynamic theory, a detailed account has been given by Roozeboom (1901). Some of these rules are considered here.

The metastable extension of a univariant isobaric curve should extend into a two-phase region in a binary diagram. In Fig. II-23 a, the two metastable extensions a and b extend into two-phase regions, while c and d extend into one-phase regions, and the shape of the phase boundary curves c and d are incorrect.

Gibbs phase rule states that $F + N = C + 2$, so for two components one gets at isobaric conditions: $F + N = 3$. The phases are, thus, at equilibrium at a given temperature, and the tie line that joins the compositions of the three phases must be a horizontal line in the T, x diagram (Fig. II-23 b). Also there cannot be four phases in equilibrium at a given temperature, and the diagram shown in Fig. II-23 c must be incorrect, point e should coincide with point f.

In binary systems with solid solution and a temperature maximum or minimum, the solidus and liquidus curves meet at a point. The tangent to the two curves at this point must be a horizontal line. The phase diagram shown in Fig. II-23 d is therefore incorrect.

A useful rule was demonstrated in detail by Rhines (1956), who stated that: "A field of the phase diagram, representing equilibrium among a given number of phases, can be bounded only by regions representing equilibria among one more or less than the given number of phases." This rule is quite general and applies to any multicomponent system as well as to pseudo-systems and P-T diagrams. In applying the rule a univariant equilibrium should be considered a region. As an example, consider the eutectic system with solid solution shown in the topmost part of Fig. II-24. There are seven different regions, three one phase regions, three two phase regions, and one three phase region represented by the invariant equilibrium between a, b, and l. The region (l, a) is bounded by region (a) and region (l), which both have one phase less than region (l, a). The third boundary is the univariant equilibrium which contains one phase more, i.e., (a, b, l).

The different basic types of binary systems are summarized in Fig. II-24, and Rhines' phase rule as well as the other rules might be applied as an exercise on these diagrams.

III Ternary Systems

1 Introduction

There is a variety of ternary phase diagrams. However, the analysis of the crystallization behavior by equilibrium crystallization and fractional crystallization as well as by partial melting can be done with the aid of a few basic principles, which will be considered here. Most helpful in this connection are the crystallization vectors dealt with in this chapter.

While there are some differences in the treatment of binary and ternary systems, there are only minor differences between the ternary diagrams and the multicomponent phase diagrams. Further, the ternary phase diagrams will provide a good approximation to the phase relations of natural rocks and magmas, since most rocks contain three to four important solid mineral phases. The ternary systems, therefore, constitute the most fundamental tool for the analysis of the phase relations of magmas.

While the principles of the ternary phase diagrams are introduced here, their application will be demonstrated in the chapters on fractional crystallization and partial melting.

For further reading, the detailed treatment of Rhines (1956), and of Ricci (1951) who has dealt with the more complicated ternary systems containing gas and two liquids are recommended. The phase diagrams of petrological relevance have been considered by Bowen (1928), Ehlers (1972), Yoder (1976, 1979), and Morse (1980).

2 Gibbs' Triangle

The phase relations of ternary systems are displayed using equilateral triangles. The use of triangles was first proposed by Gibbs (1876), and they are sometimes called Gibbs triangles. The triangular representation is used rather than a Cartesian coordinate system because a rectangular coordinate system is inconvenient (Ricci 1951).

A ternary composition is given by three variables, which have a constant sum like 1 mol, 100 g, or 100%. Using the mol unit we thus have:

$$x_1 + x_2 + x_3 = 1 \tag{1}$$

so that one variable is given when the two others are known:

$$x_3 = 1 - (x_2 + x_3) \tag{2}$$

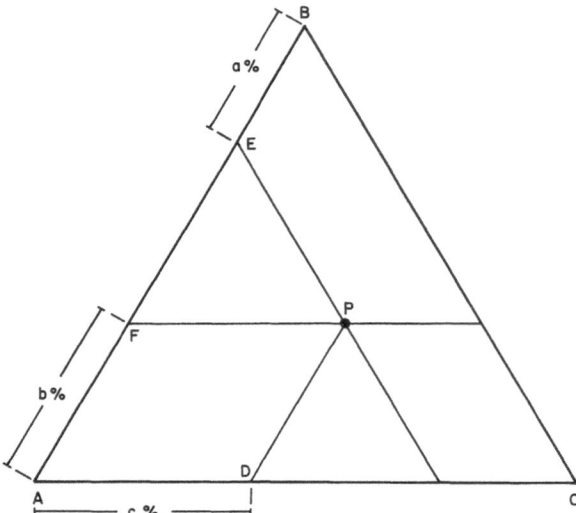

Fig. III-1. How to plot a composition a, b, c in the ternary diagram

Two variables are consequently sufficient for defining a given composition, and the ternary compositions can, therefore, be plotted on a plane. A mixture of three components may be plotted within an equilateral triangle, where the three sides are divided into mole fractions or weight percentages. Consider a mixture consisting of three components, A, B, and C, and let the weight percentages of these three components be a%, b%, and c%, respectively. This composition is plotted in the equilateral triangle by drawing a straight line parallel to the BC side that intersects the AC side at a% (Fig. III-1). Next, another line is drawn parallel to the AC side that intersects the AB side at b%, and the intersection between these two lines at the point P indicates the composition of (a, b, c).

A composition which add up to 100% can always be plotted in the equilateral triangle as is evident from Fig. III-1. The line segment AD is equal to FP, and since the triangle PEF is equilateral, it follows that AD equals FE. The sum of AF, FE, and EB, thus, always equals AB, which is defined as 100%.

3 Coordinate Transformations

Most of the data required from the ternary systems can be obtained using graphical methods, as these methods have sufficient accuracy for most applications. When the accuracy should be better than about 2% or when complicated calculations are involved, it may be necessary to make exact estimates of the coordinates, and these are performed using Cartesian coordinates. The triangular coordinates should, therefore, be transformed into Cartesian ones, and vice versa.

Let the side of the equilateral triangle be divided into 100 U, and let the origin of the Cartesian coordinate system (0, 0) be at the very A corner (Fig. III-2). Further, let the units of the x axis and y axis be the same as the units of the AC side. We

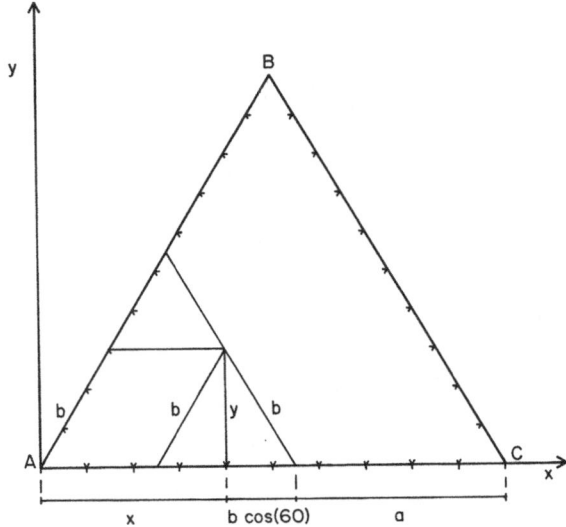

Fig. III-2. The conversion from ternary to Cartesian coordinates and vice versa

then have:

$$y = b \sin (60) \tag{3}$$
$$x = 100 - (a + b \cos (60)) \tag{4}$$

The triangular coordinates a and b are then obtained from x and y by these equations:

$$b = \frac{y}{\sin (60)} \tag{5}$$

$$a = 100 - \left(x + \frac{y}{\tan (60)}\right) \tag{6}$$

Example

An α phase consists of 80% A, 10% B, and 10% C, and a β phase consists of 20% A, 30% B, and 50% C. We want to find the exact triangular coordinates of a γ phase that consists of 40% α and 60% β (Fig. III-3).

Using Eqs. (2) and (4) we get:

$$\alpha: x_1 = 15, \quad y_1 = 8.66$$
$$\beta: x_2 = 65, \quad y_2 = 25.98$$

The length of the line joining α and β is then given by:

$$L = ((x_2 - x_1)^2 + (y_2 - y_1))^{1/2} = 52.91$$

Hence, we obtain:

$$0.6 \, L = 31.756$$

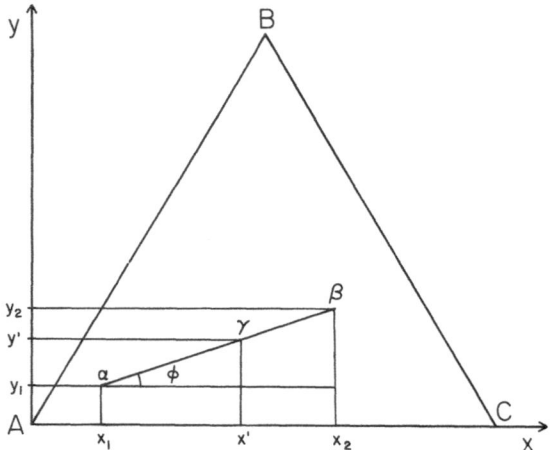

Fig. III-3. The calculation of the ternary coordinates of γ consisting of 60% β and 40% α

The angle φ is then given by:

$$\sin (\varphi) = (x_2 - x_1)/L = 0.3273,$$
$$\varphi = 19.107\,°$$

Thus, we get:

$$x' = x_1 + 31.746 \cos (19.107)$$
$$y' = y_1 + 31.746 \sin (19.107)$$

Using Eqs. (5) and (6) we finally obtain the triangular coordinates:

$$A = 44.00, \quad B = 22.00$$

4 Space Diagram and Projection

As mentioned above, the composition of a ternary mixture can be plotted on a plane. However, if a third variable like the temperature has to be shown, then we must use a space diagram. Such a space diagram is shown in Fig. III-4a using the temperature as ordinate for a simple eutectic system. It consists of three liquidus surfaces that intersect along three eutectic curves, e_1-e_t, e_2-e_t, and e_3-e_t. The three eutectic curves meet at the eutectic point e_t, which has the minimum temperature of the system. The space diagram is projected onto the plane which shows the position of the three eutectic curves (Fig. III-4a and b). The projected phase diagram allows a full analysis of all the compositional variations of liquids and solids. The temperature variations can also be shown in a projection using isotherms (Fig. III-5a and b). The temperature gradients within the system will then be evident from the distance between the isotherms, as a small distance indicates a steep gradient of the liquidus surface. Approximate estimates of temperatures can be made by interpolations between the isotherms.

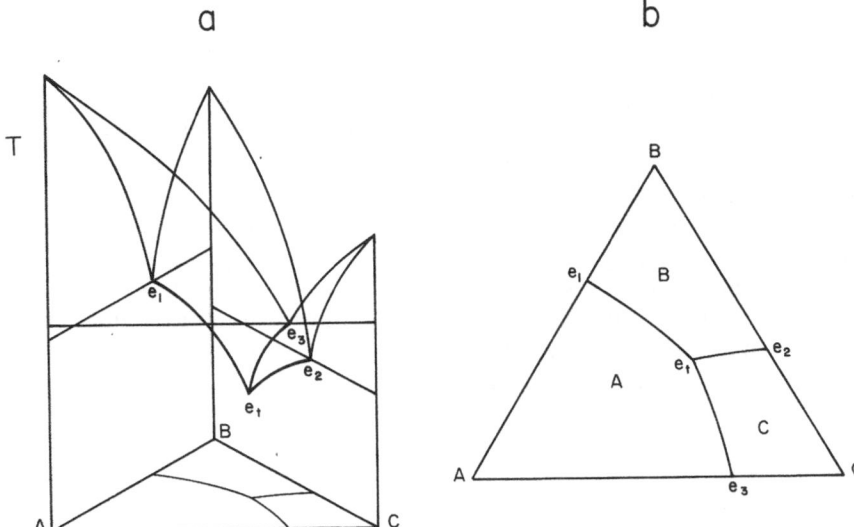

Fig. III-4 a, b. The space diagram of a ternary eutectic system (a), and its projection on the composition plane (b)

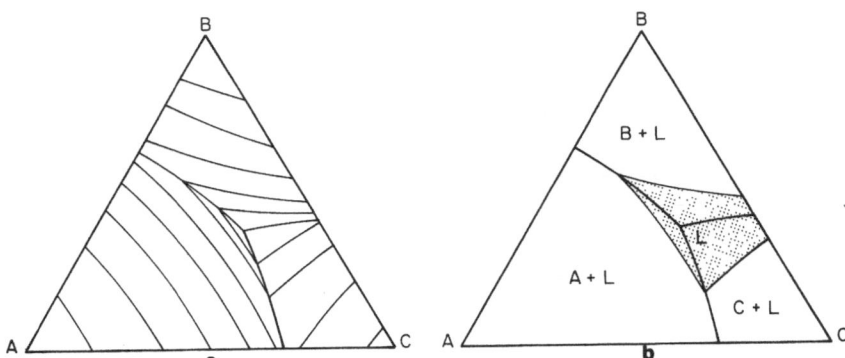

Fig. III-5 a, b. The projection of the isotherms of a ternary eutectic system (a) (Fig. III-4), and an isothermal section (b). The shaded areas only contain liquid, while at least one solid is in equilibrium with liquid in the rest of the system at this temperature. Two solids are in equilibrium with liquid along the eutectic curves

5 Ternary Eutectic System

The ternary eutectic system without any solid solution is the most simple of the ternary systems. Nevertheless some important petrogenetic features will be evident from this system, and we will also use this system further for an introduction to the quantitative analyses of ternary systems.

6 Crystallization

There is no solid solution in the system we are going to consider, and the composition of each of the solid phases will be those of the components of the system, A, B, and C (Fig. III-6). The result is that equilibrium crystallization and fractional crystallization are identical for this system. These two types of crystallization will be different when the solid phases have solid solution.

The liquid with composition M is situated within the liquidus field of A, and the liquid will first crystallize A (Fig. III-7). With decreasing temperature more A will form, and the composition of the liquid will move along a straight line away from the A apex. The crystallization of A alone will proceed until the liquid reaches the eutectic curve e_3e at l, where C begins to crystallize together with A. The liquid will thereafter move along the eutectic curve towards the eutectic minimum e. When

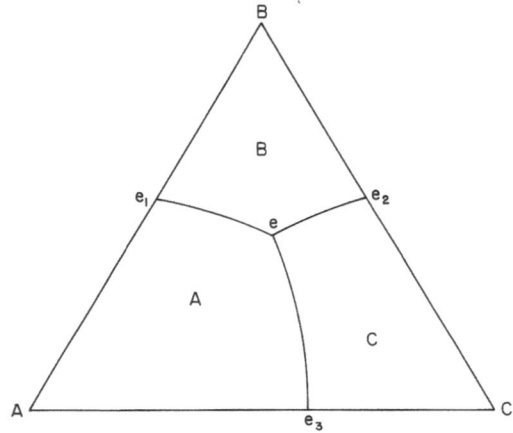

Fig. III-6. A ternary eutectic system consisting of the three solid phases *A, B,* and *C.* The minimum temperature of the system is at *e*

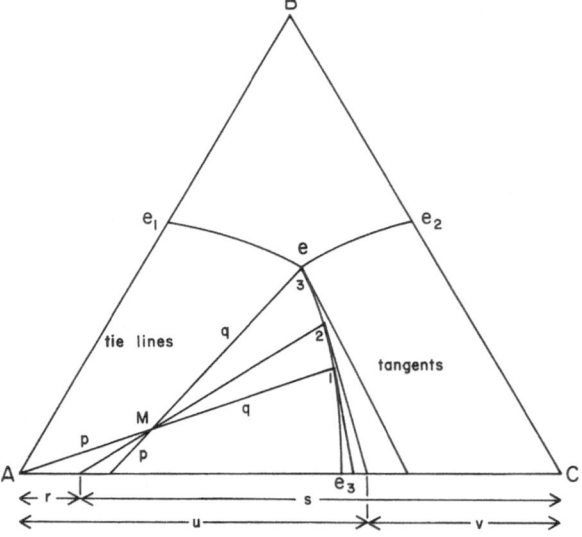

Fig. III-7. The crystallization of composition *M,* and the definition of the three different lever rules for ternary systems

Fig. III-8. The six sectors of the ternary eutectic system

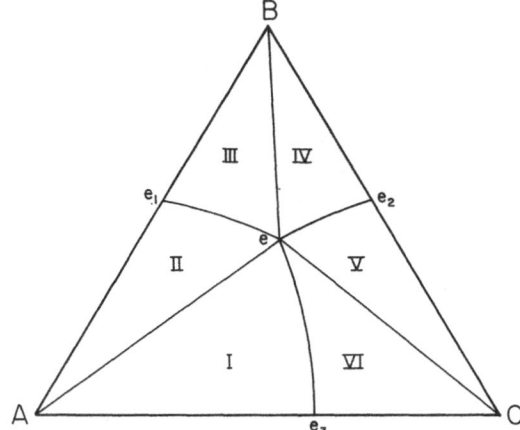

the liquid reaches e, solid B begins to crystallize, and four phases, three solids and one liquid will be in equilibrium with each other at the eutectic point. The liquid will remain at e until the last liquid disappears, and the system consists of three solids.

From Gibbs law it will be evident that the crystallization at e will occur at constant temperature. For N = 3 we get at constant pressure:

$$F + N = 3 + 2 - 1$$
$$F + 4 = 4$$
$$F = 0$$

There are, thus, no degrees of freedom or variance and the temperature must be constant at e. The sequence of phase assemblages for the crystallization just considered is as follows:

L
A + L
A + C + L
A + B + C + L
A + B + C

The crystallization sequences will be different for different compositions, and the ternary diagrams can be divided into sectors that each have the same crystallization sequence. The sectors of this ternary diagram are shown in Fig. III-8, and the various sequences are listed in Table 1.

7 Partial Melting

Partial melting will for this particular system be exactly the opposite of crystallization. A solid mixture of composition M will first form melt at the eutectic point e. The liquid will then move from e to l along the eutectic curve, and finally from l to M. The composition of the ternary mixture is insignificant for the initial composi-

Table III-1. Crystallization sequences for the ternary eutectic system

I	II	III	IV	V	VI
L	L	L	L	L	L
A+L	A+L	B+L	B+L	C+L	C+L
A+C+L	A+B+L	B+A+L	B+C+L	C+B+L	C+A+L
A+C+B+L	A+B+C+L	B+A+C+L	B+C+A+L	C+B+A+L	C+A+B+L
A+C+B	A+B+C	B+A+C	B+C+A	C+B+A	C+A+B

tion of the melt, as all liquids will start at e. By analogy the compositions of magmas formed by partial melting of rocks will have similar compositions as long as the rocks consist of the same minerals. It should also be noted that the initial addition of heat will result in the generation of more melt with the same composition. The partial melt will not change its composition before one of the solid phases has become completely melted. The rocks undergoing partial melting will, therefore, initially produce melts with a fairly constant composition, because a substantial amount of heat is required before the partial melt changes its composition.

8 Ternary Lever Rules

The lever rule introduced for the binary systems also applies to the ternary systems, and in fact for systems of any order. In the binary systems we needed only to specify one quantity, either the amount of solid or the amount of liquid. In the ternary systems there are three quantities that define the change in compositions, and they can all be estimated from three different lever rules.

The lever rules for the ternary systems can be proven by solving three linear equations with three unknowns, but they can also be proven using geometrical methods. The proofs for the lever rules were first given by Swanson (1924) who used geometric methods. The lever rules will be called Swanson's I and II rule and the tangent rule.

The amount of liquid is estimated from the line that passes through the initial composition M and the composition of the liquid (Fig. III-7). Let us estimate the amount of liquid when it has attained composition 1 on the eutectic curve. Let the line segments AM and ML_1 be p and q, respectively. The ratio $L/(L+S)$ is then given by:

$$\frac{L}{L+S} = \frac{p}{p+q} \quad \text{(Swanson I)}. \tag{7}$$

This equation defines the amount of liquid at any stage during the crystallization, and is called Swanson's I rule. When the liquid reaches L_2, the amount of liquid is again given by Eq. (7). The line through M and L_2 now intersects the AC join in agreement with the fact that the fractionate now consists of both A and C. The p and q values of the line that passes through M and e will define the amount of liquid present just when the liquid has reached e. The amount of liquid present afterwards will depend on how much heat the system has given off, and this amount can not be estimated from the lever rule.

9 Swanson's II Rule

During the crystallization the composition of the total amount of fractionate will change. Initially it will consist of only A, and then of both A and C. The relative amounts of A and C at stage 2 is given by the following ratio:

$$\frac{A}{A+C} = \frac{s}{s+r} \quad \text{(Swanson II)}. \tag{8}$$

As r initially equals zero we get $A = 1$. After the liquid has reached the eutectic curve at 1 both A and C crystallize, and the line through M and L intersects the AC join, and r becomes larger than zero.

10 The Tangent Rule

Most eutectic curves are curved and the relative proportions of A and C that form at any given temperature along the eutectic curve will, therefore, vary. At stage 2 the relative proportion between A and C is given by:

$$\frac{A}{A+C} = \frac{v}{v+u} \quad \text{(tangent rule)}. \tag{9}$$

The line segments u and v are given by the intersection between the tangent to the eutectic curve and the join AC (Fig. III-7).

Swanson's I rule, thus, affords a determination of the total amount of fractionate. Swanson's II rule estimates the relative amounts of the solid phases in the fractionate, while the tangent rule determines the relative proportions of the solid phases that form at any given instance.

Example

The point M has the composition 70% A, 10% B, and 20% C. The eutectic point e is situated at 25% A, 45% B, and 30% C, and the liquid that crystallizes from M reaches the eutectic curve at 30% A, 23.5% B, and 46.5% C.

When the liquid reaches the eutectic curve from M the amount of liquid is estimated from Swanson's I rule:

$$\frac{L}{L+S} = 0.4286$$

The amount of liquid is, thus, 42.86%, and the amount of solid consisting of A only is 57.14.

At stage 2 we get:

$$\frac{L}{L+S} = 0.3052$$

Using Swanson's II rule we obtain:

$$\frac{A}{A+C} = 0.88$$

and the tangent rule gives the following result:

$$\frac{A}{A+C} = 0.36$$

The amount of liquid left is then 30.52%, and the amount of A in the fractionate is 88%. The composition of the fractionate that forms just at stage 2 is 36% A, and 64% C.

11 Quantitative Analysis – Trend Estimate

We have now considered the procedure required for quantitative calculations, and will now analyze the compositional variations in detail. The triangular coordinates of the system is shown in Table III-2.

During the crystallization of the liquid there is a continuous change in its composition. However, this variation is not monotonous, instead the trend curve will display a kink. The compositional variation of the liquid is shown in Fig. III-9 using the B component as abcissa. From M to stage 1 the trend curve is a straight line as the fractionate consists of A only. When the liquid reaches 1 on the eutectic curve there is a sudden change in the slope of the trend, as C now enters the fractionate. The trend curve for C displays a maximum because the amount of C increases as long as A crystallizes alone, and the amount of C decreases after C enters the fractionate. From stage 1 to e the liquid follows the eutectic curve, and the trend curves display some curvature related to the curvature of the eutectic curve.

The variation in the crystallizate or fractionate that forms at any given moment is shown in Fig. III-9 as the cumulate. This variation is estimated using the tangent rule. Initially only A crystallizes, at 1 solid C suddenly enters the fractionate. When the liquid reaches e there is again an abrupt change as B now begins to crystallize.

Table III-2. Coordinates of the ternary eutectic system

	A%	B%	C%	Tangent intersection A%
M	70	10	20	
e	25	45	30	
1	30	23.5	46.5	38.5
2	27	33	40	36.0
3	25	45	30	33.0
e_1	45	55		
e_2		55	45	
e_3	40		60	

Fig. III-9. The variation in the composition of the liquid by fractional crystallization of composition M, using component B as the independent variable. When a new solid phase, like C, enters the fractionate, the trend curve displays a kink. The variation in the composition of the fractionate or cumulate is also shown. Note that there is a jump in the composition of the cumulate when C enters the fractionate and at the eutectic point

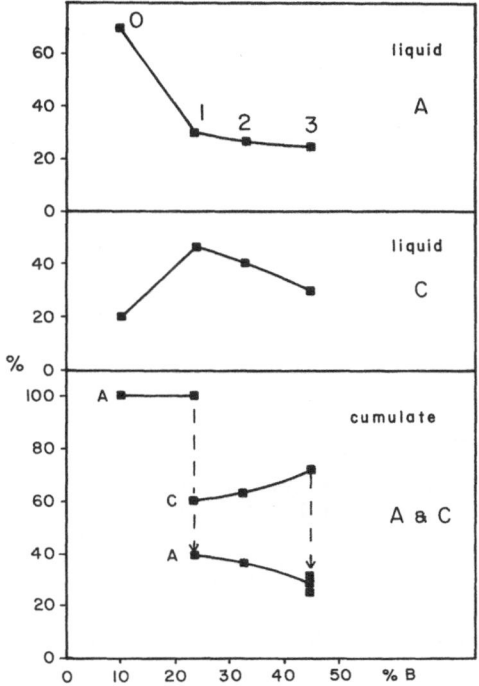

Just before the eutectic point the composition of the instantaneous fractionate is given by the tangent rule. However, at e the composition of the fractionate is given by the requirement that the composition of the liquid should be constant, and the composition of the fractionate is, thus, identical to that of the eutectic point e.

It is important to note here that each time a new solid phase enters the crystallizate there will be an abrupt change in the trend slope of the liquid. Thus, in a quaternary system there will be two kinks in the trend slope. The appearance of a new solid phase will result in a kink of the trend slope for the liquid and a jump in the composition of the fractionate or cumulate that form from the liquid.

The trends of basaltic magma suites display such kinks (Chap. X). By estimating the trend slope one can determine the composition of the fractionate and the mineral assemblage that controls the fractionation. The mineral assemblage itself can then be compared with the phase relations of the PT diagram for the rock whereafter the pressure and temperature of crystallization may be estimated.

The crystallization behavior of large magma chambers have been studied in a number of the layered intrusions, which have formed by the crystallization of basaltic, syenitic, and granitic magmas. The major part of these intrusions consist of a sequence of layered rocks which are called cumulates or layered series. The main features of the layered series fit nicely with the known phase relations (Fig. III-10), and the layered series afford an invaluable insight into the crystallization processes that occur in magma chambers.

The cumulate that would form by the crystallization of composition M is shown in Fig. III-11. It has been assumed for simplicity that the interstitial liquid leaves

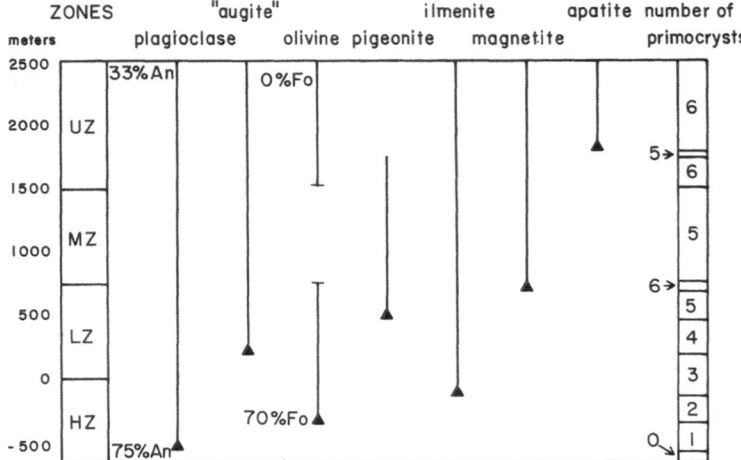

Fig. III-10. The variation in the composition of the cumulate of the layered series of the Skaergaard intrusion, East Greenland (Wager and Brown 1967). According to Rhines phase rule (Chap. V), the number of phases should change with ± 1 phase. This is the case as the number of primocrysts change with ± 1. Olivine ceases to crystallize in *zone MZ,* and begins to crystallize again in *zone UZ.* The diagram displays the general feature of magmas, the number of fractionating minerals increases with decreasing temperature. This is just what one would expect from the eutectic crystallization behavior

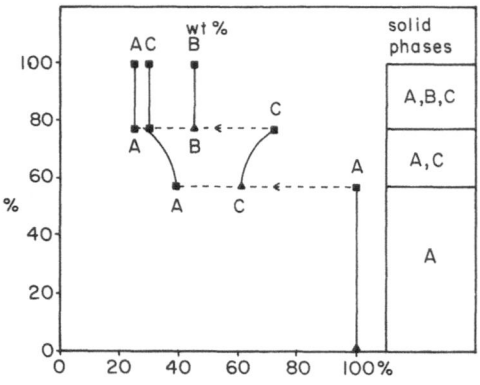

Fig. III-11. The variation in the cumulate formed by crystallization of composition M related to the amount crystallized. The diagram shows the "layered intrusion" that would form from a magma with composition M

the cumulate. The interstitial liquid may change its composition by a diffusive exchange of material with the magma situated above the layers. When this diffusive exchange is limited the interstial liquid retains its composition and the cumulates are called orthocumulates. When the diffusive exchange is intensive, cumulates may form that consist only of the primocryst minerals, and these cumulates are called adcumulates. The diagram, thus, shows the variation for an adcumulate.

12 Binary and Ternary Congruent Compounds

Both binary and ternary compounds may form within a ternary system. A system with a binary congruent compound is shown in Fig. III-12. The line from the com-

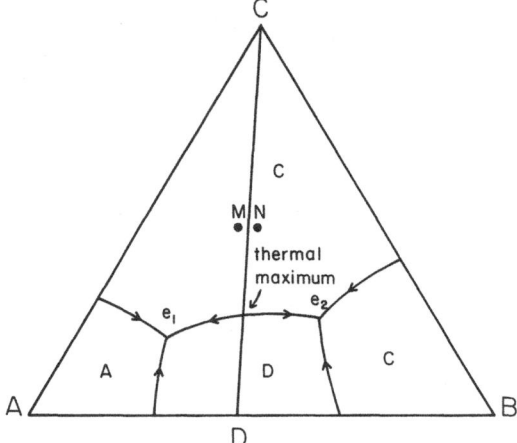

Fig. III-12. A ternary system with the binary congruent compound D. The eutectic curve e_1e_2 has a thermal maximum where it intersects the tie line between D and C

position of the compound, D, to the opposite apex, C, divides the system into two eutectic systems where the crystallization and melting occur completely isolated from each other. The line defines temperature maxima on the liquidus of C and D, and also on the eutectic curve between e_1 and e_2. The temperature maximum on the eutectic curve is called a thermal maximum or a thermal barrier. The presence of these thermal maxima is not evident from the phase rule, but can be proven from thermodynamics.

By crystallization all liquids to the left of the division line ends up in e_1, and the liquids to the right of this line end up in e_2. It should be noted that small differences in composition might cause different crystallization sequences. The two liquids M and N have similar compositions, but are situated on each side of the division line. Solid A will enter the fractionate of M, while B will enter the fractionate of N.

Basaltic rocks may be divided into silica normative and nepheline normative compositions. It was considered by some petrologists that nephelinites might form by fractionation from tholeiites at low pressures. It was shown by Yoder and Tilley (1962) on the basis of the $Fo-Ne-SiO_2$ system that the proposed fractionation scheme was impossible at low pressures. The system is shown in Fig. III-13. Albite forms a thermal barrier in the system and it will not be possible for liquids to move from the silica normative part into the nepheline normative part or vice versa. A transition from tholeiite to nephelinite might be possible at higher pressures where plagioclase becomes unstable, and the thermal barrier disappears.

A system with a congruent ternary compound is shown in Fig. III-14. This system has three thermal barriers and is divided into three isolated ternary systems with the eutectic points e_1, e_2, and e_3.

13 Crystallization Vectors

The analysis of the crystallization behavior of phase diagrams is facilitated by the use of crystallization vectors. These vectors are not of any significance for simple

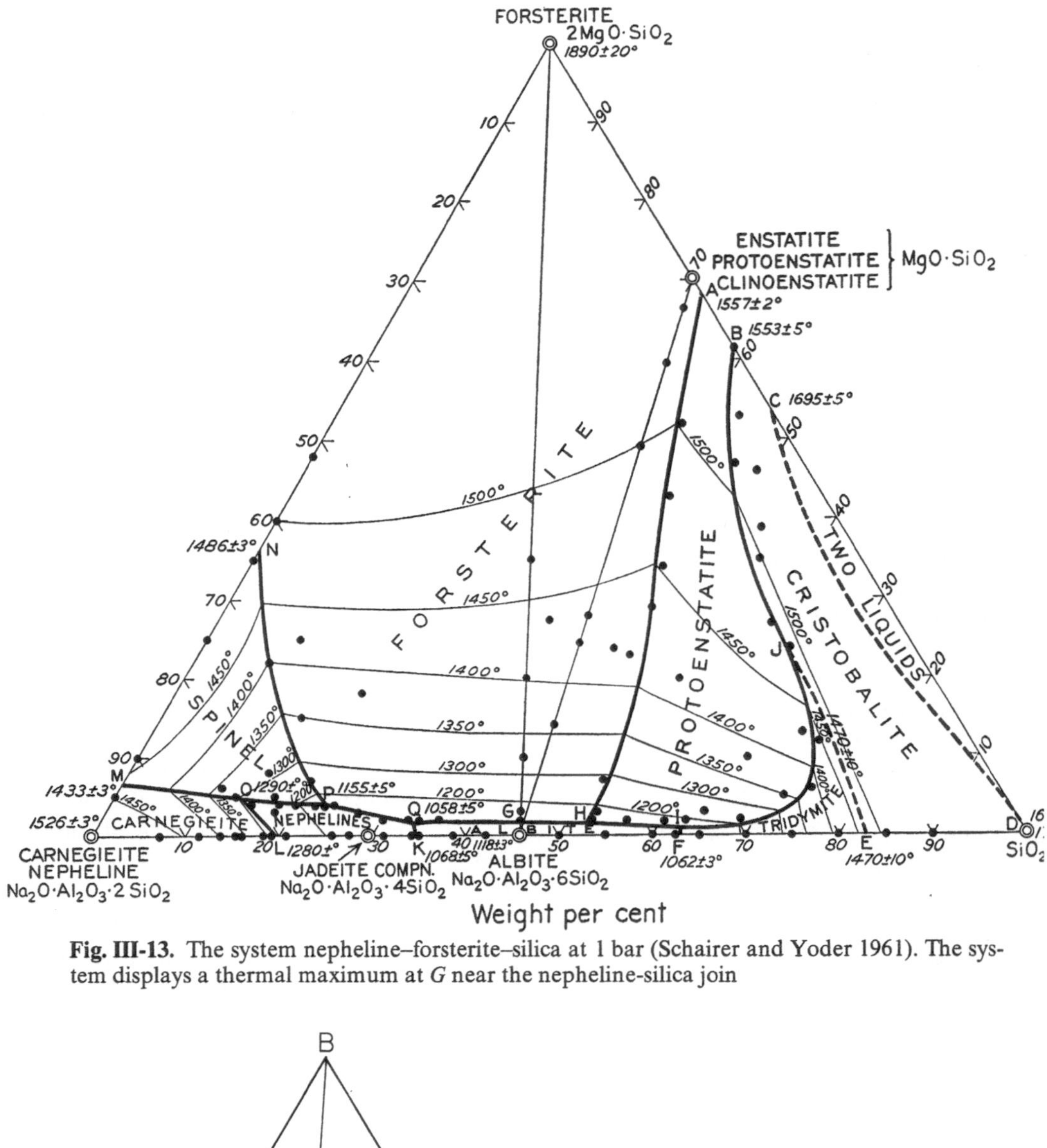

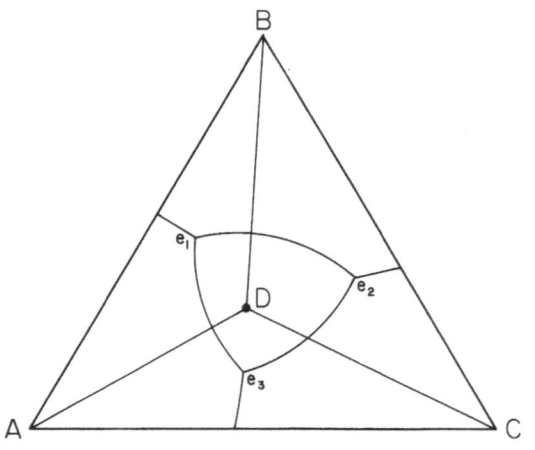

Fig. III-13. The system nepheline–forsterite–silica at 1 bar (Schairer and Yoder 1961). The system displays a thermal maximum at *G* near the nepheline-silica join

Fig. III-14. A ternary system with a ternary congruent compound *D*

Fig. III-15. The crystallization vector *v*,
and the two solid vectors *c* and *a*

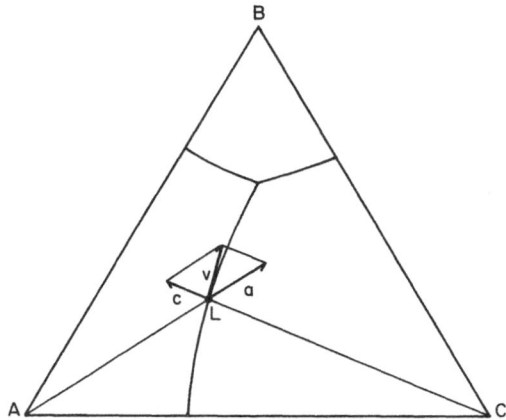

eutectic systems, but they are highly convenient for the analysis of peritectic systems and systems with solid solution.

The crystallization vector can be defined as follows:

The crystallization vector originates at the point defined by the composition of the liquid. The direction of the vector is the direction in which the liquid moves at that point.

By crystallization along a univariant curve, like a eutectic curve, the crystallization vector is coinciding with the tangent to the curve. The crystallization vector is the resultant of the solid vectors, which are defined as follows:

The solid vectors extent from the point defined by the composition of the liquid, and their directions are in the opposite directions of the solids that they represent.

The significance of these definitions is demonstrated in Fig. III-15. The solid vectors $\bar{a}$ and $\bar{c}$ originate at the same compositional point as the liquid, and their resultant is the crystallization vector $\bar{v}$. The length of the crystallization vector is not given by the above definitions, and the length may be chosen so that it indicates the amount of solid that forms by a certain decrease in temperature.

14 Ternary Peritectic System

The ternary peritectic system displays some of the fundamental phase relations of the partial melting of the herzolitic mantle because enstatite has a reaction relationship with the liquid. The reaction relationships observed in igneous rocks – Bowens' reaction series – are also explained by the peritectic system.

The space diagram for a ternary peritectic system without solid solution is shown in Fig. III-16, and the projection onto the composition plane is shown in Fig. III-17. The univariant curves that are eutectic curves may be indicated with a single arrow, while a peritectic curve, generally called a reaction curve, is indicated by two arrows (Fig. III-17). Along the reaction curve the following reaction occurs:

$$C + L = D$$

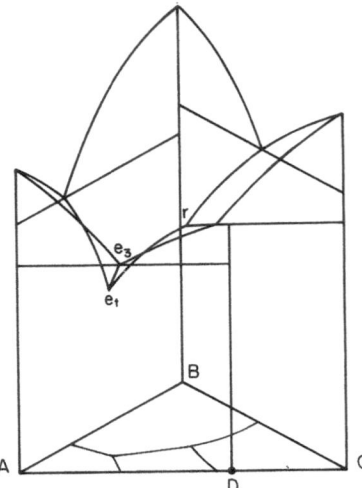

Fig. III-16. The space diagram of a ternary peritectic system. The reaction point or peritectic point is at *r*

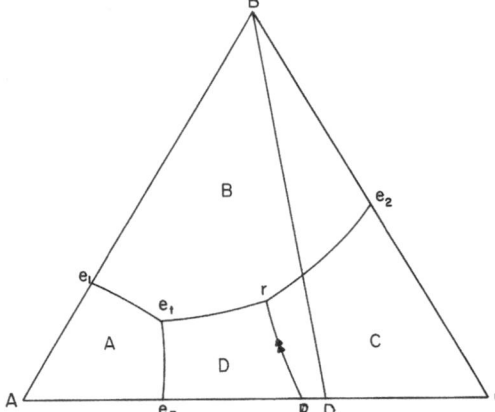

Fig. III-17. A ternary peritectic system with the reaction $C + L = D$. The *point r* ist the reaction point and the *curve p r* is the reaction curve where the reaction $C + 1 = D$ occurs. The other curves are all eutectic, and e_I is the ternary eutectic minimum point of the system

Solid phase C, thus, reacts with the liquid forming D. The point r is a reaction point, while e is an eutectic point. Equilibrium crystallization and fractional crystallization will be diffeent for this system and will be treated separately.

15 Equilibrium Crystallization

The crystallization in the peritectic system is more complicated than crystallization in the eutectic system, and the peritectic system has, therefore, been divided into different sectors with identical crystallization sequences (Fig. III-18). Before we deal ·with the detailed aspects of the crystallization it is appropriate to have an overview of the major features.

The ternary peritectic system is essentially divided into two parts. All liquids within the subtriangle ABD will end up at e by equilibrium crystallization. All

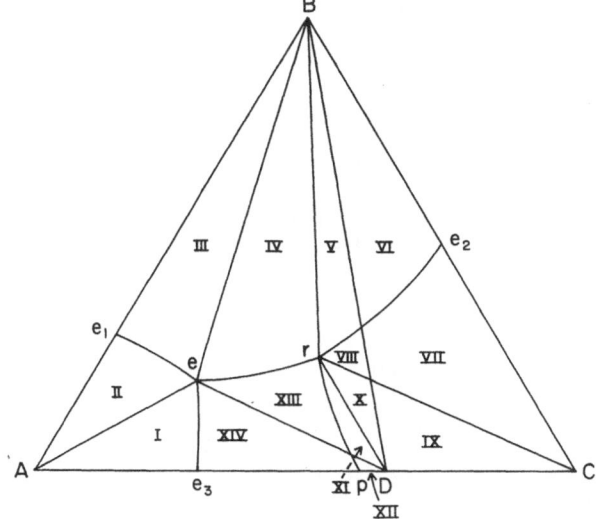

Fig. III-18. The 14 different crystallization sectors of the peritectic system that occur by equilibrium crystallization. Each of the sectors has a different crystallization sequence as listed in Table III-3

liquids within the other triangle, BCD, will end up at r by equilibrium crystallization. By partial melting all mixtures within ABD will form liquids at e, and within the triangle BCD all mixtures initially form liquids at r. By fractional crystallization this division disappears and all fractionating liquids end as usual in the eutectic point e. Liquids on the very join BD will by equilibrium crystallization move to r. By partial melting on this join the first liquid will form at r as there is no solid A in the mixture. We will now consider the different sectors.

Sectors I to IV

Crystallization within the sectors I, II, III, and IV will be simple eutectic crystallization. The crystallization sequences for these and the other sectors are shown in Table III-3.

Sector V

A liquid within sector V will a first form B and then reach the eutectic curve e_2r. Here C will begin to crystallize together with B. B and C will crystallize together, while the liquid moves towards r. At r the liquid will begin to react with C forming D. Since its total composition is to the left of the join BD there will not be enough C present to react with all the liquid. There will be some liquid left after C has reacted out, and the liquid will leave r and move towards e, crystallizing B and D. At e these two solids are joined with A.

Sectors VI and VII

A liquid within sector VI will first form B, and thereafter C when the liquid reaches curve e_2r. Thereafter the liquid will move to r, where C will begin to react with the

Table III-3. Crystallization sequences by equilibrium crystallization for the ternary peritectic system

I	II	III	IV	V
L	L	L	L	L
A+L	A+L	B+L	B+L	B+L
A+D+L	A+B+L	B+A+L	B+D+L	B+C+L
A+D+B+L	A+B+D+L	B+A+D+L	B+D+A+L	B+C+D+L
A+D+B	A+B+D	B+A+D	B+D+A	B+D+L
				B+D+A+L
				B+D+A

VI	VII	VIII	IX	X
L	L	L	L	L
B+L	C+L	C+L	C+L	C+L
B+C+L	C+B+L	C+B+L	C+D+L	C+D+L
B+C+D+L	C+B+D+L	C+B+D+L	C+D+B+L	C+D+B+L
B+C+D	C+B+D	B+D+L	C+D+B	L+D+B
		B+D+A+L		A+B+D+L
		B+D+A		A+B+D

XI	XII	XIII	XIV
L	L	L	L
C+L	C+L	D+L	D+L
C+D+L	C+D+L	D+B+L	D+A+L
D+L	D+L	D+B+A+L	D+A+B+L
D+B+L	D+A+L	D+B+A	D+A+B
D+B+A+L	D+A+B+L		
D+B+A	D+A+B		

liquid forming D. In this case the total composition is situated to the right of the division line BD, and there will consequently be more solid C present than is required for total reaction with the liquid. The liquid will, therefore, disappear first, and there will be some solid C left after the reaction. The crystallization will stop at r, and the crystallizate will consist of B, C, and D (Table III-3).

The crystallization within sector VII is the same except that C forms before B.

Sector VIII

The crystallization within this sector is the same as that of V, except that C forms before B.

Sector IX

A liquid within this sector will initially form C and then move to the reaction curve pr. The process that will occur at the reaction curve can be estimated by consid-

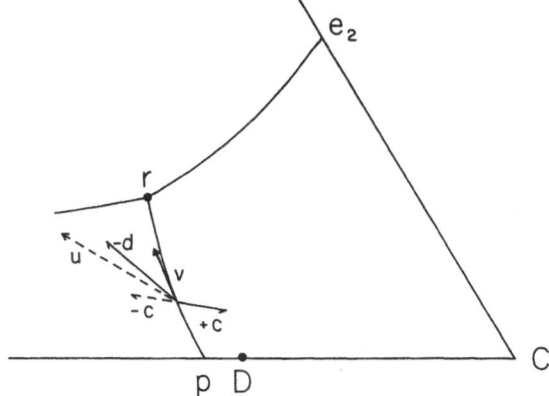

Fig. III-19. The analysis of the crystallization along the reaction curve using the crystallization vector. The liquid can only stay on the reaction curve as long as it can react with *C*. The resulting crystallization vector is then *v*. If *C* also crystallizes, the resulting crystallization vector is *u*, and the liquid would in that case leave the curve

ering the crystallization vector (Fig. III-19). The reaction curve is an isobaric invariant curve, and the liquid will move along the curve as long as both C and D are present in the liquid. The crystallization of D will tend to move the liquid away from the curve (–d̄). If C also crystallizes from the liquid (–c̄), then the liquid must clearly leave the curve as the resultant crystallization vector ū is directed away from the curve. The liquid can only move along the curve if C is added to the liquid (+c̄), and the resultant vector is v̄. Thus, the subtraction of D and addition of C will maintain the liquid on the curve.

The difference between peritectic and eutectic crystallization can be demonstrated here by considering different positions of the composition of the solid phase D. If D is situated to the left of p, the crystallization will be eutectic as both solids then can be subtracted from the liquid. If D is to the right of p then one solid, C, must be added, and the crystallization is peritectic.

The mixtures within sector IX are situated to the right of the division line BD and the liquids will, therefore, reach r where the liquid will react further with C until it disappears, leaving three solids, B, C, and D.

Sector X

The liquids will first form C, and then D when they reach pr. The liquid will move along pr, while some C reacts with it forming D. At r all C reacts out of the liquid, and the liquid thereafter moves to e, and the subsolidus assemblage will consist of B, D, and A.

Sector XI

The equilibrium crystallization of the liquids within this sector will differ from that of sectors IX and X, as the liquids not will reach r, but will instead leave the reaction curve.

A liquid of composition N will form C and after a small amount of C has formed it will reach pr at l (Fig. III-20). As the liquid now moves along pr, C reacts

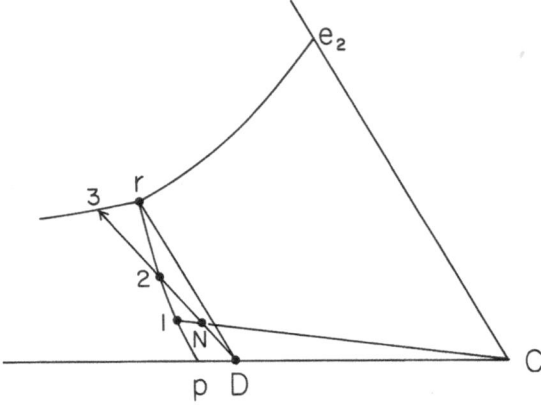

Fig. III-20. The crystallization within sector XI

with the liquid forming D. Let us consider the crystallization along pr using Swanson's II rule. The line joining the composition of the liquid and N will intersect the join AC between D and C, after the liquid reaches 1. From 1 to 2 the intersection will move towards D, and when the liquid reaches 2 the line will pass through D (Fig. III-20). This means that the crystallizate will consist of D only, there is no solid C left at this stage. The liquid must, therefore, leave the reaction curve, and will begin to cross the liquidus field of D, and move towards the eutectic curve pe, which it will reach at 3. The liquid will, thus, follow the track N–1–2–3–e.

Sector XII

C is also on the liquidus in the small sector XII, and C will crystallize until the liquid reaches the reaction curve where D begins to form. The liquid will then move over the liquidus field of D to the eutectic curve e_3e where D is joined with A.

Sectors XIII and XIV

The crystallization in these two sectors will be eutectic with the modification that the composition of D is situated outside its own liquidus field. However, this will not influence the crystallization.

16 Fractional Crystallization

The fractional crystallization will be simpler than the equilibrium crystallization, by fractional crystallization all liquids will terminate at e.

Sectors I to IV

The fractional crystallizations within these sectors are identical to equilibrium crystallization, because the solids have a constant composition. The sectors are shown in Fig. III-21, and the crystallization sequences are listed in Table III-4.

Fig. III-21. The sectors of the peritectic system by fractional crystallization

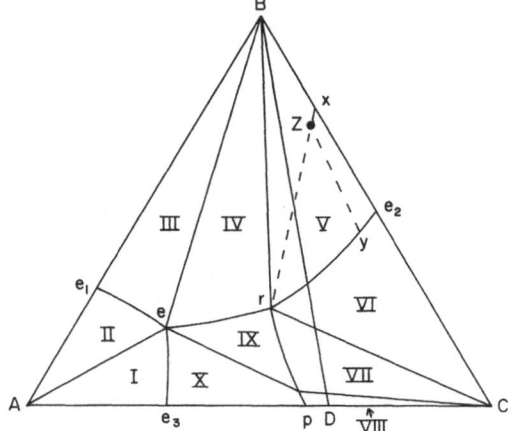

Table III-4. Crystallization sequences by fractional crystallization within a ternary peritectic system

I	II	III	IV	V
L	L	L	L	L
A+L	A+L	B+L	B+L	B+L
A+D+L	A+B+L	B+A+L	B+D+L	B+C+L
A+D+B+L	A+B+D+L	B+A+D+L	B+D+A+L	B+D+L
A+D+B	A+B+D	B+A+D	B+D+A	B+D+A+L
				B+D+A

VI	VII	VIII	IX	X
L	L	L	L	L
C+L	C+L	C+L	D+L	D+L
C+B+L	D+L	D+L	D+B+L	D+A+L
B+D+L	D+B+L	D+A+L	D+B+A+L	D+A+B+L
B+D+A+L	D+B+A+L	D+A+B+L	D+B+A	D+A+B
B+D+A	D+B+A	D+A+B		

Sectors V and VI

The fractional crystallization within these two sectors are similar except that B forms first in V and C forms first in VI. Within sector V solid B will crystallize first and the liquid will reach the eutectic curve e_2r. Thereafter C and B will fractionate together and the liquid will reach r. At r no crystals will be present as they are fractionated out of the liquid. The liquid will pass r while C stops its crystallization and D is formed instead. The presence of the reaction point will, thus, be evident from a sudden change in the solid phase assemblage which will change from L, S (B, C) to L, S (B, D). Such a change is not possible by eutectic crystallization where a new phase is added to the phase assemblage. At a reaction point one solid is replaced by another solid. The liquid will move along r–e, while B and D crystallize. Finally, the liquid will end in e, where A will begin to crystallize.

Sectors VII and VIII

Within these sectors C forms first and the liquid subsequently moves to pr. At the reaction curve the liquid will start to form D instead of C, and it will move across the liquidus field of D until it reaches the pe curve, and then D and B will fractionate out of the liquid. At e solid A will form part of the fractionate.

Within sector VIII the fractionation is essentially the same except that the liquid will reach the eutectic curve e_3e.

We have hitherto assumed perfect fractional crystallization, and that the crystals leave the liquid as soon as they are formed. In reality it will take some time before the crystals sink out of the liquid. The crystals of C that are present when the liquid reaches the reaction curve may react with the liquid so that a corona of D is formed around cores of C. Also it is likely that crystals of C may form nucleation centers for D so that new crystals of D form by heterogeneous nucleation on crystals of C. It may be noted that corona formation also may happen by eutectic crystallization, so that corona formation itself does not hint at a reaction relationship.

Sectors IX and X

The crystallization within this part of the system is simple eutectic crystallization. The liquids will move away from the composition of solid D and reach either curve re or curve e_3e.

17 Partial Melting of Composition Z

The lherzolitic mantle consists mainly of olivine and enstatite with minor amounts of diopside and garnet or spinel. The phase relations of the system considered here may elucidate aspects of the partial melting process that occur in the mantle. Let B be olivine, C enstatite, and D diopside. The composition Z will then be a "lherzolitic" composition.

The composition of Z is within the liquidus field of B, and to the right side of the division line BD. The first melt will form at r. The liquid will remain at r until solid D has become completely melted. When this happens the system will consist of B, C, and liquid. The degree of partial melting required for complete melting of D can be determined by drawing a line through r and Z (Fig. III-21). This line intersects the BC join at x. Using the lever rules the amount of liquid is calculated as 16% and the solid residuum contains 76% B and 24% C. The liquid will by increasing temperature move towards e_2, and it will leave the eutectic curve re_2 at y. Here there is 50% liquid present. It is apparent from Fig. III-21 that the intersection between the line through Z and liquid will intersect the BC join progressively nearer to B, as the liquid moves from r to y. At y there is no C left in the system. Finally, the liquid will move from y to Z.

The minerals of lherzolite display solid solution and the reaction point r will be replaced by a reaction curve. However, the variation in composition of the first formed melts will be small if the melting occurs at dry conditions.

18 Solid Solution

Most minerals display solid solution, and many igneous minerals are solid solutions of three or more components. The composition of magmas will both be controlled by the eutectic and peritectic phase relations, as well as the phase relations of solid solution. By granitic and andesitic rocks plagioclase and augite display the largest range of solid solution. By basaltic rocks extensive solid solution is displayed by olivine, pyroxenes, and plagioclase. The minerals of peridotitic rocks like lherzolite and wehrlite have a limited range of solid solution with $MgO/(MgO + FeO)$ ratios of about 0.9. However, spinel and garnet may vary in composition.

19 One Solid Phase

The liquidus and solidus surfaces of ternary systems form curved surfaces which either are monotoneous or have a maximum or a minimum. The most simple ternary system with solid solution has a monotonous increase in temperature and only one solid phase (Fig. III-22). The liquidus surface is situated above the solidus surface, and the two surfaces meet at the three corners at the melting points of the three end members. A solid of a given composition can only be in equilibrium with a liquid with a specific composition, so that tie lines between solid and liquid do not intersect.

A characteristic feature of equilibrium crystallization is the constant composition of the entire crystallizing system. Equilibrium crystallization is a closed

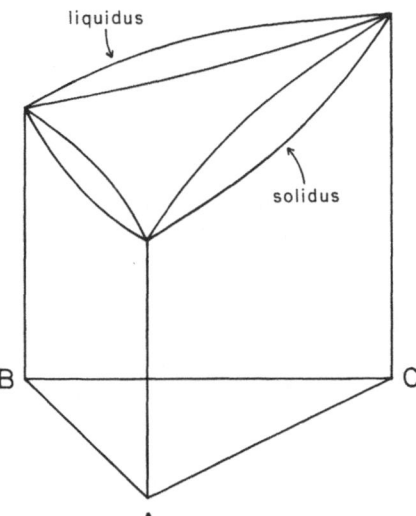

Fig. III-22. The space diagram for a ternary system with total solid solution. The system only contains two phases, a solid phase and a liquid phase

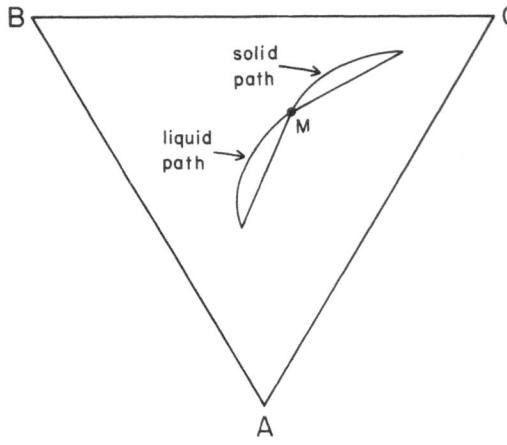

Fig. III-23. The solid and liquid paths by equilibrium crystallization. The detailed relationships of the crystallization are shown in Fig. III-24

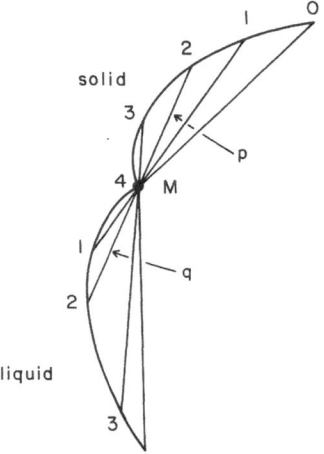

Fig. III-24. The determination of the crystallization path by equilibrium crystallization. All tie lines between solid and liquid have to pass through the original composition M at any temperature

system process, while fractional crystallization is an open system process. The constant composition criterion defines the crystallization path by equilibrium crystallization. The tie lines between solid and liquid should always pass through the initial composition of the liquid.

We will now consider the crystallization of the mixture M shown in Fig. III-23. In this system C has the highest melting point, and A the lowest one. The first formed solid will, therefore, contain mostly C, and more B than A. The composition of the first formed solid is shown at 0 in Fig. III-24. After some equilibrium crystallization has occurred the liquid will have lost more of component C than of A and B, and the liquid will mainly have moved away from the C corner. During the crystallization the solid will move along a curve like s_0-s_4, while the liquid moves along $M-l_4$. The tie lines will at any temperature pass through M. The relative

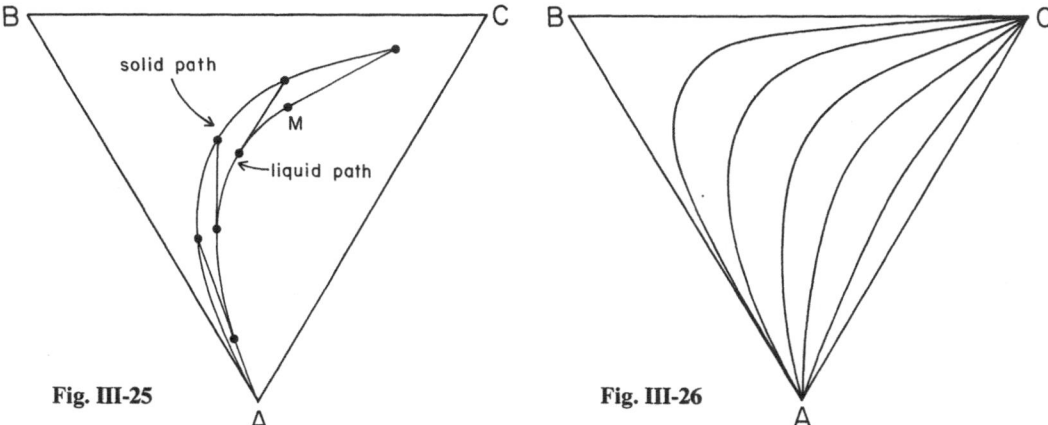

Fig. III-25

Fig. III-26

Fig. III-25. The solid and liquid paths by fractional crystallization. The tie lines between liquid and solid form tangents to the liquid curve

Fig. III-26. The general fractionation curves for a ternary system with total solid solution. C has the highest melting temperature and A has the lowest melting point

amounts of solid and liquid can be estimated using the lever rule. At stage 2 the amount of liquid will be given by the ratio $p/(p+q)$.

The crystallization paths of the liquid and the solid can only be estimated when the liquidus and solidus surfaces have been estimated by experiments. As yet only a very few ternary systems with solid solution have been estimated completely, and the position of tie lines can only be determined approximately. If silicate melts were ideal solutions one could calculate the equilibrium compositions, but silicate melts are generally not ideal solutions. An excellent polymerization model has been proposed by Burnham (1979) which will give accurate results for some silicate melts. We will later see how one can obtain an approximate estimate for the crystallization path by extrapolation of the binary phase relations.

During fractional crystallization the composition of the liquid will always move away from the composition of the solid. The tie lines between the liquid and solid will form tangents to the curve which display liquid compositions (Fig. III-25). The curves for the solid and liquid compositions of mixture M are shown in Fig. III-25. The liquid will move from M to the A apex as solid A has the lowest melting point within the system. The composition of the solid will vary from s_0 to A, and the solid curve will be situated on the B side of the liquid curve, as B is more refractory than A.

By fractional crystallization it is possible to estimate general fractionation curves that are valid within the entire system (Fig. III-26). The fractionation paths will be independent on how much liquid is present, and a liquid of a given composition will always follow the same path. In this respect fractional crystallization is simpler than equilibrium crystallization where each liquid composition has its own path, because the initial composition of the liquid controls the variation in compositions.

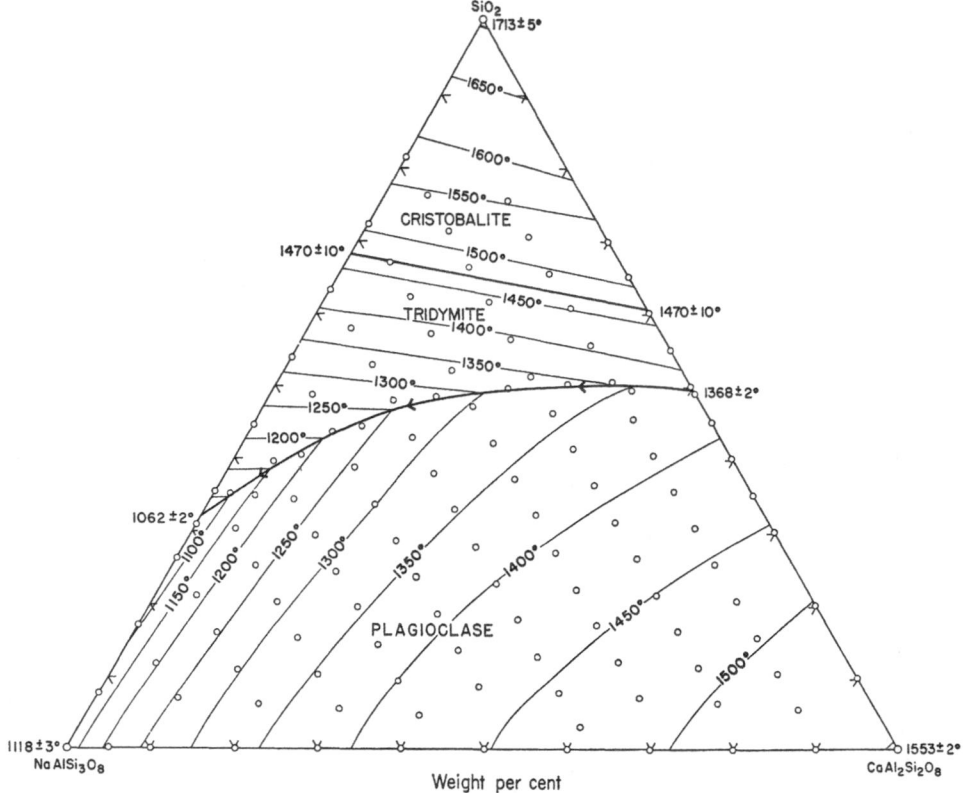

Fig. III-27. The albite-anorthite-silica system at 1 bar (Schairer 1957, Yoder 1967)

20 Albite–Anorthite–Silica System

The crystallization behavior of a system where only one solid displays solid solution may be demonstrated from the Ab–An–SiO₂ system. The phase relations of this system at 1 bar is shown in Fig. III-27 (Schairer 1957; Yoder 1967). Albite and anorthite display solid solution, while both form eutectic systems with silica.

The determination of crystallization paths within the liquidus field of plagioclase can only be made if we know the equilibrium compositions within the entire field. A systematical investigation of the equilibrium compositions of the ternary compositions has not yet been made, but we can obtain approximate estimates of the crystallization paths if we assume ideal solution between plagioclase and silica liquids. Plagioclase and silica liquids do not form ideal solutions, but the deviation from ideal solution is minor. Accurate estimates can be done using the polymerization model by Burnham (1979). Here we will assume ideal solution for simplicity, but it is emphasized that this simplifying approach cannot be used generally.

21 Equilibrium Crystallization

Let us consider the equilibrium crystallization of a mixture consisting of 40% Ab, 40% An, and 20% SiO₂. The weight percentage of anorthite in the plagioclase part

Fig. III-28. Equilibrium crystallization of composition M
within the system albite–anorthite–silica

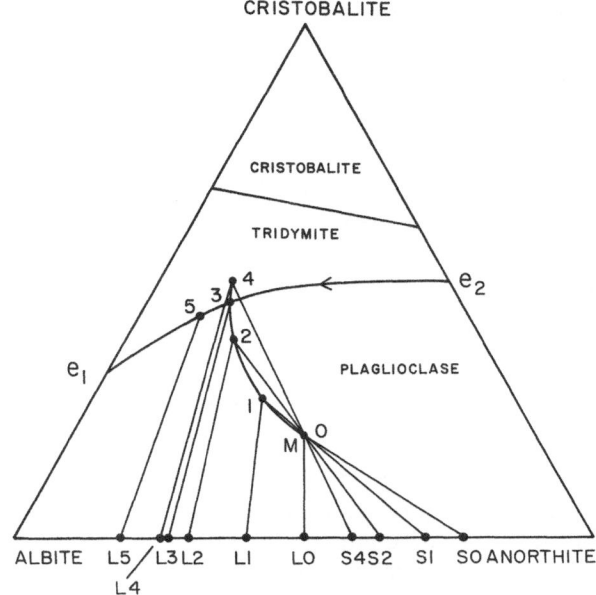

is then 50% An (Fig. III-28). The phase relations of the binary Ab–An system will
now be used for an estimate of the crystallization path. The phase diagram could be
used, but it is convenient to use the following equation which was estimated from
the phase relations:

$$\frac{x_1}{x_s} = a\,x_1 + b\,\sqrt{x_1} + c \qquad (7)$$

where:

a = 0.37
b = 0.58
c = 0.05

Here x_1 and x_s are in mole fractions. We must, therefore, recalculate the mole fractions to weight fractions and vice versa, as the phase diagram is in % (Fig. III-27). A
plagioclase with 50 wt% An contains 48.52 mol% An. Using Eq. (7) the anorthite
content of the first formed solid is calculated as 76.59 mol% An and 77.63 wt% An.

After some crystallization the anorthite content of the liquid will be 40 wt% An,
and the position of the liquid within the ternary system can be estimated from the
intersection between two lines. The liquid L1 with 40 wt% An is in equilibrium with
plagioclase containing 71.01 wt% An (S1) (Table III-5). The tie line between liquid
and solid has to pass through M, so we draw a line through M and S1. We also
draw a line between the silica apex and the point indicating 40% An on the Ab–An
join as all liquids containing 40 wt% An will fall on this line. The composition of L1
is then given by the intersection between these two lines at 1. At stage 2 the liquid
contains 30 wt% An. Using Eq. (7) we can again estimate the composition of
plagioclase, and by repeating the above construction the composition of L2 may be

Table III-5. Compositions obtained by equilibrium crystallization within the Ab–An–SiO$_2$ system

Stage	Wt%		Mol%		Liquid SiO$_2$ wt%	Liquid left (g)
	Liquid $\dfrac{An}{An+Ab}$	Solid $\dfrac{An}{An+Ab}$	Liquid $\dfrac{An}{An+Ab}$	Solid $\dfrac{An}{An+Ab}$		
0	50	77.63	48.52	76.59	20.0	100
1	40	71.01	38.59	69.77	27.1	74
2	30	62.92	28.77	61.33	38.5	52
3	26.6	59.71	25.46	58.28	45.8	44
4	25	58.11	23.91	56.65	50.3	
5	18	50.00	17.17	48.52	43.1	0

estimated. This procedure may be repeated until stage 4, where the estimated composition is situated within the liquidus field of tridymite. The crystallization path is given by the curve through M–1–2–3–4, and the curve intersects the eutectic curve at 3, so that crystallization within the liquidus field of plagioclase stops here. The anorthite content of the liquid at stage 3 is estimated by drawing a line from the SiO$_2$ apex through the intersection point to the Ab–An join.

After the liquid has reached the eutectic curve it will move from 3 towards e_1. The crystallization will come to an end when the plagioclase contains 50 wt% An, and this happens at stage 5 where the liquid contains 18.0 wt% An.

22 Fractional Crystallization

The path of the liquid can only be estimated from solutions of the Rayleigh equation when the crystallization is fractional. The solution for the binary Ab–An system is given in Chap. X and we will apply this equation here, assuming that it also is valid for the ternary liquids. Using the solution to the Rayleigh equation one first calculates the amount of plagioclase component left. The initial amount of this component is here 80 g, and the fractionation equation will show the amount left of this initial amount (Table III-6). Having estimated the amount of plagioclase left we can also estimate the wt% of SiO$_2$ in the liquid, as the absolute amount of the silica component remains constant (20 g). The wt% An in the plagioclase part of the liquid was our starting point for the calculations, and we now also know the silica content. The position of the liquids within the ternary system is then estimated by drawing lines from the SiO$_2$ apex to the various anorthite contents of the liquid on the Ab–An join (Fig. III-29). The positions of the liquids are given by the intersections between these lines and the silica wt% listed in Table III-6. The curve through these points is the fractionation curve of the liquid. This curve intersects the eutectic curve, and the fractionation of plagioclase alone will proceed until this intersection. Thereafter both plagioclase and tridymite will fractionate, while the liquid moves to e_1. The intersection occurs when the plagioclase part of the liquid contains 16.1 wt% An, while the intersection by equilibrium crystallization occurred

Table III-6. The variations in composition of liquid and plagioclase by fractional crystallization

An/An + Ab				Mol pl. in liq.	Gram pl. in liq.	Wt% SiO₂ in liq.
Mol%		Wt%				
x_l	x_s	x_l	x_s			
48.52	76.59	50.00	77.63	0.2963	80.00	20.00
40.00	70.82	41.43	72.03	0.2220	59.63	25.12
30.00	62.67	31.26	64.05	0.1622	43.31	31.59
20.00	52.17	20.96	53.64	0.1194	31.70	38.68
10.00	36.98	10.55	38.37	0.0866	22.60	·46.95

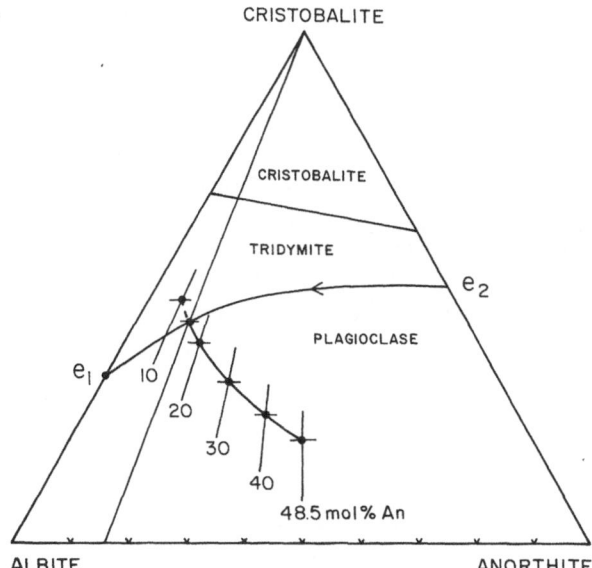

Fig. III-29. Fractional crystallization of composition *M,* containing 40% albite, 40% anorthite, and 20% silica

at 18.0 wt% An. The difference between the equilibrium path and the fractional path is, thus, small in the present case.

23 Solid Solution – Two Solids

When two solids display limited solid solution in a ternary system, the phase relations become similar to those of a binary eutectic system (Fig. III-30). When two solids are in equilibrium with liquid there will be three phases present, and the variance of the system will be 1 according to Gibbs' phase rule. There will, therefore, not be an eutectic point within the system, instead the liquids of the three phase equilibria will be situated along an eutectic curve.

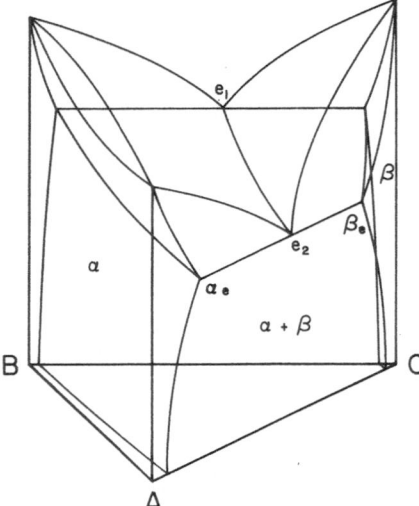

Fig. III-30. The space diagram of a ternary system with two solid phases

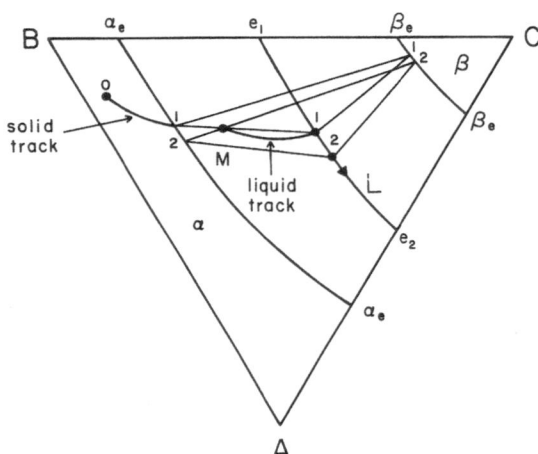

Fig. III-31. Equilibrium crystallization within the ternary system with two solid phases. The initial liquid composition is M, and the composition of the first formed solid phase is at 0

In the system shown in Fig. III-30, there is continuous solid solution between A and B, and limited solid solution between A and C, and B and C. The eutectic point of the binary system BC is e_1, and that of AC is e_2. The temperature decreases from e_1 to e_2, and e_2 is the minimum point for the entire system. Liquids along $e_1 e_2$ are in equilibrium with α_e and β_e, and the equilibrium compositions of the liquid are situated on the low temperature side of the join α_e–β_e. Thus, the three phase equilibria form triangles within the system (Fig. III-31).

24 Equilibrium Crystallization

Composition M is situated within the liquidus field of α, and the first formed α may have a composition at 0. The liquid will move towards L1, while the composition of

Fig. III-32. The relationship between the tie lines between the two solid phases α_e and β_e and the initial composition M. When the tie line intersects the initial composition the crystallization is completed (at *3*)

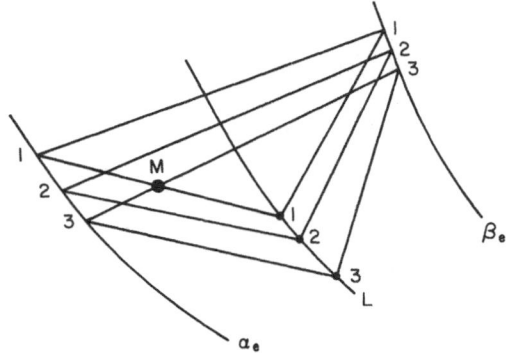

α moves towards $\alpha 1$. When the liquid reaches the eutectic curve at L1, β will begin to crystallize, and the liquid will thereafter move towards L2. The crystallization is completed when the tie line between α and β intersects the initial composition M, and the final liquid has composition L2.

This crystallization behavior is shown in more detail in Fig. III-32 which depicts three stages during the crystallization. When α attains a composition of the α_e curve at stage 1, β begins to crystallize. At stage two the system contains α, β, and liquid. With decreasing temperature the tie line between α and β moves towards M, and the amount of liquid decreases. When the tie line intersects the composition of M, the phase assemblage will consist of only α and β, and there will be no liquid left. The line that defines the amount of liquid passes through the composition of the liquid and M, and intersects the tie line between the two solids.

By partial melting the liquid will as usual follow exactly the same path as by equilibrium crystallization, but in the opposite direction.

It may be noted that the composition of the three phases fall on a line within the binary system, while the assemblage forms a triangle within the ternary system. As the binary system BC and AC is approached, the equilibrium triangles become more and more narrow until they form a line at e_1 and e_2, respectively.

25 Fractional Crystallization

By fractional crystallization of a liquid with composition M the liquid will move from M to L1 while the composition of α moves from α_0 to α_1. The tangent to the liquid curve will pass through the composition of α (Fig. III-33). At stage 1 β begins to crystallize and the liquid will thereafter move towards e_2. When the liquid reaches e_2 both α and β will be mixtures of only A and C.

26 Solid Solution – Peritectic System

A ternary peritectic system with solid solution is shown in Fig. III-34. By comparison with the binary system (Figs. III-2, III-17), it is evident that the phase relations are similar. The projection of the system is shown in Fig. III-35, and we will now consider the equilibrium crystallization of composition M.

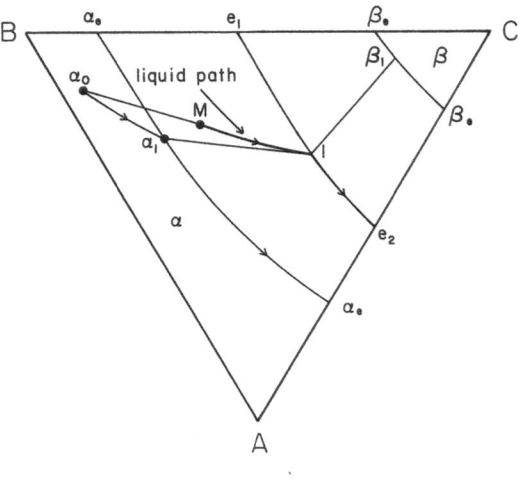

Fig. III-33. Fractional crystallization of composition M. The composition of the liquid moves from M to e_2, and the composition of the solid phase α moves from α_0 to α_e, and that of β moves from β_1 to β_e

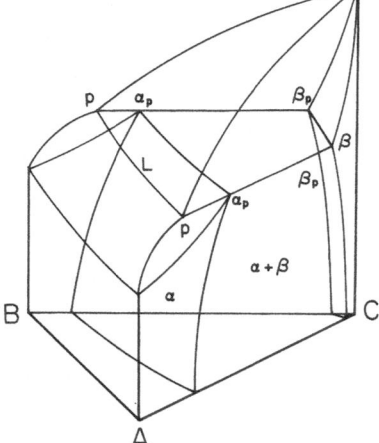

Fig. III-34. The space diagram of a ternary peritectic system with two solid phases. The *curve pp* is the reaction curve that displays the compositions of the liquid

A liquid with composition M is situated within the liquidus field of β, and the first formed solid phase is therefore β, which may have a composition like β_0. The liquid will by decreasing temperatures move towards the reaction curve p_1p_2, which it reaches at L_1, where α begins to crystallize. The liquid will then move along the reaction curve, while α and β crystallize. At stage 3 the tie line between α_3 and β_3 intersects the total composition M, and the last trace of liquid must disappear at this stage. The lever rules may be applied in the usual manner for an estimate of the amount of liquid and the composition of the crystallizate.

The peritectic system with solid solution may elucidate some of the phase relationships of the mantle undergoing partial melting. The lherzolitic mantle consists of olivine, enstatite, diopside, and spinel or garnet. Enstatite displays a reaction relationship with the liquid at high pressures: $En + L = Di$. From the binary phase relations we would expect that liquids generated at the peritectic curve will not contain a phase like enstatite, and that is in fact the case.

Fig. III-35. The crystallization of composition M within the ternary peritectic system

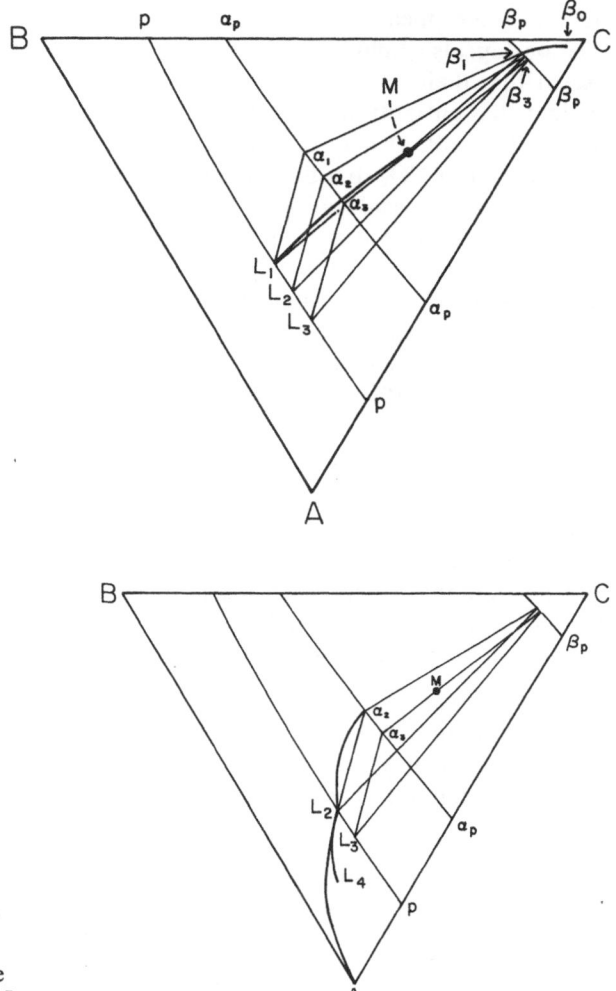

Fig. III-36. The partial melting of composition M. The first formed liquid has composition L_3, and the liquid will move from L_3 to M via L_2

Consider that a solid phase assemblage consists of α and β and has composition M (Fig. III-36). The first formed melt must have composition L_3 (cf. Fig. III-35). By about 28% melting the liquid will have attained composition L_2. This melt is now removed from its source, in the same manner as a primary magma is removed from its source when it begins to ascend. If the melt L_2 begins to crystallize by equilibrium crystallization, then the liquid will move from L_2 to L_4 where it becomes completely solidified. The composition of α simultaneously moves from α_2 to L_2. By fractional crystallization the liquid L_2 will move from L_2 to the A apex, and the final liquid will consist of pure A.

By the experimental investigations of a supposed primary magma composition, modelled here by L_2, one will observe the following phase relations. The subsolidus phase assemblage will consist of α. When α begins to melt the first formed liquid will have composition L_4. The liquid will thereafter move to L_2, where the last trace of α disappears, and there will be no evidence whatsoever of a β phase. This relationship also holds for liquids that have undergone fractional crystallization, except that their liquidus temperatures will be below that of L_2. By analogy it is evident that enstatite will be absent from the melting interval of primary magmas formed by small degrees of partial melting. However, it should be mentioned that the addition of CO_2 to a lherzolitic composition may change the reaction relationship to eutectic crystallization.

From Fig. III-35 it is evident that all liquids generated at higher temperatures than L_1 must contain the β phase on their liquidi. The phase with a reaction relationship will, thus, enter the melting interval when the partial melting has become sufficiently high. So far only komatiites have enstatite within their melting interval, and the komatiitic magmas must have formed at high degrees of partial melting of about 20% to 50%.

27 Types of Univariant Curves in Ternary Systems

For far the majority of ternary systems the univariant curves maintain their eutectic or peritectic character along their entire expiration. However, exceptions exist and it is possible that some phase reactions in magmas may be related to a change in the character of the univariant curves. In general, there are four types of univariant curves in ternary systems (Fig. III-37):

1. eutectic only;
2. transition from eutectic to peritectic;
3. peritectic only;
4. transition from peritectic to eutectic.

In Fig. III-37 the s-curves show the variation in composition of the two solids α and γ which are in equilibrium with liquid. The arrows show the direction of falling temperature, and one arrow indicates eutectic crystallization and two arrows a peritectic crystallization.

28 Eutectic Crystallization

The two solid phases α and γ crystallize from the liquid, and the liquid will move away from both α and γ as indicated by the two solid vectors $\bar{a}$ and $\bar{c}$. The reaction occurring along the univariant curve is:

$$l = \alpha + \gamma$$

When both solids crystallize from the liquid the crystallization is called positive. When the sign is the same, the crystallization is even, when the sign is + and – the crystallization is odd.

29 Eutectic–Peritectic Transition

Along the first part of the univariant curve the crystallization is eutectic and positive for both phases. At the point Q the crystallization changes from eutectic to

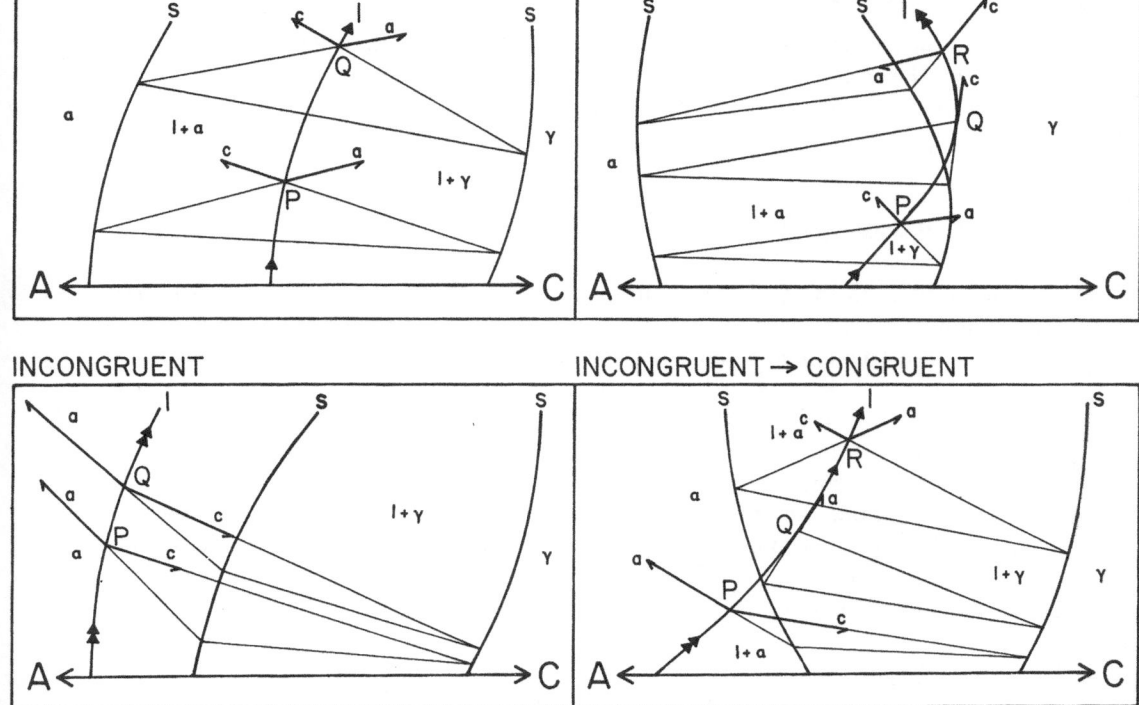

Fig. III-37. The four different types of isobaric univariant curves in a ternary system

peritectic, and from Q to R the curve is peritectic. So along the first part of the curve the following reaction takes place:

$$l = \alpha + \gamma$$

and from Q to R this reaction occurs:

$$l + \alpha = \gamma$$

Let us consider the crystallization along the curve in some detail. At P both α and γ crystallize and the tangent to the univariant curve intersects the α–γ join between α and γ. As the liquid moves away from P towards Q the tangent intersects the join nearer to γ. This means that progressively less α crystallizes from the melt, while the fraction of γ is large and nearly l. Just before Q, almost only γ crystallizes while a diminutive amount of α is formed. The decrease in the amount of α is caused by the curvature of the univariant curve. At the very point Q, the tangent to the univariant curve coincides with the tie line between the liquid and the γ phase, with the result that only γ crystallizes. Just after Q the crystallization becomes peritectic, and the solid α now begins to react with the liquid, and the crystallization of α is, therefore, called negative. Summarizing, the crystallization of γ is positive along the whole part of the curve, while the crystallization of α first is positive, and becomes negative after Q.

30 Peritectic Crystallization

The solid phase γ is reacting with the liquid forming α, so that this reaction occurs: $\gamma + L = \alpha$. The solid phase γ is resorbed along the univariant curve, while α crystallizes from the liquid. Thus, moving from P to Q, γ is resorbed while α is crystallizing.

31 Peritectic–Eutectic Transition

At the point P on the univariant curve, α is crystallizing while γ is resorbed. The resorption of γ will proceed until the point Q where γ begins to crystallize. This time it is also convenient to consider the behavior of the tangent to the univariant curve.

At P the tangent to the univariant curve intersects the tie line between α and γ to the left of the composition of α. As the liquid moves from P to Q this intersection moves towards α, and at the point Q the tangent intersects the composition of α. The crystallization, therefore, changes from eutectic to peritectic at Q, and the crystallization of γ becomes positive instead of negative. The crystallization of α remains positive from P to R. From P to Q the reaction is $\gamma + L = \alpha$, and from Q to R the reaction changes to $L = \gamma + \alpha$.

IV Pseudobinary Systems

1 Introduction

The pseudobinary systems differ from the true binary ones by the presence of one or more phases with compositions lying outside those of the binary join, i.e., the compositions of all the phases cannot be expressed as a linear combination of the two end member components. In the system albite–orthoclase, sanidine melts incongruently to leucite plus liquid, but the leucite composition cannot be expressed from those of the two feldspars. Similarly spinel occurs in the binary join anorthite–forsterite, but has a composition outside the join. The pseudobinary systems occur frequently in silicate systems and interpretation of these systems is required for an evaluation of important phase relationships. The pseudobinary systems also bear a relationship to P–T diagrams of rock compositions, since these P–T diagrams may be regarded as distorted pseudobinary systems.

2 Rhines Phase Rule

The phase rule of Gibbs is also valid for pseudobinary systems, however, the shape of the equilibrium fields for pseudobinary phase assemblages may have much more varied shapes than for the true binary systems, and their experimental determination, therefore, becomes more difficult. A theorem explicitly stated by Rhines (1956) is of general validity, but is very useful considering the pseudosystems:

Rhines' theorem: A field of the phase diagram, representing equilibrium among a given number N of phases, can be bounded only by regions representing equilibria among one or more or one less, i.e., $N \pm 1$, than the given number of phases N. In applying this rule, a univariant equilibrium curve should be considered a region.

3 Ternary Eutectic System

The three principal pseudobinary sections of a ternary eutectic system is shown in Fig. IV-1. Pseudobinary joins will normally be between two mineral compositions, and not arbitrarily between compositions as here, but the ternary eutectic system is convenient to start with.

The pseudobinary phase relations for the join I–C is shown in Fig. IV-Ia, and is constructed from the ternary system as follows. First, the compositional range for the liquidus fields is estimated. Here, there are two, one for A and one for C, and

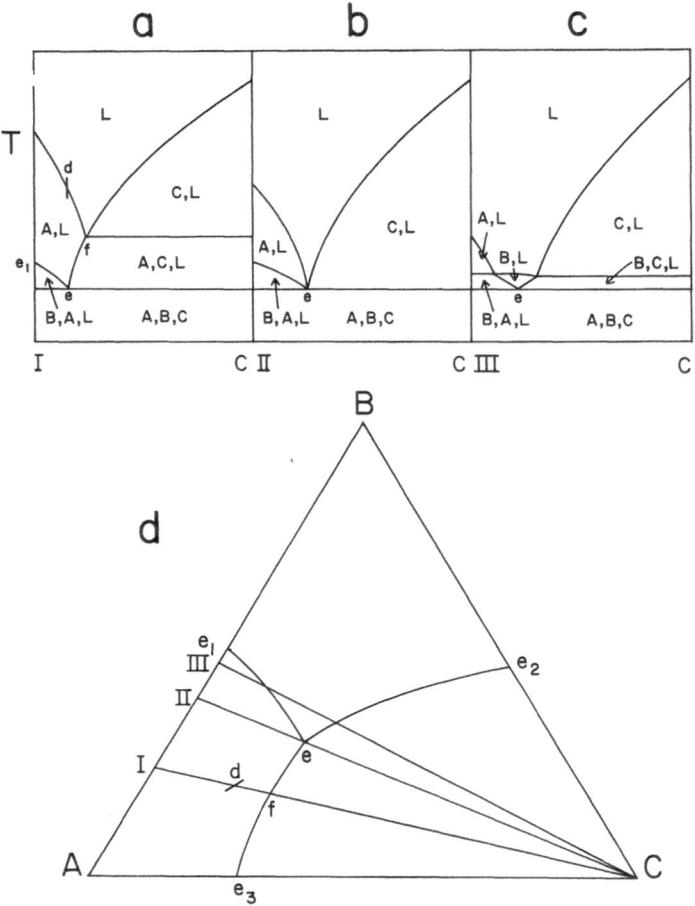

Fig. IV-1 a–d. The three principal pseudobinary sections of a ternary system. The positions of the sections *I-C, II-C,* and *III-C* are shown in diagram **d**

they meet at f where the join line I-C intersect the univariant curve e_3–e. The next step is an estimate of the composition ranges for which a second phase will crystallize. All compositions along f–C will first form C and then A. The sequence of crystallization for compositions between I and f will vary with composition. Compositions along I–d will first form A and then B, while compositions between d and f will display the sequence A–A, C. The actual shape of the boundary curves will depend on the particular phase diagram studied. The curvature shown in Fig. IV-1 a, will be typical for a system with ideal solution in the liquid state. The pseudobinary diagram for the join II-C passing through the ternary eutectic point is shown in Fig. IV-1 b. The join III-C cross three liquidus fields and the pseudobinary diagram is shown in Fig. IV-1 c.

In the true binary systems the composition of the liquid will always be moving away from one of the end components. This is not generally the case for pseudosystems. If a composition along f–C is crystallizing, it will at first move directly away

Fig. IV-2a, b. The pseudobinary section of a ternary system with congruent compound D

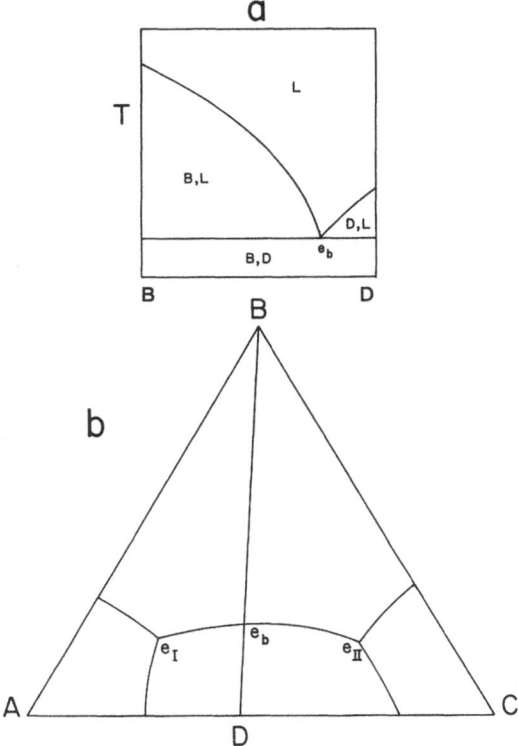

from C, however, during crystallization along f–e, the composition of the liquid will move towards B, and the content of A compared to C will be increasing as evident from Fig. IV-1 d. The curve f–e in Fig. IV-1 a do not show the composition of the liquid from f to e, but rather the boundary between the stability fields of L, S(A) and L, S(A, C). Because univariant curves generally have a course not being parallel to the join, and because solid phases with compositions outside the join, like B, might crystallize, it is evident that the course of the liquid cannot be estimated from the diagram itself.

4 Ternary System with a Congruent Compound

A ternary system with a congruent compound D is shown in Fig. IV-2 b. The join B–D is not a pseudobinary system as all compositions are situated on the join, but it is the most simple join within a ternary system.

5 Ternary System with an Incongruent Compound

A ternary system with an incongruent compound D is shown in Fig. IV-3 c. The crystallization along r–e is even, while the crystallization is odd along p–r because of the reaction $C + L = D$. All liquids within the triangle ABD will end up in e both by

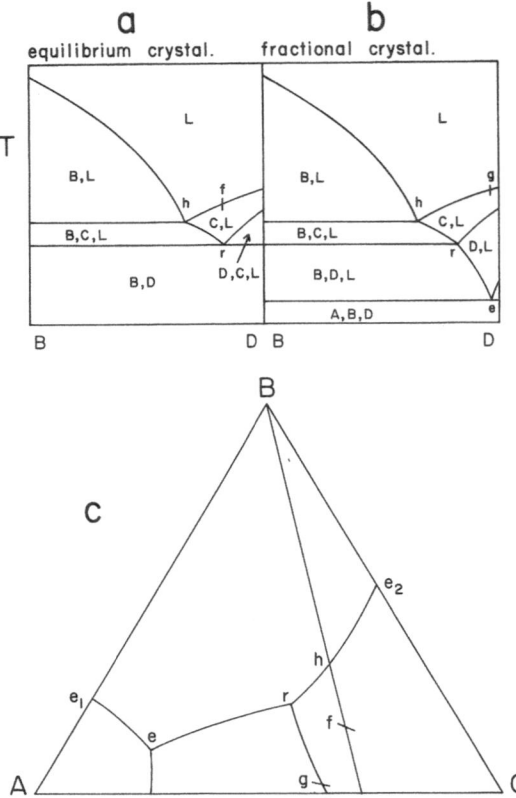

Fig. IV-3a–c. Diagrams **a** and **b** show the pseudobinary sections by equilibrium and fractional crystallization, respectively, for a ternary system with an incongruent compound *D*. In **c** the eutectic points are labeled *e*, and *f, g*, and *h* are auxillary points for the construction of the pseudobinary sections .

equilibrium crystallization and fractional crystallization. Liquids with initial compositions on the join BD will end up in r by equilibrium crystallization and in e by fractional crystallization. Liquids within the triangle BCD will end in the reaction point r by equilibrium crystallization and in e by fractional crystallization.

The pseudobinary join for BD is shown in Fig. IV-3a for equilibrium crystallization and is estimated as follows. By equilibrium crystallization the liquids between B and h will first crystallize B and then (B + C) from h to r. At r all C is resorbed in the liquid, and D is formed. Liquids between h and f will first crystallize C and then (C + B). At r all C will again be completely resorbed whereby D is formed. The liquid with composition f will only crystallize C from f to r, where all C is resorbed. Liquids between f and D will crystallize C and then (C + D) along p–r. When the liquid reaches r, B will begin to form and C will react out of the liquid.

Applying Rhines' theorem it will be evident that the phase field differs by ±1 phase. The field for 1, S(B, C) is surrounded by the fields L, S(B), L, S(C), and the isobaric invariant equilibria L, S(B, C, D).

The pseudobinary diagram for fractional crystallization is shown in Fig. IV-3b. The main difference is that all liquids will end up in e, and the D-rich liquids will pass over the liquidus field for D.

Fig. IV-4. A ternary system with two incongruent compounds D and E. The point r is a double reaction point, i.e., r is the reaction point for both D and E. The liquidus field for D is to the left of $r-e_1$, and the liquidus field for E is to the right of this curve. There is a thermal maximum between m between e_1 and e_2. The pseudobinary sections for $A-E$ and $D-C$ are shown in Fig. IV-5

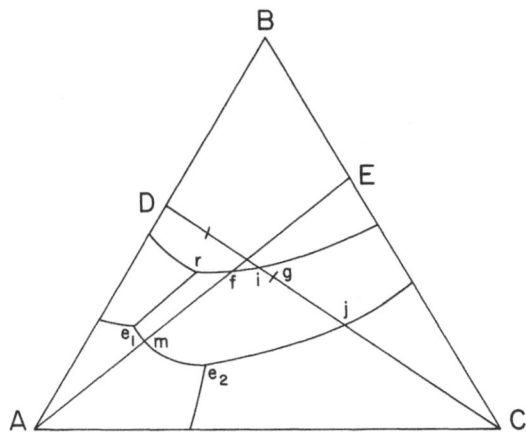

6 Ternary System with Two Incongruent Compounds

The pseudobinary systems may be quite complicated, and their experimental evaluation difficult. The liquidus projection for the system shown in Fig. IV-4 appears quite simple, but one of the joins is nevertheless a rather complicated pseudobinary system. The pseudobinary join A–E is simple, while the join D–C displays several phase fields. By the experimental determinations of the phase relations of silicate systems with reaction relationships, true thermodynamic equilibrium might be difficult to obtain. An experimental determination of a join like D–C might result in a confusing mixture of equilibrium and fractional crystallization (Fig. IV- 5b and c). The evaluation of the pseudobinary joins shown in Fig. IV-5 is left as an interesting exercise for the reader.

7 Pseudobinary Joins Within the Ternary System Wollastonite–Enstatite–Alumina

As an example of a pseudobinary system, two joins within the system wollastonite (wo)–enstatite (en)–alumina (al) will be considered. This system has bearings on the phase relations of the mantle and the generation of eclogite.

The join diopside–pyrope is pseudobinary at all pressures up to at least 40 kbar, and is shown at 30 kbar in Fig. IV-6 (O'Hara 1963). The system has also been estimated at 0 kbar (O'Hara and Schairer 1963) and 40 kbar (Davis 1965). There are in all four solid phases present, diopside, enstatite, garnet, and spinel. Obviously enstatite and spinel do not have compositions on the join (Fig. IV-7), and the system must, therefore, be pseudobinary. Whether or not diopside and garnet have compositions on the join cannot be evaluated from the phase relations of the join, but in general the compositions of minerals displaying solid solutions will not be on the joins, when these are pseudobinary. The compositions of both diopside and garnet solid solutions are outside the join. As the mineral compositions are outside the join, and the total composition of the phase assemblages are on the join, the compo-

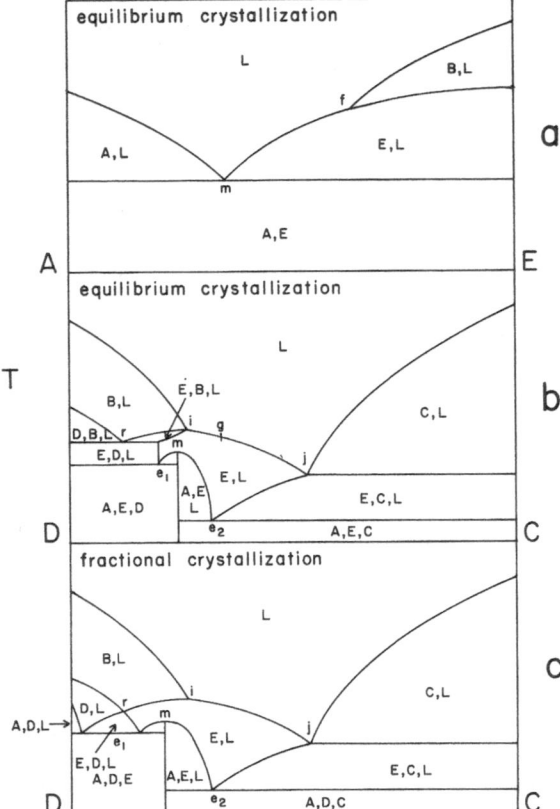

Fig. IV-5 a–c. The pseudobinary sections for the ternary system shown in Fig. IV-4. **a** and **b** Equilibrium crystallization; and **c** fractional crystallization

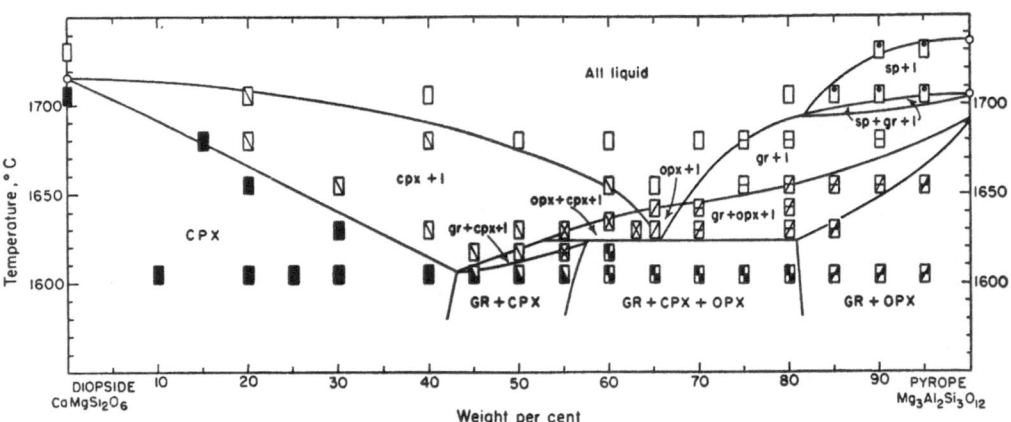

Fig. IV-6. The pseudobinary section for the system diopside–pyrope at 30 kbar (O'Hara and Schairer 1963). The abbreviations are as follows: *cpx* clinopyroxene; *gr* garnet; *opx* orthopyroxene; *sp* spinel; and *l* liquid

Fig. IV-7. The positions of different mineral compositions within the CMAS-system. Gehlenite and anorthite have compositions situated in the plane for CaO-Al$_2$O$_3$-SiO$_2$

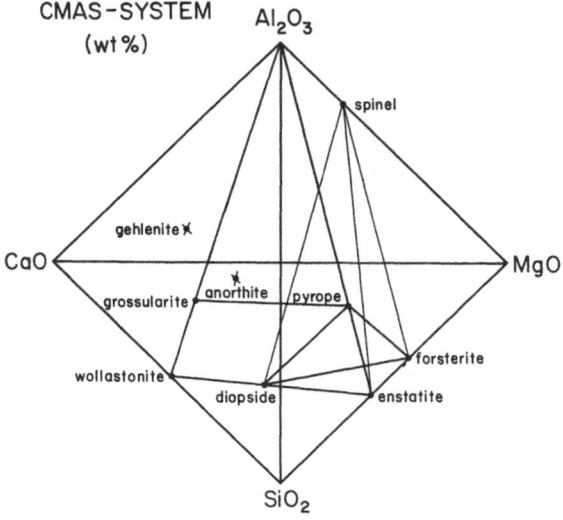

CMAS-SYSTEM
(wt %)

Al$_2$O$_3$

spinel

gehlenite ✕

CaO — MgO

grossularite — anorthite — pyrope

forsterite

wollastonite

diopside — enstatite

SiO$_2$

sitions of the liquids involved must also be outside the join. The boundary curve separating the phase fields for cpx and cpx + l is, therefore, not a solidus curve in the general sense, and the same applies for the curve separating the fields for gar + opx + l and gar + opx.

Applying Rhines' phase rule it will be evident that this rule is fulfilled for all the phase fields. Apparently an exception is the horizontal line separating the fields gar + opx + l and gar + cpx + opx. At this line the four phases gar, cpx, opx, l are in equilibrium. The compositions of these and spinel may be defined within the CMAS-system (CaO–MgO–Al$_2$O$_3$–SiO$_2$), which is a four component system. An isobaric univariant equilibria in a four component system involves five phases according to Gibbs' phase rule. The horizontal line may, thus, appear in error, however, this is not the case. The compositions of the four phases may all be defined within the ternary system wo–en–al. The diopside–pyrope system is true ternary in this part of the system, and the horizontal line represents a univariant equilibria. As spinel also is stable within the join, the whole join becomes a pseudobinary join within a quaternary system, but the order of the system may be lower in parts of the system. This is a general feature of a pseudobinary system, the order of the system may be different in various parts of the system.

Spinel is stable at high temperature in the pyrope-rich part of the system, but is absent in the subsolidus assemblage. Spinel, thus, displays a reaction relationship, the reaction being sp + l = gar. It is evident from the central part of the system that enstatite also has a reaction relationship, the phase field for opx, cpx, l extends over the field for gar, cpx, and the reaction is opx + l = cpx.

The join intersects four liquidus fields. The point where the liquidus fields for spinel and garnet meet must be a piercing point for a reaction curve. The point where the liquidus fields for diopside and enstatite meet must also be a piercing point for a reaction curve. The liquidus field for enstatite is quite small and the ternary invariant reaction point for the equilibria L, S(gar, opx, cpx) must be situated near the join.

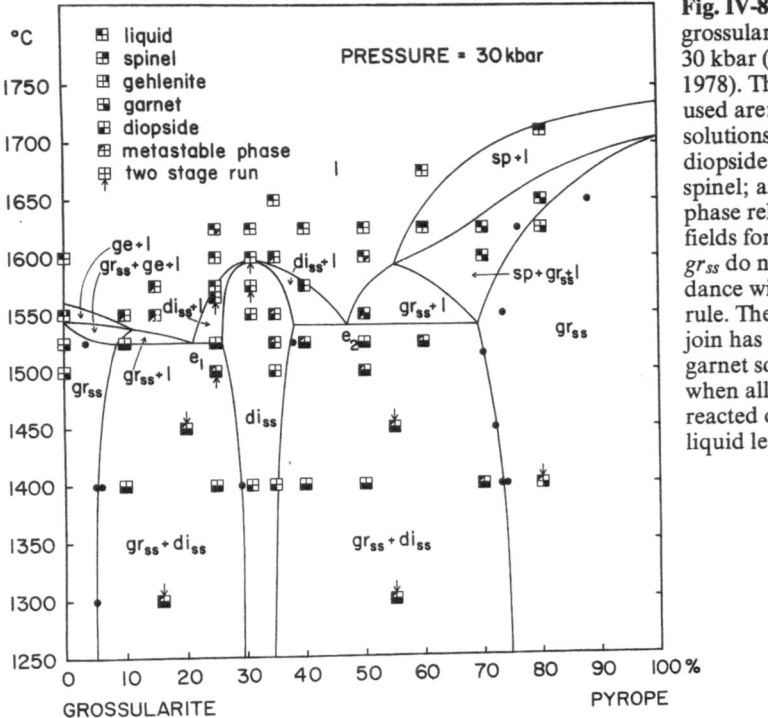

Fig. IV-8. The system grossularite–pyrope at 30 kbar (Maaløe and Wyllie 1978). The abbreviations used are: *gr_ss* garnet solid solutions; *ge* gehlenite; *di_ss* diopside solid solutions; *sp* spinel; and *l* liquid. The phase relations between the fields for *sp* + *gr_ss* + *l* and *gr_ss* do not appear in accordance with Rhines' phase rule. The reason is that the join has the composition of garnet solid solutions, i.e., when all the spinel has reacted out then there is no liquid left

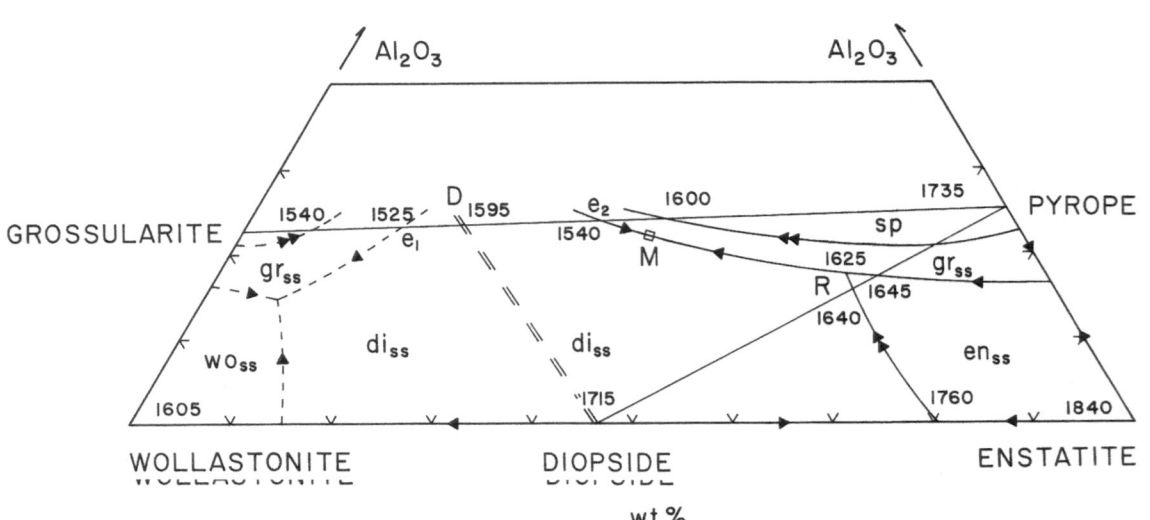

Fig. IV-9. The liquidus projection for part of the system CaSiO₃–Al₂O₃–MgSiO₃ at 30 kbar (Maaløe and Wyllie 1978). *R* is a reaction point, and *M* is a thermal minimum point. *Double arrows* indicate curves of odd reaction, and *single arrows* curves of even reaction. The stippled line diopside–D is a thermal maximum curve

Further evaluation of the phase relations for the system requires a determination of the binary joins wollastonite–diopside, pyrope–enstatite, wollastonite–grossularite, and grossularite–pyrope. The latter join is also pseudobinary, and is shown in Fig. IV-8. The run data for this system show that complete experimental equilibrium was not obtained in all parts of the system, the phase relations for the join was worked out using Rhines' phase rule and the boundary rule. The compositions of gehlenite and spinel are not situated on the join, and the compositions of all the solid phases have to be defined within the quaternary CMAS-system. The join grossularite-pyrope is, therefore, a pseudobinary join within a quaternary system. This join differs in one important respect from the diopside–pyrope join, the compositions of both diopside and garnet are situated on the join. The two eutectics at e_1 and e_2 are, therefore, thermal maxima on the univariant curves $L, S(gr_{ss}, di_{ss})$ and $L, S(py_{ss}, di_{ss})$ within the ternary system wollastonite–enstatite–alumina. Combining available experimental data, the liquidus projection for the ternary system might be worked out (Fig. IV-9). Some of the phase relations for this system were actually worked out by O'Hara (1963), on the basis of the diopside–pyrope system alone. The liquidus projection is based on the piercing points of the pseudobinary joins, and the liquidus relations of the binary joins (Maaløe and Wyllie 1978). The present example demonstrates how the phase relations of petrological systems are built up from binary systems, considering the number of experiments required for the determination of these binary systems, it may be understood that the determination of phase relations is a slow and tedious process.

V P–T Diagrams

1 Pressure–Temperature Diagrams

The pressure–temperature diagrams of rock compositions display the stability fields for the different phase assemblages at various pressures and temperatures. These diagrams afford information about the possible natural phase relations for rocks and magmas, and indicate the possible combinations of phenocrysts that might constitute a fractionate, and might as well show the P–T regions for the generation of a particular composition by partial melting. The disadvantage of the P–T diagrams for natural compositions is that the particular phase relations estimated, is only valid for the composition investigated. If an incongruent reaction occurs for a given composition, it is not granted that the reaction also occurs for a similar but slightly different composition. The influence of a small difference in composition on the phase relations is especially evident from systems with congruent compounds and thermal divides. The synthetic systems with pure end member compositions will clarify the fundamental phase relations, and indicate the compositional ranges for which a particular phase relation is valid. On the other hand, the synthetic systems do not define accurate P–T values for the natural magmas. Thus, the two types of systems supplement each other, the fundamental phase relations are evident from synthetic systems, while accurate estimates of pressure and temperatures may be obtained from experiments on natural compositions at different P–T values.

The rationale of P–T diagrams may be understood by comparison with the pseudosystems. A pseudosystem shows the phase relations for a join at various temperatures and compositions. In a P–T diagram the composition is constant, while the phase relations vary around the compositional point as temperature and pressure changes.

2 Eutectic System

The eutectic system with solid phases A and B shown in Fig. V-1b, has the eutectic point e_0 at 0 kbar, and the eutectic point shifts continuously with encreasing pressure to e_{40} at 40 kbar. The P–T diagram for composition M within the system is shown in Fig. V-1a. The P–T diagram is constructed by observing the crystallization sequence for composition M at different pressures. At low pressures solid A is on the liquidus. A liquid with composition M has the very eutectic composition at 20 kbar, and the melting interval is consequently zero at this pressure. At pressures above 20 kbar solid B is the liquidus.

Fig. V-1 a, b. The P-T diagram for a binary
eutectic system. The diagram applies for
composition *M* shown in **b**

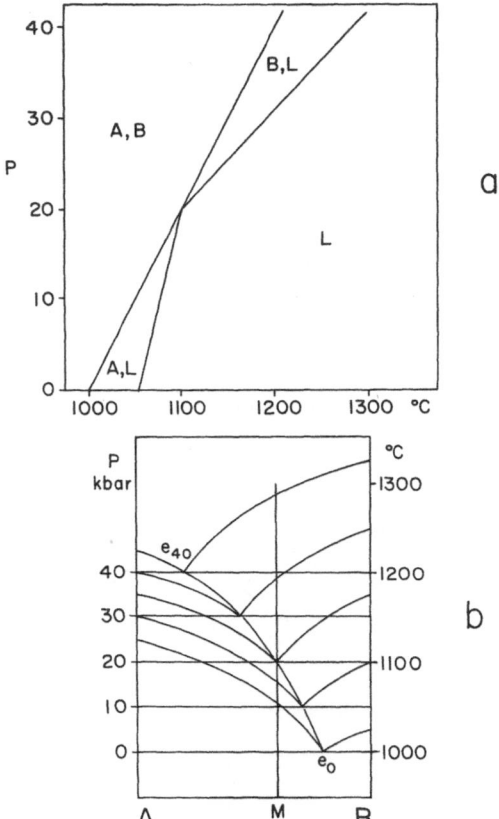

Fig. V-1 a, b. The P-T diagram for a binary
eutectic system. The diagram applies for
composition *M* shown in **b**

The P–T diagram for an eutectic system with the solid phases displaying limited solid solution is shown in Fig. V-2. Again the eutectic point crosses the selected composition M by increasing pressure, and the melting interval becomes zero.

The similarity between the phase fields in the binary T–x diagrams and the P–T diagrams is quite obvious from the two eutectic systems considered here. Comparison between the binary systems shown in Figs. II-11 and II-12 and the P–T diagrams shown here in Fig. V-1 and V-2, shows that the phase fields have the same arrangement, while the shape of the fields are different. As long as the phase relations remain the same at various pressures the phase fields in the P–T diagrams will have the same arrangement as in the binary systems. The P–T diagrams may be regarded as distorted binary diagrams.

3 Congruent–Incongruent Transition

By varying pressures the relative stability of the different solid phases changes, and an incongruent solid phase may become congruent or vice versa. A transition from

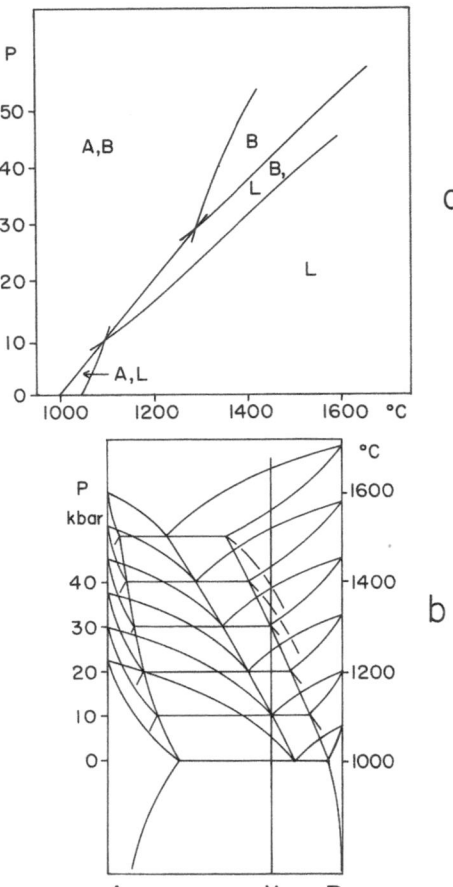

Fig. V-2a, b. The P-T diagram for a binary eutectic system with solid solution. The diagram is for composition *M* shown in **b**

incongruence to congruence with increasing pressure is shown in Fig. V-3b. The mixture M is situated to the right of D, and the liquids formed by crystallization of an initial liquid of composition M will always end in the eutectic point between D and B. If M had a composition between A and D there would have been an abrupt shift in the solidus temperature by equilibrium crystallization with increasing pressure. A liquid with composition M will first crystallize A at pressures between 0 and 11 kbar. By decreasing temperature the liquid will reach the peritectic point where A reacts with the liquid forming D. The curve between the two phase fields (A,L) and (D,L) is, thus, an univariant curve with the equilibria A+L=D.

4 Congruent Compound

A characteristic feature for the P–T diagrams of rock compositions is the appearance of new solid phases in different pressure regimes. A most notable example con-

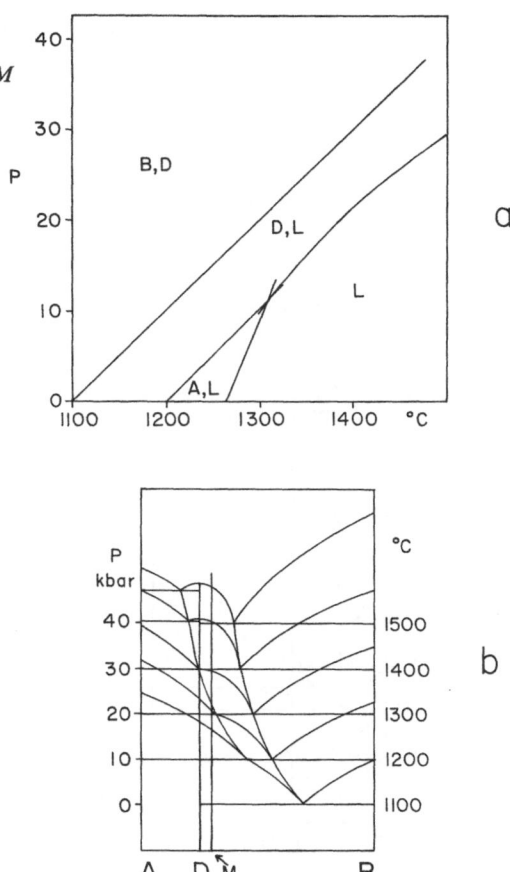

Fig. V-3a, b. The P-T diagram for a binary system with an incongruent compound *D*. The reaction point crosses the composition *M* with increasing pressure

sidering the mantle is the appearance of garnet in lherzolitic compositions around 20 kbar. The stability of garnet is higher than that of spinel at 20 kbar and higher pressures. The P–T diagram for a mixture M in a system A–B where a new solid phase D becomes stable by increasing pressures is shown in Fig. V-4. At pressures below 15 kbar liquids of composition M will change in composition towards the eutectic point. However, after the liquidus of D has intersected the liquidus curve for B, the liquids will change their compositions in the opposite directions towards the new eutectic point. At low pressures the crystallization occurs within the system A–B, and at high pressures within the system D–B. Because the liquids of composition M crystallize within different systems at different pressures, the solidus temperatures for composition M change abruptly as evident from Fig. V-4a. This abrupt change has not been reported in P–T diagrams for rock compositions. It is probably very difficult to detect small amounts of glass in the experimental changes, or perhaps this feature has been neglected, with the result that solidus curves just have been drawn smoothly to give a continuous curve.

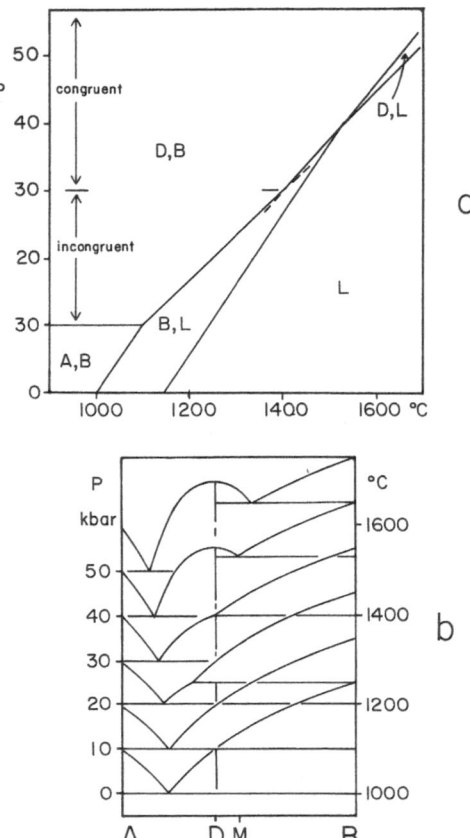

Fig. V-4a, b. The appearance of a new phase with increasing pressure. The new phase first melts incongruently, and then above 30 kbar congruently

5 Ternary Eutectic System

The P–T diagram for a ternary eutectic system is shown in Fig. V-5a. The diagram applies for the composition M shown in Fig. V-5b, and is constructed from this diagram in the following manner.

The solid triangles show the positions of the eutectic point in intervals of 10 kbar from 0 to 50 kbar. The temperature of the eutectic point increases with 100 °C for each 10 kbar starting at 1000 °C at 0 kbar. The isotherms around the eutectic points are only for the eutectic point at 0 kbar. They are drawn at 100 °C intervals and are assumed linear for simplicity. The isotherms for the eutectic points at higher pressures have the same configuration. At 0 kbar the liquidus phase for composition M is C, and above 15 kbar the liquidus phase is A. By decreasing temperatures at 0 kbar the liquid starting at M will reach the curve L, S(A, L) at 1180 °C, whereafter the liquids move towards the eutectic point. Thus, the sequence of phase assemblages at 0 kbar will be L – L, S(C) – L, S(C, A) – L, S(C, A, B) – S(C, A, B). The crystallization sequence is estimated in a similar manner at higher pressures for each 10 kbar, whereafter the P–T diagram may be constructed; the result is shown in Fig. V-5a.

Fig. V-5a, b. The P-T diagram
for a ternary eutectic system
(a). The composition for the
P-T diagram is shown as *M* in
diagram *b*. The *solid triangles*
show the variation in composition of the eutectic point with
pressure

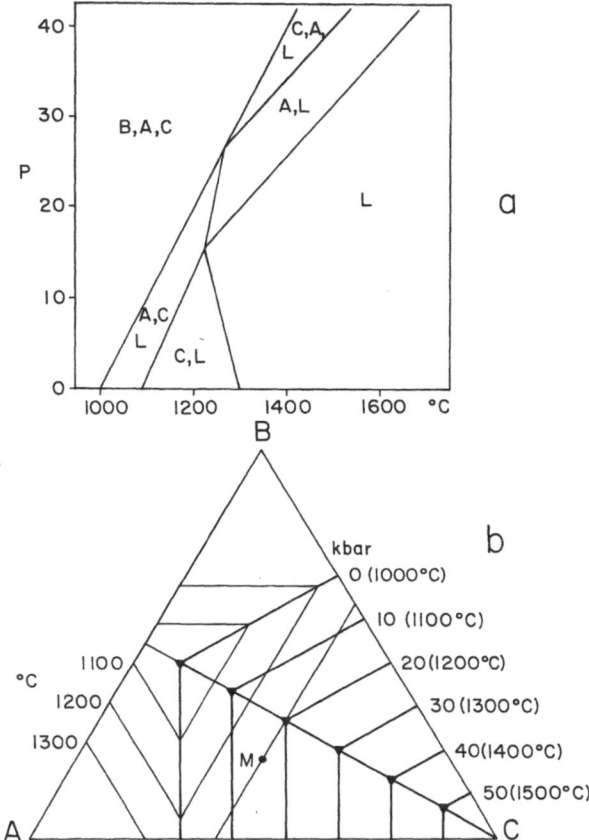

At 15 kbar the univariant curve L,S(A,C) will intersect composition M. The
crystallization behavior of composition M is, therefore, equivalent to the crystallization along the join I–C shown in Fig. V-1. Comparison between Fig. V-1a and
Fig. V-5a, shows that the pseudobinary diagram is similar to the P–T diagram. The
general appearance of the two other types of P–T diagrams for a ternary eutectic
system will be evident from Fig. V-1b and c.

6 Lherzolite

The P–T diagram for a garnet lherzolite was first estimated by Ito and Kennedy
(1967), and their diagram is shown in Fig. V-6 with minor modifications. The crystallization curve for spinel was not terminated towards higher pressures, but stopped
at s. Using Rhines' phase rule it is obvious that this curve should terminate at t,
whereby all the phase fields become in accordance with this rule.

The phase relations in the P–T diagram show that spinel lherzolite is stable near
the solidus at pressures lower than 23 kbar, while the garnet lherzolitic assemblage
is stable above this pressure. It is known from binary systems that enstatite reacts

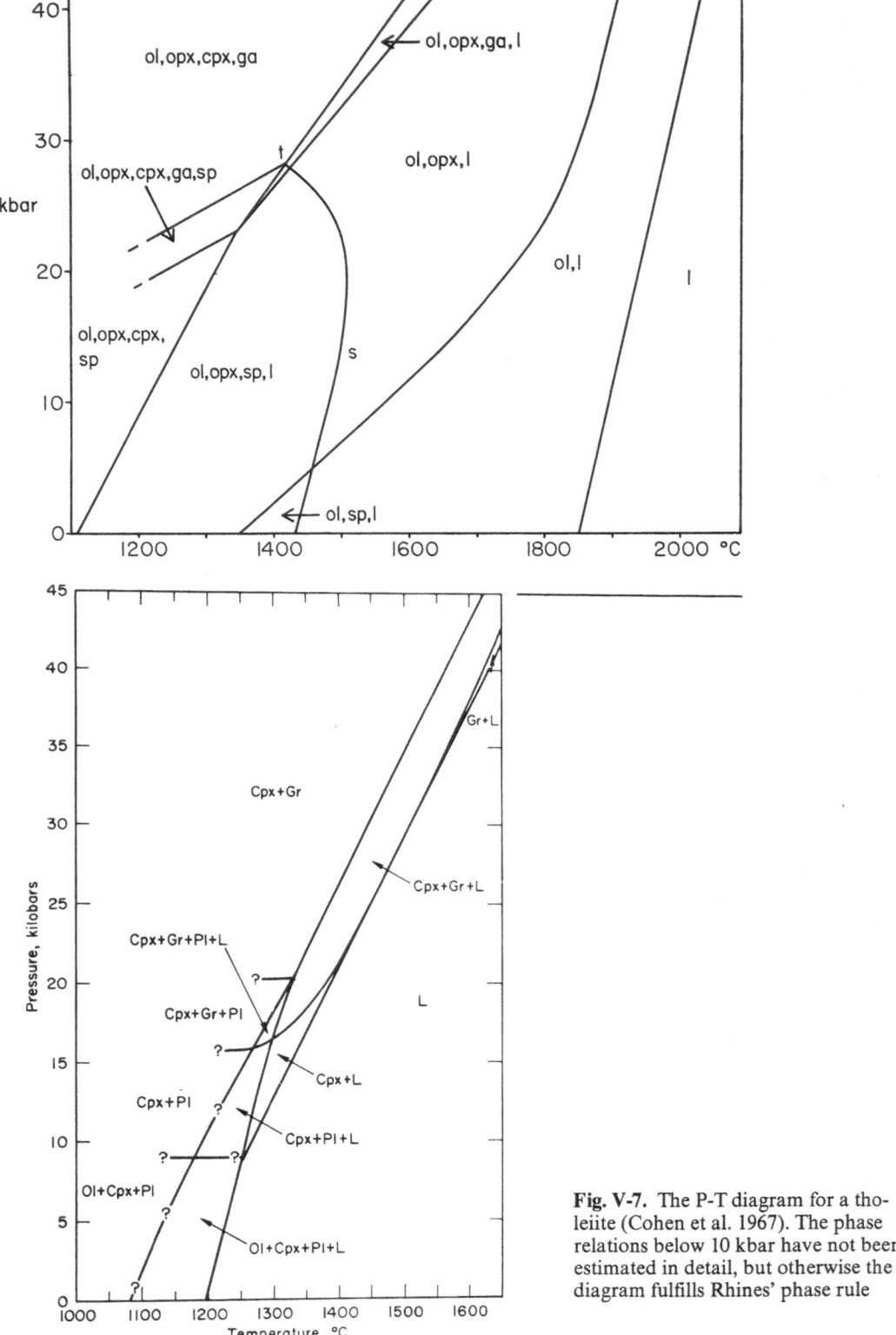

Fig. V-7. The P-T diagram for a tholeiite (Cohen et al. 1967). The phase relations below 10 kbar have not been estimated in detail, but otherwise the diagram fulfills Rhines' phase rule

Fig. V-6. The P-T diagram for a lherzolite estimated by Ito and Kennedy (1967). The diagram is slightly modified from their diagram, as the melting curve for spinel has been prolongated from s to t. The diagram shown is in accordance with Rhines' phase rule

◄───

with liquid to form diopside at all pressures up to at least 40 kbar. This reaction relationship is not evident from the P–T diagram, the reason is that the lherzolite contains too much enstatite for the reaction to be completed.

7 Tholeiite

The P–T diagram for a tholeiite with 7.6% MgO is shown in Fig. V-7. The phase relations deduced from experimental results are in agreement with Rhines' phase rule. Olivine is only stable up to about 10 kbar, while plagioclase is stable up to 20 kbar. All the solid phases occurring within the melting interval are also stable at subsolidus conditions, and there is no evidence for reaction relationships, although this does not exclude a reaction relationship as was mentioned above. Olivine, clinopyroxene, and plagioclase appears nearly simultaneously at pressures below 8 kbar, and the tholeiite might, therefore, have been formed by fractionation at this pressure.

VI Schreinemakers' Phase Theory

1 Introduction

The classic thermodynamic theory was worked out by Gibbs (1876), who also considered various phase relations. The thermodynamic theory of Gibbs was applied in detailed studies of phase relations by Roozeboom (1893, 1901), and one of Roozeboom's co-workers, F. A. H. Schreinemakers evaluated the methods and theorems required for an estimation of the systematic relationships of univariant equilibria and P–T diagrams. The works of Schreinemakers (1912–1925) are highly original and his methods very powerful, as only a limited amount of information lead to an estimate of fundamental phase relations. Schreinemakers largely applied geometrical methods, while the analytical treatment of heterogeneous phase equilibria was worked out by Morey and Williamson (1918) and Morey (1936). The petrological application of Schreinemakers' theorems have been considered by Niggli (1937), and their experimental applications have been demonstrated in a series of fundamental phase studies by Wyllie (1966, 1976, 1977).

Schreinemakers' theorems lead to an estimate of the relative position of univariant curves and bivariant regions in a P–T diagram. The relative position of the univariant curves are estimated from the univariant reactions, which will be known when the compositions of the reacting phases are known. The theorems do not allow an estimation of the slopes of the univariant curves in the P–T diagram, but the slopes may in many cases be approximately estimated using the Clausius-Clapeyron equation. A special advantage of the P–T diagrams is that the phase relations may be plotted in a plane, irrespective of the number of components involved. The detailed evaluation of phase relationships requires experimental work, and Schreinemakers' theorems cannot evaluate all the desirable phase relations. Most notably, it is not possible to state absolute values for pressures and temperatures for univariant curves, and the more accurate dependence of phase assemblages on compositions cannot be estimated from the theorems.

The present chapter will consider the fundamental aspects of Schreinemakers' phase theory. For a more detailed account and wider aspects, the original works of Schreinemakers' (1912–1925) should be consulted, as well as Rhines (1956), Zen (1966), and Mohr and Stout (1980).

2 Variance

The thermodynamic phase rule of Gibbs (1876) states:

$$N + P = C + 2 \tag{1}$$

where N is the degree of freedom, P the number of phases present, and C is the number of components. In a binary system there will be four phases in equilibrium at the invariant point, neither temperature, pressure, or composition can be changed if all four phases have to be present. The invariant point, therefore, has unique coordinates in the P–T diagram. Similarly the projection of the invariant point for ternary and quaternary systems on the P–T plane occurs at a specific point. In the binary system there are four phases in equilibrium at the invariant point, and three phases along the univariant curves. Univariance connotes that a given phase assemblage may be retained if only one variable is changed arbitrarily. Thus, if the temperature is changed then the other two variables, pressure and composition, will have to attain specific values, i.e., they are both functions of temperature. The result is that univariant equilibria forms curves in the P-T-x space, quite independently of the number of components.

From the invariant point in a binary system a total of four univariant curves originate, as four three phase assemblages are possible. Each of these univariant curves is labeled by the phase that is absent from the equilibrium. Thus, (3) signifies that the three other phases, 1, 2, and 4 are in equilibrium along the curve labeled (3). At the univariant curve (3) three bivariant equilibria originate: $1+2$, $1+4$, and $2+4$. Considering all univariant and bivariant equilibria the following result is obtained:

(1) $2+3+4$ $2+3$, $2+4$, $3+4$
(2) $1+3+4$ $1+3$, $1+4$, $3+4$
(3) $1+2+4$ $1+2$, $1+4$, $2+4$
(4) $1+2+3$ $1+2$, $1+3$, $2+3$

Half of the bivariant equilibria are the same, and there are, thus, six different bivariant equilibria in all around an invariant point in a binary system.

In a ternary system there will be five phases in equilibrium at the invariant point, and consequently five different univariant curves originate at the invariant point. The P–T diagram for a ternary system depends on the type of reactions occurring. One example of a ternary system is shown in Fig. VI-1. The univariant equilibria are defined by the following reactions:

(1) $2+3=4+5$
(2) $1+5=3+4$
(3) $1+2=4+5$
(4) $1+5=2+3$
(5) $1+2=3+4$

From the univariant curve (5) the following bivariant phase assemblages originate:

$1+2+4$
$1+4+3$
$2+4+3$
$1+2+3$

From each of the five univariant curves, four bivariant equilibria originate. As half of these are identical there are in all ten different bivariant equilibria around a ternary invariant point. These are for the diagram shown in Fig. VI-1: 123, 134, 145, 345, 234, 245, 135, 152, 523, and 124.

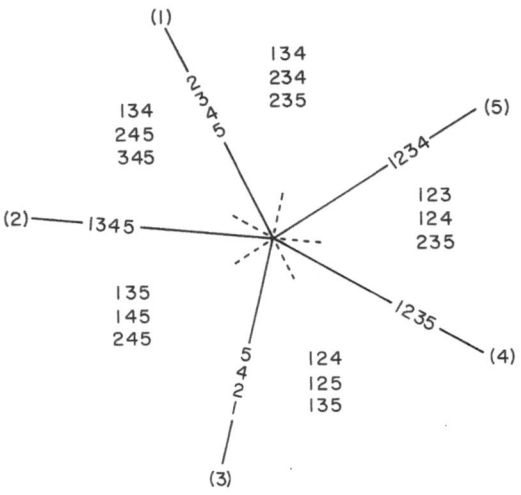

Fig. VI-1. The chemograph for a ternary system of type I. The curves for univariant equilibria are labeled with the phase being absent in parentheses. The bivariant assemblages between the univariant curves are also shown. The configuration of the bivariant surfaces originating at (5) is shown in Fig. VI-8

Generally for an N-component system there are the following quantitative relationships:

N + 2 phases at the invariant point;
N + 1 phases in equilibrium along univariant curves;
N phases in bivariant equilibria;
½ (N + 2) (N + 1) bivariant equilibria.

The phase regions between the univariant curves are called sectors. As an example, the phase region between (1) and (5) in Fig. VI-1 is a sector. The bivariant assemblage 234 is unique for this sector. However, a bivariant assemblage may overlap a univariant curve, and need not be confined to the region between two univariant curves. The sector for 235 occurring between (1) and (4) is such an overlapping sector. The diagram shown in Fig. VI-1 is called a P–T projection or a chemograph, the latter notation is convenient as the P–T diagrams for rock compositions are essentially different.

3 The P-T-x Diagram for a Binary Eutectic System

The nature and phenomena of Schreinemakers' theorems may best be understood by considering a simple P-T-x diagram, and the phase relations for a eutectic system will, therefore, be considered here. The theorems may be deduced without considering the detailed phase relations of a P-T-x diagram, as was done by Zen (1966), however, the theory becomes rather abstract, and difficult to perceive.

The binary eutectic system shown in Fig. VI-2 consists of the two components, A and B. The two solids have the compositions A and B, while the compositions of the gas and liquid phases are mixtures of A and B. The P–T diagram for component A is shown in the hindmost plane, and that for B in the frontal P–T plane. At invariant conditions there are four phases in equilibrium, two solids consisting of pure A and

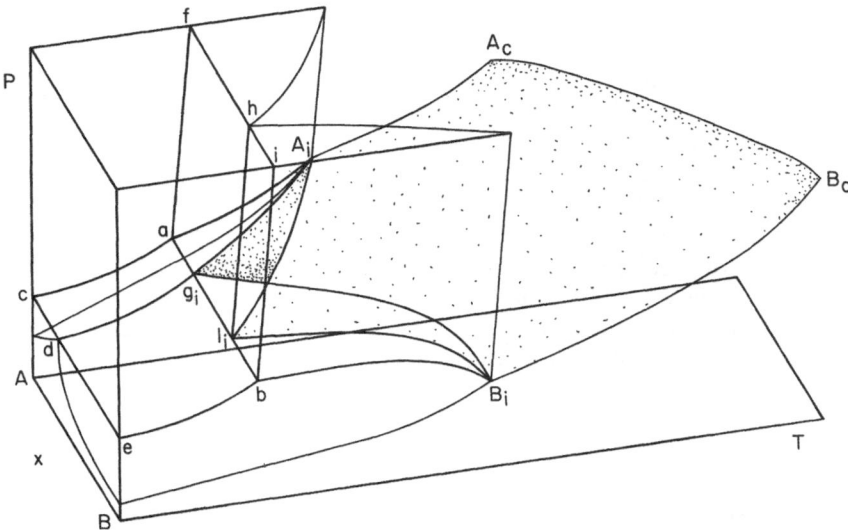

Fig. VI-2. The P-T diagram for a binary eutectic system. A_i and B_i are the invariant points for pure A and B. A_c and B_c are the critical points for A and B, respectively

B, and a gas and liquid phase, being different mixtures of A and B. These four phases have compositions along the line a–b, the compositions of the gas and liquid phases being g_i and l_i, respectively. Four univariant curves branch off from the univariant equilibrium. The univariant curves involve equilibrium between three phases:

(A)   BLG
(B)   ALG
(L)   ABG
(G)   ABL

The univariant curve (A) extends from l_i to B_i and (B) extends from g_i to A_i. The univariant curve (L) extends from g_i to d, and (G) extends from l_i to h. The phases in univariant equilibrium at (L) are A, B, and G. The gas phase G has compositions between g_i and d. Since the equilibrium condition is univariant, only one variable may be changed arbitrarily, and at constant temperature the equilibria becomes invariant. For given values of P and T the composition of the gas phase will be fixed. The tine lines between A, G, and B will, therefore, be parallel with the composition line A–B, and the surface c-a-b-e is, therefore, a cylindrical evolution surface. The same applies for the other surfaces defined by univariant tie lines. Above the plane c-a-b-e the two solids A and B will be in equilibrium, below this surface there are three stable phase assemblages, A + G, G, and B + G. As evident from Fig. VI-2 the phase assemblage present will depend on the composition of the mixture, for A-rich mixtures the phase assemblage will be A + G, and for B-rich mixtures the phase assemblage is B + G. For intermediate compositions and at low pressures only the gas phase is stable. The univariant equilibrium (G) implies the reaction A + B = L, and extends from l_i towards h. To the left of the plane a-f-i-b A+B is stable, and to the right of this plane either A+L, L, or B+L is the stable phase assemblage. The univariant equilibrium (B) is for the reaction A+L=G, and is displayed by the plane al_iA$_i$. Above this plane A+L is stable and below there are three possible assemblages, A+G, G, and L+G. Finally, the univariant equilibrium (A) is for the reaction G+B=L, and is displayed by the plane g$_i$B$_i$b. Below this plane G+B is stable, and above there are three possible assemblages, L+G, L, and L+B. It is evident from the equilibrium relationships of the eutectic system that there might be either one or three stable phase assemblages on one of the sides of a univariant equilibrium curve. Which one is stable of the three possible assemblages will depend on the composition of the mixture. Schreinemakers' theorems define the possible bivariant assemblages between univariant curves, but the theorems cannot define which one actually is present.

The four univariant equilibria for the eutectic system is shown in Fig. VI-3a, and their projection onto the P–T plane in Fig. VI-3b. As demonstrated above there are different bivariant phase assemblages possible between the univariant curves, and these are indicated in Fig. VI-3b. The univariant and bivariant phase relations for the P–T projection have not been estimated from the binary P–T diagram. This procedure will always be possible for a binary system, as the phase relations are three-dimensional. The phase relations for a ternary system is four-dimensional, and the present procedure cannot be applied. The theorems of Schreinemakers' allow an estimation of the P–T projection not only of binary systems, but also of ternary, quaternary, and higher ordered systems.

The P–T projection of univariant and bivariant phase relations shown in Fig. VI-3b is called the chemograph for the binary eutectic system. The purpose of the present treatment has been to demonstrate directly some of the general features of a

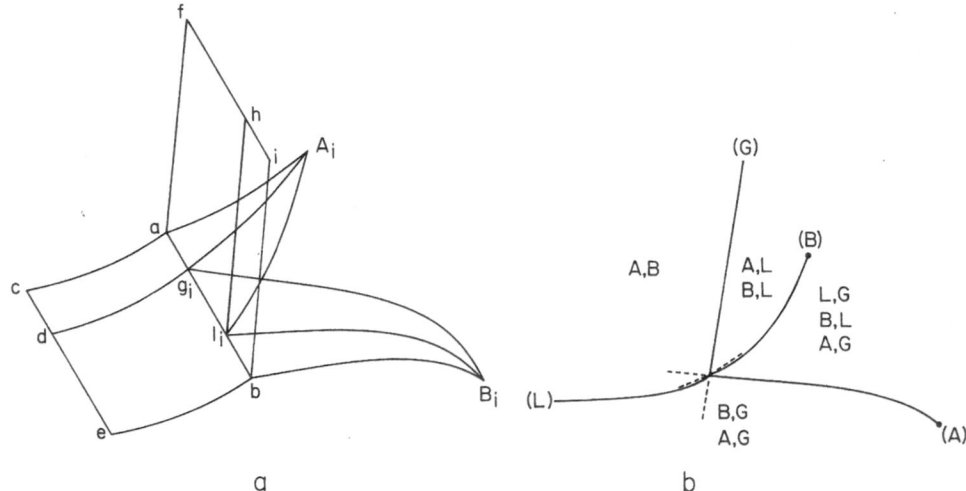

| a | b |

Fig. VI-3a, b. The four univariant equilibria for the eutectic system shown in Fig. IV-2. **a** The surfaces for the univariant tie lines, and **b** the projections of the univariant curves onto the P-T plane. The bivariant phase assemblages are also shown

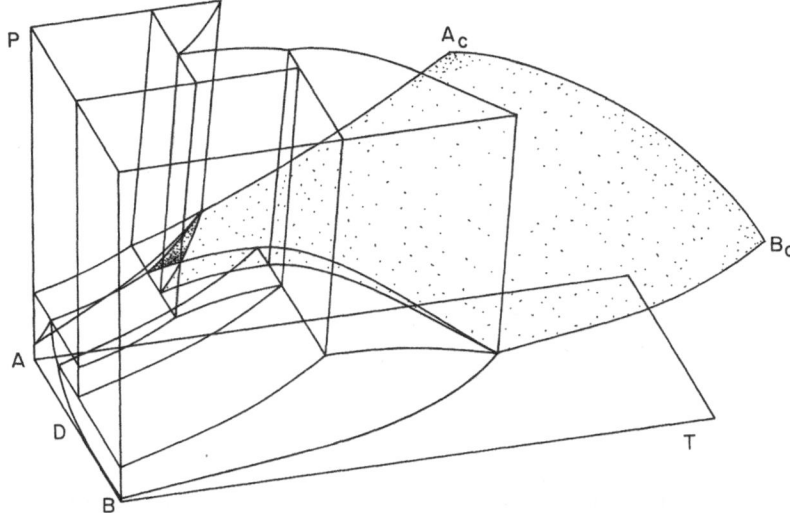

Fig. VI-4. The P-T-x diagram for a peritectic system. The incongruently melting phase is labeled *D*

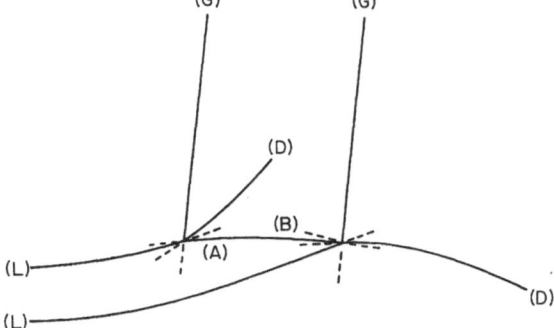

Fig. VI-5. The chemograph for the peritectic system shown in Fig. VI-4. The univariant curve (*A*) at the invariant point to the left is identical to (*B*) of the invariant point to the right, the univariant reaction is in both cases $G + D = L$

chemograph: (1) the univariant equilibria are projected onto the P–T plane as curves: (2) the bivariant equilibria will be stable in sectors defined by the univariant equilibrium curves.

The chemograph for a eutectic system involves only one binary invariant point. The chemograph for a binary peritectic system has two invariant points, the P-T-x diagram for such a system is shown in Fig. VI-4, and the resulting chemograph is shown in Fig. VI-5.

4 The Bounding Theorem

In a system consisting of C components, a total of C + 1 bivariant phase assemblages originates at each univariant curve. These bivariant equilibria have a limited extension in the chemograph, and the bounding theorem defines their region of stability.

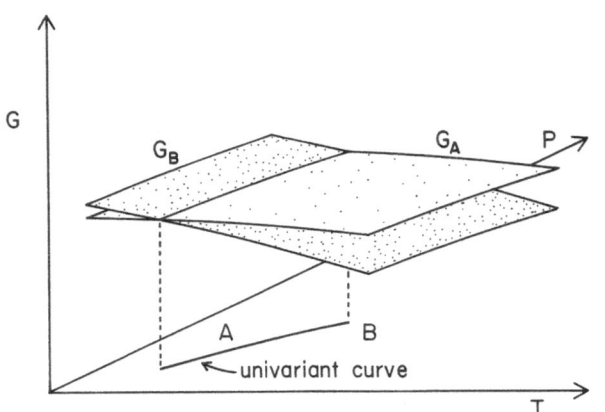

Fig. VI-6. Gibbs' free energy surfaces for two phase assemblages *A* and *B*. The two assemblages may consist of any number of phases > 1

The bounding theorem: A bivariant region originating at a univariant curve will be stable only on one side of the univariant curve, and is metastable on the other.

Proof: The free energy of a phase assemblage varies with pressure and temperature, and the variation may be estimated from:

$$\delta G = V\,\delta P - S\,\delta T$$

$$
\begin{aligned}
G &= \text{Gibbs free energy} \\
V &= \text{Volume} \\
P &= \text{pressure} \\
S &= \text{entropy} \\
T &= \text{temperature}
\end{aligned}
$$

(2)

According to Eq. (2):

$$\frac{(\delta G)}{(\delta P)_T} = V \tag{3}$$

$$\frac{(\delta G)}{(\delta T)_p} = -S \tag{4}$$

Both V and S are monotoneous functions of pressure and temperature for a given phase assemblage as long as the number of phases remains unchanged. The values of V and S will be different for two different phase assemblages, and the G-surfaces for the two phase assemblages will have different slopes in a GTP coordinate system. The free energy surfaces for two bivariant phase assemblages, say A and B, will therefore have different slopes, and intersect each other along a monotoneous curve, which will be a univariant curve (Fig. VI-6). The projection of the univariant curve onto the P–T plane represents the univariant curve in a chemograph. It follows from the first and second laws of thermodynamics that the phase assemblage with the lowest free energy is the stable one. Due to the different slopes of the free energy surfaces, one assemblage will be stable on one side of the curve of intersection, and the other assemblage will be stable on the other side. As evident from Fig. VI-6 assemblage A is stable to the left of the univariant curve, and assemblage B is the stable assemblage on the right side. The projection of the univariant curve in the P–T plane divides this plane into two regions, one region in which A is stable and another one in which B is stable. Phase assemblage A is metastable with respect to B on the right side of the univariant curve where B is stable and vice versa. A bivariant assemblage like A is consequently bounded by univariant curves. If no intersection occurs between two phase assemblages, the one with the lowest free energy will be stable throughout the region considered.

Schreinemakers' other theorems will be deductions from the bounding theorem. Two features allow the derivation of the bounding theorem, first the fact that univariant curves are a function of only one variable, which means that they form curves in any coordinate system, let it be a GTP or a P-T-x diagram. Second, the equilibrium conditions derived from classical thermodynamics define regions of stability bounded by the univariant curves.

5 Morey-Schreinemakers' Theorem

The phase rule states that $C+2$ phases are in equilibrium at the invariant point, and further the rule implies that $C+2$ univariant curves radiate from the invariant point. The latter feature has been shown for a binary and ternary system in Fig. VI-3 and Fig. VI-1, respectively. Morey–Schreinemakers' theorem defines the relative positions of the univariant curves in the chemograph (Schreinemakers 1919, p. 121):

Morey-Schreinemakers theorem: A sector angle is always smaller than or equal to 180°.

Proof: A sector for a bivariant assemblage A with a sector angle Θ being larger and smaller than 180° is shown in Fig. VI-7. If the sector angle is larger than 180° for the stable part of assemblage A, then the stable region of the assemblage is situated on both sides of the metastable extension of the univariant curve (u_1). The bivariant assemblage is, thus, assumed stable on both sides of the univariant curve (u_1) which is impossible according to the bounding theorem, and the sector angle must consequently always be smaller than or equal to 180°.

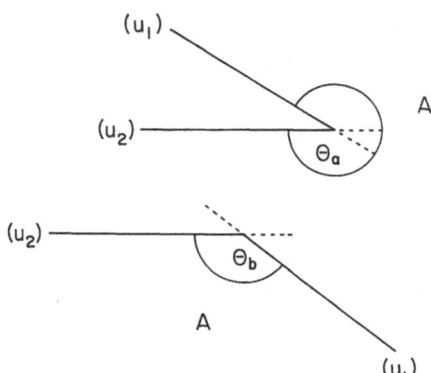

Fig. VI-7. Two sectors for assemblage A between the univariant curves u_1 and u_2. In the upper diagram θ_a is larger than 180°, and in the lower diagram θ_b is smaller than 180°

As $C + 2$ univariant curves originate from an invariant point, a direct consequence of Morey–Schreinemakers' theorem is that: The P–T regions in the neighborhood of an invariant point is divided by univariant curves in $C + 2$ sectors, each of which is less than or equal to $180°$ in angular extent (Zen 1966).

6 The Overlap Theorem

The univariant curves divide the P–T region in $C + 2$ sectors. Between the univariant curves at least one, and frequently some bivariant phase assemblages are stable. Consider as examples Fig. VI-1 and Fig. VI-3. In Fig. VI-3 the bivariant assemblages L + G, B + L, and B + G are stable between (A) and (B). In Fig. VI-1 for a ternary system, the bivariant assemblages $1 + 3 + 4$, $2 + 3 + 4$, and $2 + 3 + 5$ are stable between (1) and (5). Since the phase relations for a system consisting of several phases are projected onto a single plane, the P–T plane, several bivariant phase assemblages may be stable in a sector defined by two univariant curves. Between two univariant curves, say (1) and (2), the bivariant assemblage (1, 2) is stable. The bounding theorem states that (1, 2) is stable between (1) and (2), while the overlap theorem defines which additional bivariant assemblages are stable in the same bivariant sector of (1, 2).

The overlap theorem: Every sector which extends itself over the stable or metastable part of a univariant curve (j) contains the phase j.

Proof: Consider the bivariant assemblage (j, j + 1). This assemblage extends to (j) and (j + 1), but never crosses these two curves in the chemograph according to the bounding theorem. Further, according to Morey–Schreinmakers' theorem the sector (j, j + 1) is smaller than $180°$ in angular extent, and the assemblage (j, j + 1) can consequently extend to, but not beyond the metastable extensions of (j) and (j + 1). In conclusion, a bivariant assemblage which lacks the phases j and j + 1 has to be bounded by (j) and (j + 1). Therefore, if an assemblage extends over an univariant curve, say (j), this assemblage has to contain the phase j.

A phase assemblage containing the phase j does not necessarily extend over (j).

In the chemograph for the binary eutectic system (Fig. VI-3), the bivariant assemblage B + L consisting of the two phases B and L is stable between (G) and (B). However, this bivariant assemblage is also stable between (B) and (A), and therefore, extends over (B). According to the overlap theorem the assemblage has to contain the phase B, which is the case.

7 Principle of Opposition

Along a univariant curve $C + 1$ phases are in equilibrium, in a binary system there will be three phases in euqilibrium along a univariant curve. Specific reactions will occur between the phases when the pressure is changed or heat is added or subtracted from the univariant assemblage. Consider the ternary reaction:

$$3 + 4 = 1 + 2 \tag{5}$$

Fig. VI-8. The bivariant equilibria around the univariant curve (5) in a ternary system (cf. Fig. VI-1). The assemblage 3+4 is a low temperature assemblage and 1+2 is a high temperature assemblage. There are four bivariant surfaces, *234, 134, 123,* and *124*. Between the bivariant surfaces there are space regions with two-phase assemblages like 1+2 and 3+4

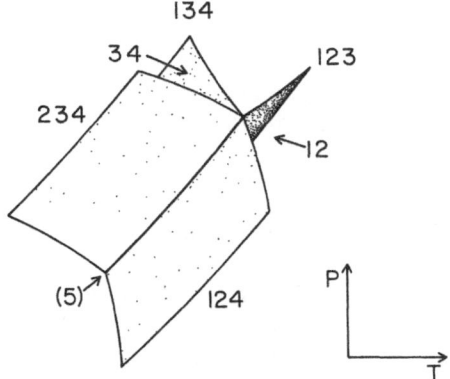

If this reaction is endothermic, then it will proceed to the right if heat is added to the system, and by completion of the reaction phases three and four will disappear. The assemblage 3+4 will represent a low temperature assemblage, being stable on the low temperature side of (5), while 1+2 is stable on the high temperature side of (5). The two assemblages may, therefore, be said to be in opposition with respect to (5). As phases three and four are stable on the low temperature side the univariant curves lacking these two phases, i.e., (3) and (4) will be situated on the high temperature side. Similarly, (1) and (2) will be situated on the low temperature side of (5) (cf. Fig. VI-1; VI-8).

8 Arrangement of Univariant Curves

The above theorems are the fundamental theorems of Schreinemakers' theory, and they allow the construction of chemographs if the compositions of reacting phases are known. Schreinemakers (1912–1925) had considered additional aspects of univariant phase relations and chemographs, but the theory represented here and by Zen (1966) is sufficient for the construction of chemographs. The practical application of the theorems will now be exemplified by using them for a ternary system.

Four bivariant equilibria will originate at a univariant curve in a ternary system. If the univariant curve is defined by the reaction just mentioned above there will be the following four bivariant equilibria:

(4)(5): 123
(3)(5): 124
(2)(5): 134
(1)(5): 234

As the assemblages 1+2 and 3+4 are in opposition according to the reaction 3+4=1+2, the assemblages 123 and 124 will be situated on one side of (5), and the assemblages 134 and 234 on the other side. Let as before, 123 and 124 be on the high temperature side of (5), and 134 and 234 on the low temperature side. In Fig. VI-1, 123 and 124 will be situated to the left of (5) and 134 and 234 to the right of

(5). Each of the four bivariant sectors of the chemograph of the ternary system is limited by two univariant curves:

123 by (4) and (5)
124 by (3) and (5)
134 by (2) and (5)
234 by (1) and (5)

Since the angular extension of each sector is less than 180°, the univariant curves (4) and (3) must be situated on the same side of (5), and opposite (1) and (2), as shown in Fig. VI-1. The relative position of (3) and (4) is not defined from the reaction mentioned above. Additional information is required. If the ternary system is peritectic another reaction will be:

(1): $2 + 3 = 4 + 5$

Thus, (2) and (3) should be in opposition to (4) and (5) with respect to (1). The relative position of the invariant curves are defined from the two reactions. The chemograph might also have been the mirror image of the chemograph shown in Fig. VI-1, however, it was assumed that the reaction $3 + 4 = 1 + 2$ was endothermic, so in the present case the chemograph will be of the type shown in Fig. VI-1. The two reactions considered above may be represented by the schemes:

(3) (4) | (5) | (1) (2)

(2) (3) | (1) | (4) (5)

which indicates that (3) and (4) should be situated on one side of (5), and (4) and (5) on the other.

The actual slopes of the univariant curves have to be estimated from the Clausius–Clapeyron equation:

$$\frac{\delta P}{\delta T} = \frac{\varDelta H}{T \varDelta V}$$

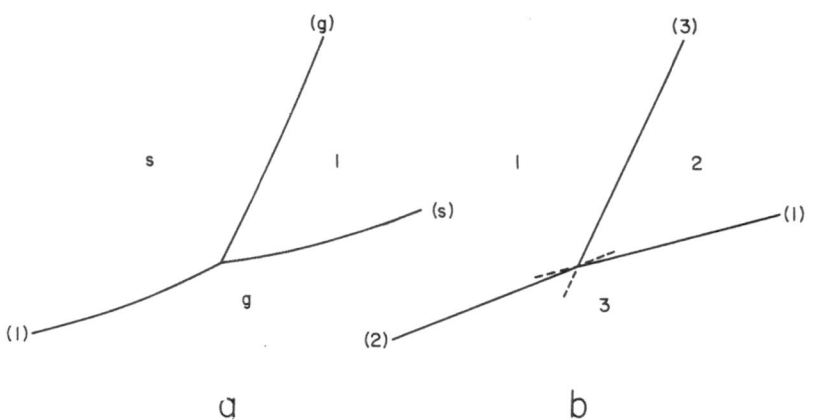

Fig. VI-9a, b. The chemographs for a monary system. **a** The familiar P-T diagram for a monary system; **b** The monary chemograph obtained from Schreinemakers' theorems

where ΔH is the heat reaction and ΔV is the volume change by the reaction. These two parameters will generally be known for some of the reactions, and the approximate chemograph may, therefore, be constructed as the sector angles have to be less than 180°. If the slopes cannot be estimated for any of the univariant curves, an approximate chemograph may be constructed from the known P–T behavior of the phases.

The basic theorems for the construction of chemographs have now been mentioned, the application of the theorems for monary, binary, ternary, and quaternary systems will be examined in the subsequent sections.

9 Monary Systems

In a monary system there are three phases in equilibrium at the invariant point, and consequently three univariant curves. If the phases are labeled 1, 2, and 3, then the following univariant equilibria occurs:

(1): $2=3$
(2): $1=3$
(3): $1=2$

The chemograph for a monary system is shown in Fig. VI-9. The univariant curve (1) has to be situated between (2) and (3), and (2) must be situated between (1) and (3). The slopes for the univariant curves cannot be estimated from the theorems, but may be determined from experimental data. The chemographs for water and the polymorphs of Al_2SiO_5 are shown in Fig. VI-10.

10 Binary Systems

The compositions of binary phases are always situated along a line, if the sequence of phases along the compositional line is 1, 2, 3, and 4, then the following univariant reactions occur:

(1): $2+4=3$
(2): $1+4=3$
(3): $1+4=2$
(4): $1+3=2$

With (1) as the reference curve, (2) and (4) should be in opposition with (3). The resulting chemograph is shown in Fig. VI-11a. However, the relative position of (2) and (4) cannot be estimated from the univariant reactions (1) alone. The univariant reaction (2) requires that (1) and (4) are in opposition with respect to (3). Thus, the chemograph shown in Fig. VI-11b fulfills the constraints of reactions (1) and (2). The chemograph in Fig. VI-11b is also in accordance with the requirements of the two univariant reactions (3) and (4).

The chemograph in Fig. VI-11b is in agreement with the bounding theorem, none of the bivariant equilibria are both stable and metastable within their sectors. Also Morey–Schreinemakers' theorem is fulfilled as all the sector angles are less

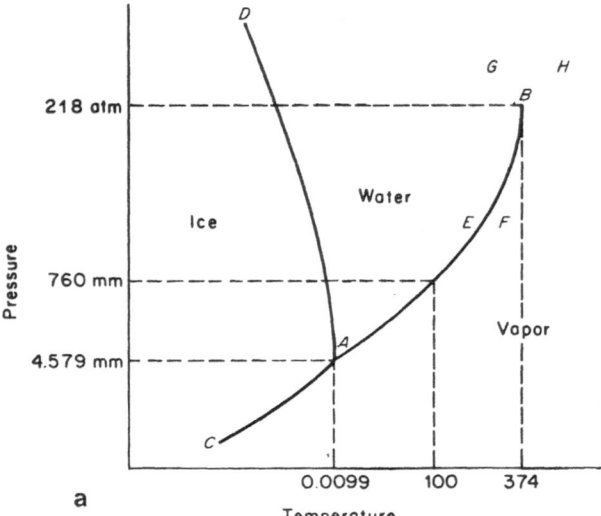

a

Fig. VI-10a, b. The phase diagrams for water **a**, and the polymorphs of Al$_2$SiO$_5$ **b** (Turner 1968)

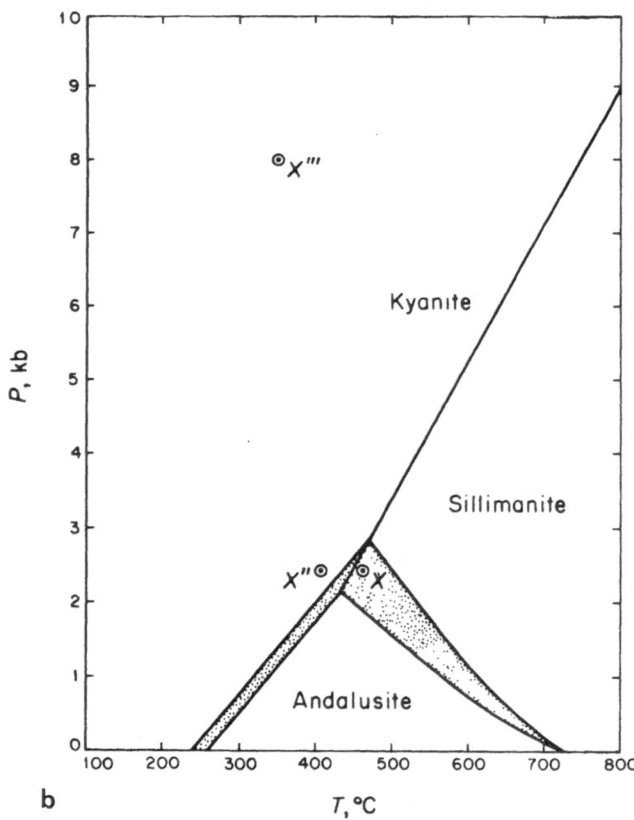

b

Fig. VI-11 a, b. The chemographs obtained from binary
univariant reactions. Diagram
a is obtained from the *reaction
(1):* 2+4=3, which results in
the two possibilities shown.
The second *reaction (2):*
1+4=3 defines the binary
chemograph shown in diagram **b**

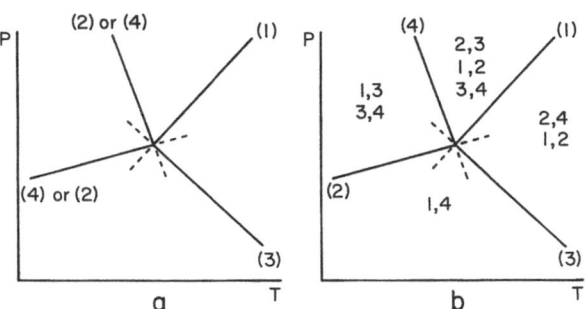

Fig. VI-12. The chemograph for
the system $CaO-CO_2$ with the four
phases, CaO, $CaCO_3$, CO_2, and
liquid (Wyllie and Boettcher 1969)

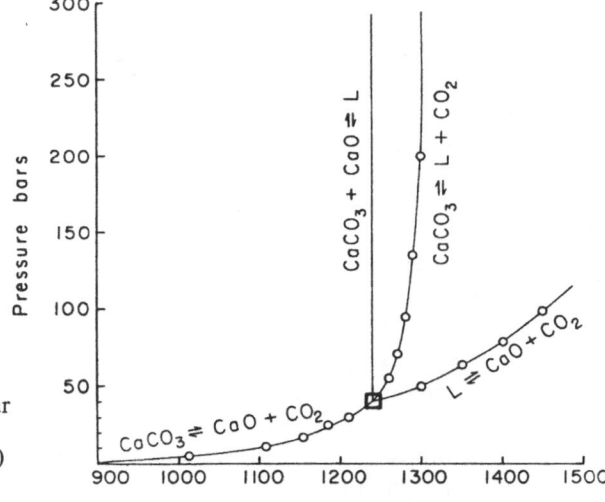

than 180°. Further, the bivariant sectors comply with the overlap theorem. The bivariant equilibrium 1+2 overlaps (1), and contains phase 1. Similarly, 3+4 overlaps (4) and contains (4). The sectors for the bivariant sectors are estimated as follows. The bivariant equilibrium 2+3 is stable between (4) and (1), 2+4 between (1)
and (3), 1+4 between (2) and (3), and 1+3 between (2) and (4). However, further
bivariant equilibria are stable in two of these sectors. Morey–Schreinemakers' theorem requires only that the sector angles should be less than 180°. The bivariant
equilibrium 1+2 is stable between (4) and (2) as its sector angle is less than 180°.
Also 3+4 is stable between (2) and (1). There are, thus, a total of four single bivariant sectors, and two double sectors.

The binary system $CaO-CO_2$ has been studied by Wyllie and Tuttle (1960) and
Baker (1962), and the chemograph shown in Fig. VI-12 was compiled from exper-

imental data by Wyllie and Boettcher (1969). The liquid has its composition between CaO and $CaCO_3$ so the four possible univariant reactions are as follows:

$$(L): \quad CaO + CO_2 = CaCO_3$$
$$(CaCO_3): \quad CaO + CO_2 = L$$
$$(CO_2): \quad CaO + CaCO_3 = L$$
$$(CaO): \quad L + CO_2 = CaCO_3$$

The invariant points have the coordinates 1230 °C and 39.5 bar. The reactions define the relative positions of the univariant curves, but the experimental data are required for an estimate of the coordinates of the invariant point and the curvature and slopes of the univariant curves. Additional chemographs of carbonate-bearing systems may be found in Wyllie and Huang (1976).

11 Degenerate Systems

In the binary system dealt with above, all four phases have different compositions. However, in some systems different phases have the same composition, and are polymorphic phases with identical composition. As a familiar example, the polymorphic forms of Al_2SiO_5 may be mentioned, kyanite, sillimanite, and andalusite are all binary compositions within the system Al_2O_3–SiO_2, but their mutual phase relations are not binary. The presence of polymorphic forms changes the chemograph, and systems where polymorphic forms occur are called degenerate systems. There are two types of degeneracy:

1. *Polymorphic Phases*. In this case two or more phases have the same composition. This type of degeneracy may occur in systems of any order. Monary systems may be considered as degenerate binary systems according to this definition.

2. *Compositional Coincidence*. If in a C-component system C or more phases are colinear in composition then the system is degenerate. This type of degeneracy applies to systems with more than two components, i.e., for C equal to or larger than 3. In binary systems the compositions are always colinear, and the binary systems may be considered degenerate ternary systems. If C=4 and four or more phases are coplanar then the system is degenerate.

Generally, if in a C-component system the composition of one of the phases can be defined with less than C components then the system is degenerate. Those phases that contain less than C components are called singular phases.

Those phases that are not singular ones are grouped together and are called indifferent phases. There are two types of indifferent phases. The absolute indifferent phases do not participate in any of the univariant equilibria. Thus, graphite or carbon dioxide are absolute indifferent phases at subsolidus conditions in the system Al_2O_3–SiO_2. The relative indifferent phases participate in some of the univariant reactions. The relative indifferent phases in a degenerate system may perhaps better be described as those phases that behave as in a nondegenerate system, they are the "normal" phases. These definitions will now be clarified considering the degenerate binary systems.

The nondegenerate binary chemograph is shown in Fig. VI-13a, where it may be compared with the four degenerate binary types. In the first type the phases 1, 2,

Fig. VI-13 a–e. The non-degenerate binary chemograph, and the four degenerate binary chemographs (**b, c, d,** and **e**)

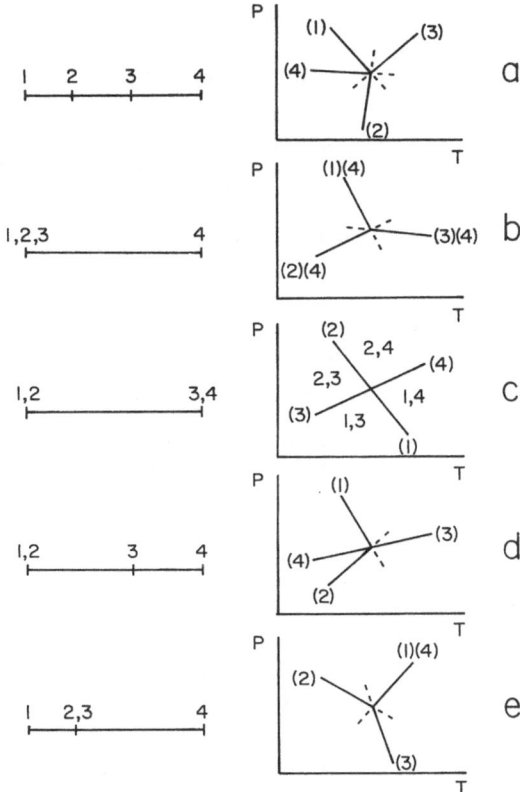

and 3 are polymorphs and singular phases (Fig. VI-13 b). Phase 4 is absolute indifferent. The following reactions may occur:

$$(1): \ 2 = 3$$
$$(2): \ 1 = 3$$
$$(3): \ 1 = 2$$
$$(1, 2, 3): \ 4 = 4$$

Phase 4 is stable throughout the P–T region and does not participate in any of the reactions between 1, 2, and 3. The resulting chemograph is the same as for a monary system with one phase added (Fig. VI-13 b).

When each pair of the phases has the same composition the following reactions may occur:

$$(1)\,(2): \ 3 = 4$$
$$(3)\,(4): \ 1 = 2$$

The resulting chemograph is shown in Fig. VI-13 c. Phases 1 and 2 will be stable on each side of a univariant curve, likewise phases 3 and 4 will be stable on each side of a univariant curve. That the two curves are univariant might appear surprising as we are dealing with a binary system, however, the equilibria between 1 and 2 is quite independent of the equilibria between 3 and 4; there are so to speak two overlapping monary and univariant equilibria. The curves may be labeled using the overlap rule, however, concerning the two curves it can only be stated that the phases are stable on each side of the curves (Fig. VI-13 c).

When only two phases are polymorphs, two types of chemographs are possible. In the chemograph shown in Fig. VI-13 d the relative indifferent phases have compositions on the same side of the singular phases, and the following reactions take place:

$$(1): \quad 2 + 4 = 3$$
$$(2): \quad 1 + 4 = 3$$
$$(3)\,(4): \qquad 1 = 2$$

The chemograph is determined by these reactions and is shown in Fig. VI-13 d.

By the second type the relative indifferent phases have compositions on each side of the singular ones, and the following reactions occur:

$$(1)\,(4): \qquad 2 = 3$$
$$(2): \quad 1 + 4 = 3$$
$$(3): \quad 1 + 4 = 2$$

The chemograph is shown in Fig. VI-13 e. The last two examples demonstrate the coincidence rule (Zen 1966): When the two indifferent phases chemographically lie on the same side with respect to the singular phases, then the univariant curves bearing the labels of the indifferent phases coincide stable to metastable; when the two indifferent phases lie on opposite sides of singular phases, then the univariant curves bearing the labels of the indifferent phases coincide stable to stable. The same rule applies for ternary and quaternary systems as well, in a ternary system the indifferent phases are situated with respect to a line through singular phases, and in a quaternary system they should be related to a plane through the singular phases.

12 Ternary Systems

In a ternary system there are five phases in equilibrium at the invariant point, and five univariant curves where four phases are in equilibrium. There are three non-degenerate and 16 degenerate types of chemographs (Zen 1966).

Only the nondegenerate types will be described here, they are shown in Fig. VI-14. The following reactions are valid for type I:

$$(1): \quad 2 + 3 = 4 + 5$$
$$(2): \quad 1 + 5 = 3 + 4$$
$$(3): \quad 1 + 2 = 4 + 5$$
$$(4): \quad 1 + 5 = 2 + 3$$
$$(5): \quad 1 + 2 = 3 + 4$$

The chemograph of type I applies for a ternary system with only peritectic reactions (Rhines 1956). At isobaric conditions the univariant equilibria becomes invariant. If

Fig. VI-14. The three nondegenerate ternary chemographs. The compositional relationships are shown in the diagrams to the left of the chemographs

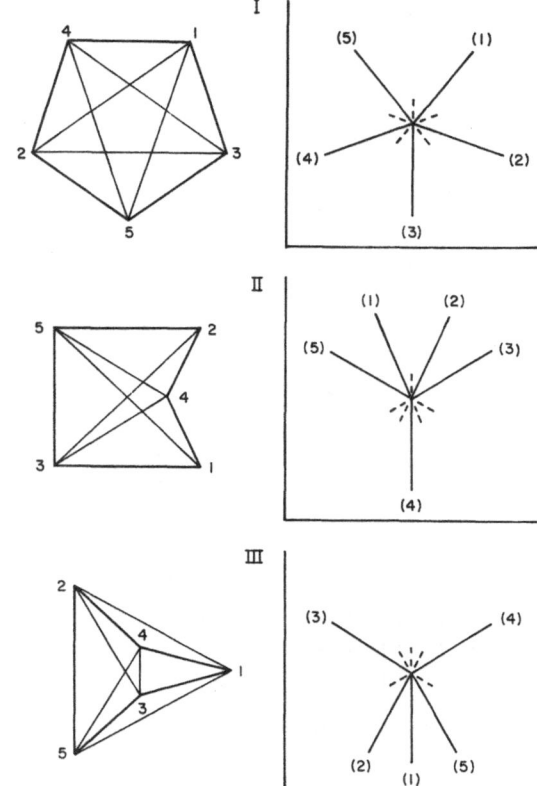

phase 4 is the liquid phase then reaction (5): $1+2=3+4$ might be the reaction occurring at the peritectic curve if 1, 2, and 3 are solid phases.

By type II one of the phases is situated inside the quadrangle formed by the other four phases (Fig. VI-14). As a ternary eutectic point is situated inside the triangle formed by the three solid phases, it might not be surprising that type II may represent a ternary eutectic system. The reactions defining the chemograph are:

(1): $2+3=4+5$
(2): $1+5=3+4$
(3): $1+2+5=4$
(4): $2+3=1+5$
(5): $1+2+3=4$

The chemograph for type II is shown in Fig. VI-14.

By type III, two of the phases are situated inside a triangle formed by the other three phases (Fig. VI-14). Type III is valid for a ternary system with only one peritectic reaction. The reactions are:

(1): $3+2=5+4$
(2): $1+4+5=3$

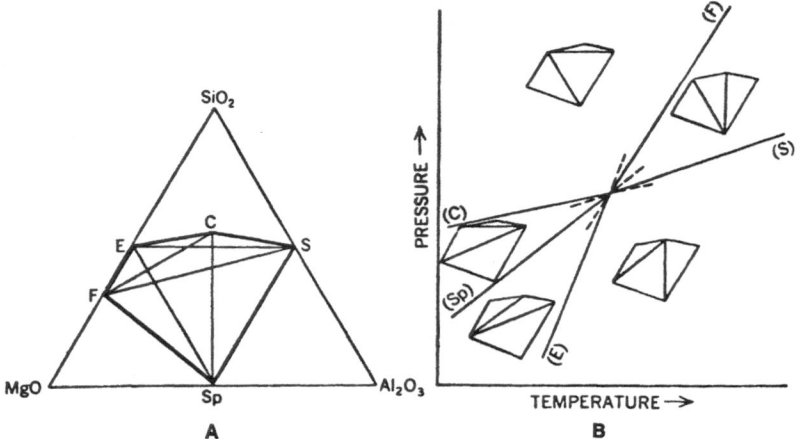

Fig. VI-15 A, B. The chemograph for the system $MgO-SiO_2-Al_2O_3$ with the five phases, enstatite, forsterite, sillimanite, cordierite, and spinel (Zen 1966)

(3): $1+2+5=4$
(4): $1+2+5=3$
(5): $1+2+3=4$

The resulting chemograph is also shown in Fig. VI-14.

13 Examples of Ternary Chemographs

Some examples of ternary chemographs will now be considered. The first example is nondegenerate, while the other two are degenerate ternary equilibria.

The five phases enstatite (E), forsterite (F), sillimanite (S), cordierite (C), and spinel (Sp) belong to the ternary system $MgO-Al_2O_3-SiO_2$ (Fig. VI-15). For these five phases the univariant reactions are (Zen 1966):

(E): $5F + 10S = 3C + 4Sp$
(F): $5E + 5S = 2C + Sp$
(S): $10E + 2Sp = 5F + C$
(C): $3E + Sp = S + 2F$
(Sp): $2S + 4E = F + C$

As evident from Fig. VI-15 A the chemograph for these reactions must be of type I. The chemograph is shown in Fig. VI-15 B.

The five phases kaolinite (K), pyrophyllite (P), andalusite (A), kyanite (Ky), and quartz (Q) belong to the ternary system $Al_2O_3-SiO_2-H_2O$. These mineral compositions are plotted within this system in Fig. VI-16 A. As andalusite and kyanite have the same composition, the invariant equilibria is of the degenerate type. If andalusite and kyanite had different compositions the equilibrium would be of the non-

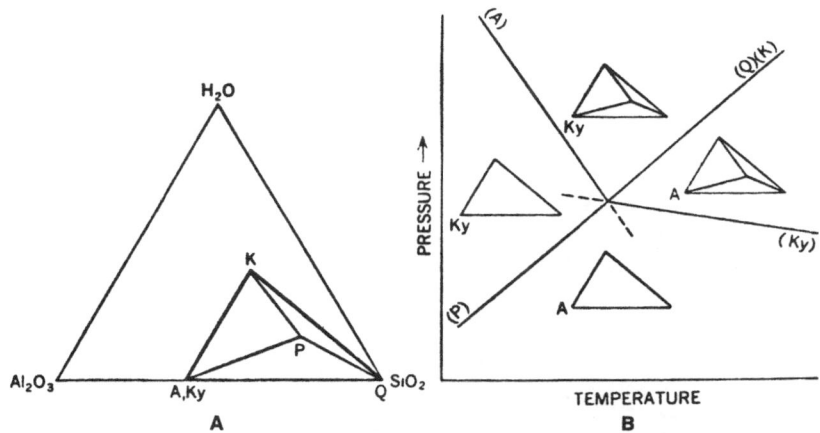

Fig. VI-16 A, B. The chemograph for the system Al_2O_3–SiO_2–H_2O with the phases kaolinite, pyrophyllite, andalusite, kyanite, and quartz (Zen 1966)

degenerate type II. The following reactions occur around the invariant point (Zen 1966):

$$(A):\ Ky + 5Q + K = 2P$$
$$(Ky):\ A\ \ + 5Q + K = 2P$$
$$(Q)\ (K)\ (P):\ A\ = Ky$$

Kaolinite, pyrophyllite, and quartz are relative indifferent phases, while kyanite and andalusite are singular phases. With respect to (A) the univariant curves (Ky), (Q), and (K) must be in opposition to (P). The univariant curve (P) is in opposition with (A), (Q), and (K) with respect to (Ky). Considering the coincidence rule then the indifferent phases K, P, and Q lie on the same side of the two singular phases A and Ky. The univariant curve (Q), (K), therefore, coincides stable to metastable with (P).

Another degenerate example may be illustrated using the system Al_2O_3–SiO_2–H_2O once more, but with some new phases. The five phases corundum (C), diaspore (D), gibbsite (G), water vapor (W), and pyrophyllite (P) have compositions within this system. The compositions of corundum, diaspore, gibbsite, and water are situated along a line (Fig. VI-17 A), and the invariant equilibrium is, therefore, degenerate. These four phases are singular, while pyrophyllite is absolute indifferent. The chemograph is estimated from these reactions (Zen 1966):

$$(C)\ (P):\ D + W = G$$
$$(D)\ (P):\ C + W = G$$
$$(G)\ (P):\ C + W = D$$
$$(W)\ (P):\ C + W = D$$

The chemograph is that for a binary four phase system C–W with a fifth phase P being present throughout (Fig. VI-17 B).

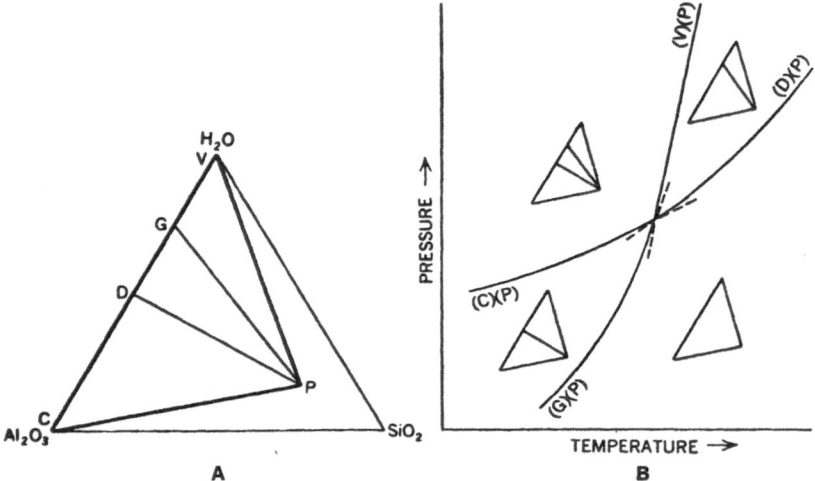

Fig. VI-17 A, B. Another chemograph for the system Al_2O_3–SiO_2–H_2O. In this diagram the five phases are corundum, diaspore, gibbsite, water, and pyrophyllite

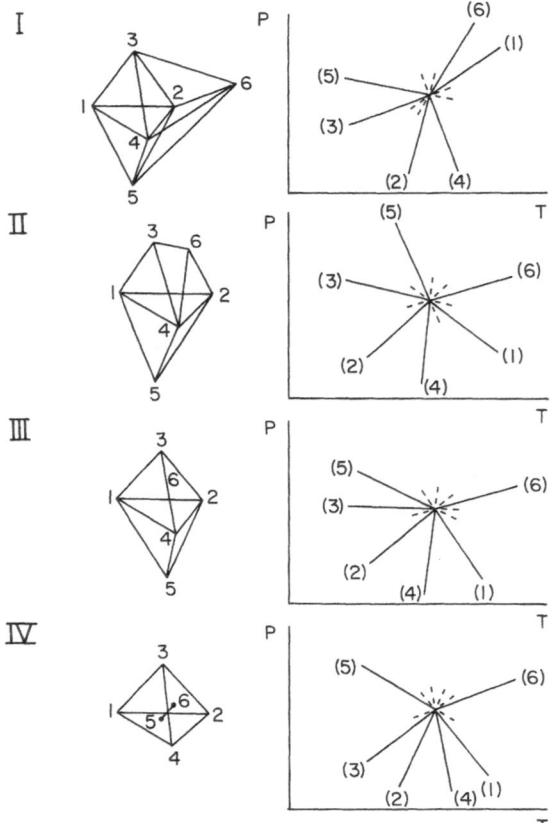

Fig. VI-18. The four possible chemographs for quaternary nondegenerate equilibria

14 Quaternary Systems

There are four different types of nondegenerate quaternary systems. The compositions of the quaternary phases may be shown in three-dimensional polyhedrons (Fig. VI-18). There are six phases in equilibrium at the invariant point, and six univariant curves originate at this point. The reactions for the four different types are listed below.

Type I: (1): $2 + 4 = 3 + 5 + 6$ Type II: (1): $5 + 6 = 2 + 3 + 4$
            (2): $3 + 5 = 1 + 4 + 6$ (2): $3 + 5 = 1 + 4 + 6$
            (3): $2 + 4 = 1 + 5 + 6$ (3): $5 + 6 = 1 + 4 + 2$
            (4): $1 + 6 = 2 + 3 + 5$ (4): $1 + 6 = 2 + 3 + 5$
            (5): $1 + 6 = 2 + 3 + 4$ (5): $1 + 6 = 2 + 3 + 4$
            (6): $3 + 5 = 1 + 2 + 4$ (6): $3 + 5 = 1 + 2 + 4$

Type III: (1): $6 = 2 + 3 + 4 + 5$ Type IV: (1): $6 = 2 + 3 + 4 + 5$
            (2): $3 + 5 = 1 + 4 + 6$ (2): $3 + 5 = 1 + 4 + 6$
            (3): $5 + 6 = 1 + 2 + 4$ (3): $5 = 1 + 2 + 4 + 6$
            (4): $1 + 6 = 2 + 3 + 5$ (4): $1 + 6 = 2 + 3 + 5$
            (5): $6 = 1 + 2 + 3 + 4$ (5): $6 = 1 + 2 + 3 + 4$
            (6): $3 + 5 = 1 + 2 + 4$ (6): $5 = 1 + 2 + 3 + 4$

The chemographs for the four different types are shown in Fig. VI-18.

The present chapter has been intended as an introduction to Schreinemakers' phase theory. The theory has in recent years been developed further by Braun and Stout (1975) and Mohr and Stout (1980).

VII Gas-Bearing Systems

1 Introduction

The addition of a gas component to silicate melts has pronounced effects for the phase relations. The melting temperature decreases, and the phase boundaries are shifted away from those of the dry systems. The composition of the melts formed in the presence of a gas phase is, therefore, different from those of the dry system. The addition of water results in melts with a higher SiO_2 content, and by the addition of carbon dioxide the melts become less silica saturated. The effect of water and carbon dioxide for the melting temperature of diopside is shown in Fig. VII-1 for vapor-saturated melts. The lowering of the melting temperature is most pronounced for water, but the addition of CO_2 also has a significant effect. The variation in the composition of melts by the addition of water and carbon dioxide is shown in Fig. VII-2 for the ternary system forsterite–diopside–silica. The melts of the four phase point $L, S (Fo, En, Di)$ will become more olivine normative by an increase in pressure at dry conditions, while the addition of water moves the point into the

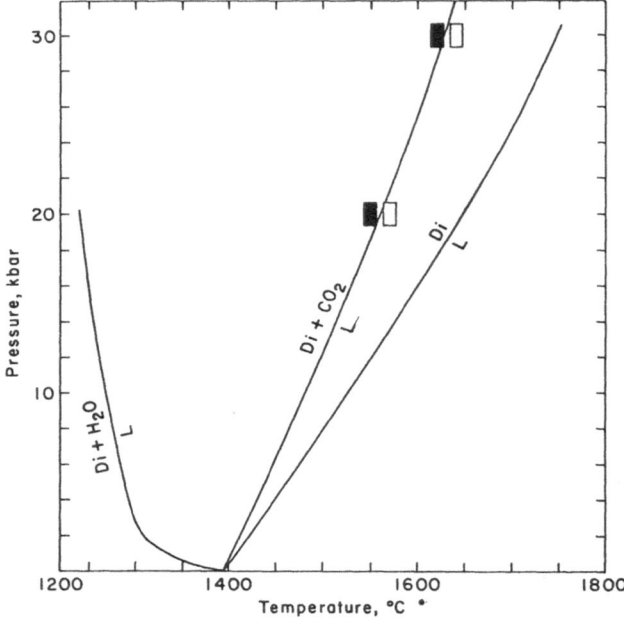

Fig. VII-1. The variation in melting temperature for diopside at vapor saturated conditions with H_2O and CO_2, at vapor absent conditions (Eggler 1973)

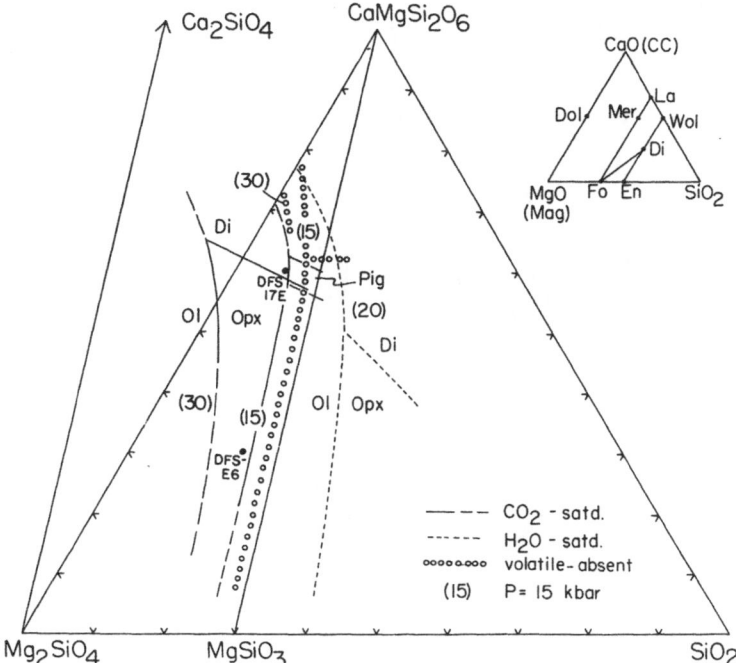

Fig. VII-2. The variation in the four phase point L, S(ol, di, en) at different pressures and gas contents (Eggler 1974). At 30 kbar and CO_2-saturated conditions the point moves into the larnite normative field, and at water-saturated conditions the point moves into the silic normative field

silica-saturated part of the system. The addition of CO_2 moves this point away from the silica apex, and the melts become more silica-undersaturated by increasing CO_2 pressure (Eggler 1974).

The phase relations just considered apply for gas-saturated systems. If the melt not is saturated with the gas component, the effect on melting temperatures and compositions will be less significant. This is demonstrated in Fig. VII-3, which shows the variation in melting temperature for albite at different water pressures. At 10 kbar total pressure the melting temperature is decreased by about 100 °C if the water pressure is 1 kbar, but if the melt is saturated so that P_{H_2O} equals 10 kbar then the melting temperature is decreased more than 600 °C.

The solidus temperature for a system or a rock will be the same by gas undersaturated and saturated conditions as long as the gas content is larger than the gas content of the solids or minerals. Even the nominal anhydrous minerals have low contents of water. They do contain some water, and the gas concentration must be larger than that of the solids if the gas should have any significant influence on the phase relations. The gas concentrations of minerals and magmas should, therefore, be known before the effect of volatiles can be evaluated. It can be generally stated that the H_2O contents of granitic, andesitic, and basaltic melts are so low, less than 1.5% H_2O, that the phase relations of these magmas are not seriously affected by the presence of water. The situation is different for the melilitic and kimberlitic magmas

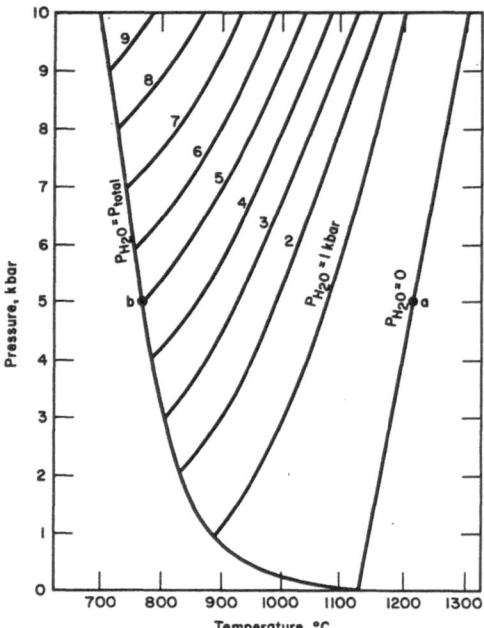

Fig. VII-3. The liquidus temperatures for albite at different water pressures up to 10 kbar. The point b is situated on the solidus curve, while point a is situated on the dry liquidus curve. At 5 kbar H_2O pressure the liquidus and solidus temperatures are coinciding at point b (Yoder 1976)

where the CO_2 concentrations are relatively high, probably about 1–5% CO_2. We will consider the available evidence for gas contents later on, here we will first deal with the effect of water.

2 Binary and Multicomponent Phase Relations

The phase relations of mineral–water systems may be illustrated from the albite–water system at 5 kbar total pressure, shown in Fig. VII-4 (Yoder 1976). The addition of water will lower the melting point of albite from T_a to T_s where the melt becomes saturated with water, and contains about 10 wt% H_2O. The decrease in melting temperature of albite is large, but it should be noted here that it is the molar fraction, and not the wt% that is controlling the decrease in melting temperature. An albite melt that contains 1 wt% H_2O contains 12.84 mol% H_2O, so that the mol% is much larger than the wt%. If the melt contains more water than required for saturation of the melt, then the melt will contain small bubbles of water vapor, and the mixture is situated within the liquid + gas field. The water vapor may dissolve some albite, so that albite is dissolved entirely at large water contents above 95 wt% H_2O.

The behavior by partial melting will depend on the water content of the mixture, the melting behavior will be different on each side of the saturation point S. Mixtures with less water than S will start to melt at T_s, and albite will melt at T_s until all water has been dissolved in the melt. The melt will thereafter move along the liquidus curve for albite, where the last crystals of albite will disappear, and the mixture is completely melted. Mixtures to the right of S will also start to melt at T_s, but all

Fig. VII-4. Estimate of the albite-H_2O system at 5 kbar (Yoder 1976). The solidus temperature is T_s, and the saturation point is S. By the addition of H_2O the liquidus temperature for albite changes from T_a to T_s

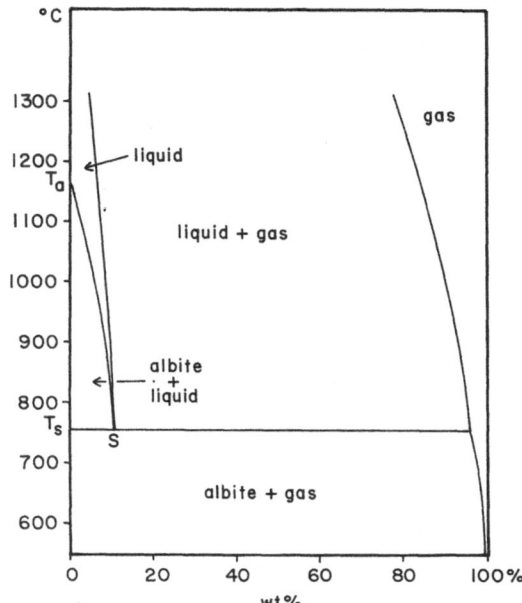

albite will in this case melt at T_s. After albite has melted the mixture will consist of a silicate melt with dissolved water and a separate gas phase forming bubbles in the melt.

The diagram shown in Fig. VII-4 is only a minor part of the albite–H_2O system. The general features of mineral–H_2O systems are shown in the isobaric diagram in Fig. VII-5. The isobaric diagrams of these systems vary with pressure, and the diagram shown is valid at pressures above the critical pressure of water and at pressures lower than the critical pressure of albite. Similar diagrams at other pressures have been shown by Ricci (1951) and Wyllie and Tuttle (1960). The melting point of the pure silicate is T_a, and the melting point at vapor saturated conditions is T_s, the water content at saturation being S. The water vapor will begin to condense and form liquid water at T_v, and ice begins to form at a temperature slightly higher than T_i. Below T_i the system consists only of solids. The vapor + melt field end at T_g because the critical point of silicate melts is substantially higher than the pressure for which the diagram applies.

The solubility of water in magmas increases with increasing pressure until a certain limit, which depends on the composition of the magma. Andesitic magmas can dissolve up to about 20–25% H_2O at pressures above 20 kbar (Wyllie et al. 1976). A decrease in total pressure may, therefore, cause vesiculation of vapor saturated melts. The vesiculation of the magma may be followed by the accumulation of the gas, in which case large volumes of gas may be formed. It is, thus, possible that some explosive volcanic activity is related to the exsolution of gas during the ascend of magma. The formation of a separate gas phase by a release in total pressure was studied by Yoder (1965), who demonstrated the principles from the anorthite–H_2O

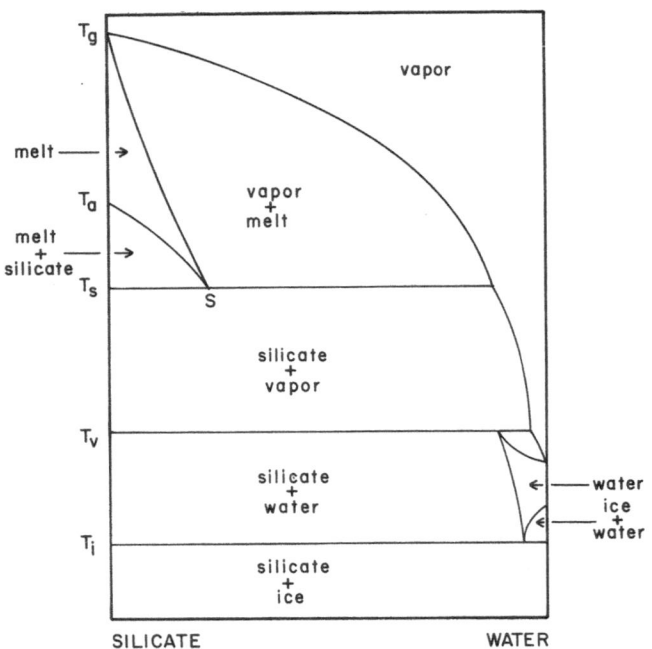

Fig. VII-5. The general phase relations for a silicate-water system. The diagram has not been drawn to scale in order to show the different phase fields

system at 5 and 10 kbar (Fig. VII-6). Consider that a melt has formed initially at A at 10 kbar, and that the pressure is reduced to 5 kbar. The liquid will now be within the An+G field and anorthite will begin to crystallize while the vapor vesiculates in the melt. After crystallization is completed the vapor will be situated between grain boundaries or in miarolitic cavities. The gas is highly mobile and might in nature form hydraulic fractures, so that a system with a separate gas phase has potential for explosive activity.

By a similar decrease in pressure the liquid at B will move into the L+G field at 5 kbar, and the mixture will now consist of silicate melt and vapor. The presence of a gas phase may again result in explosive activity. A decrease in temperature may also cause formation of a separate gas phase. Consider the liquid at point C which is on the liquidus at 10 kbar. By subsequent cooling at 10 kbar the liquid will move along the liquidus while anorthite crystallizes, and reaches the saturation point G,L,S(An) where a gas phase is formed. As anorthite proceeds to crystallize at this point more gas will be formed.

It was shown from experimental work by Burnham and Jahns (1958) that the formation of pegmatites may be explained by the presence of a separate gas phase. They observed that the crystals in a granitic melt tend to grow large when there are gas bubbles present in the melt. Pegmatites may either be of the simple type without exotic minerals, or of the complex type, containing minerals like beryl, allanite, and lepidolite. The pegmatites occur as dykes above or around granite batholiths, and it is likely that the simple pegmatites formed from a granite magma that vesiculated by a decrease in total pressure, while the complex pegmatites may have formed

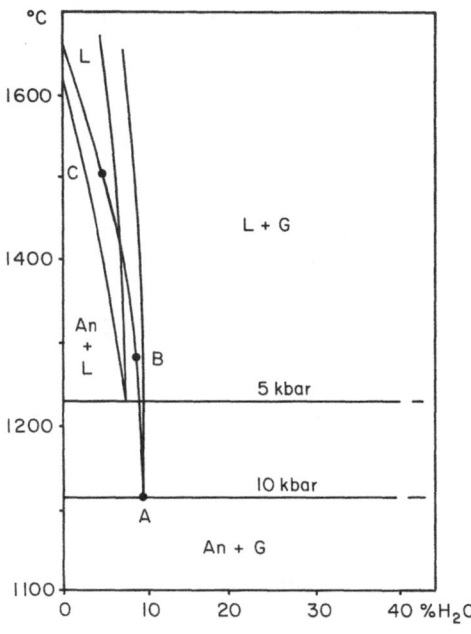

Fig. VII-6. The anorthite-H_2O system at 5 and 10 kbar (Yoder 1965)

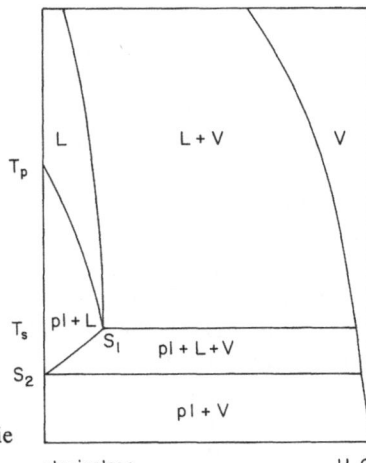

Fig. VII-7. The general phase relations of the system plagioclase–H_2O as estimated by Robertson and Wyllie (1971)

from highly fractionated granite magmas, which have vesiculated mainly due to a decrease in temperature (Maaløe 1974).

The binary systems considered so far have neglected the presence of several solid phases as well as solid solution, and we will now consider systems with more than one solid component. The general phase relations of such systems was worked out by Robertson and Wyllie (1971). The phase relations of the plagioclase–H_2O system is shown in Fig. VII-7. The liquidus curve T_p–S_1 is equivalent to the liquidus curve

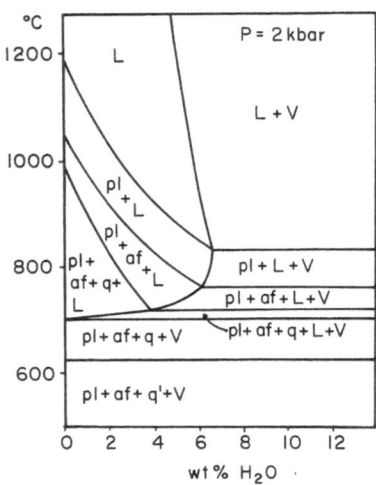

°C

Fig. VII-8. The phase relations for a granite-H_2O system at 2 kbar (Whitney 1975). Abbreviations: L liquid; pl plagioclase; af alkali feldspar; q beta-quartz, q' alfa-quartz; and V H_2O vapor

for albite except that plagioclase now changes its composition along the curve. When a crystallizing melt reaches S_1, saturation occurs and a separate gas phase is formed. During further decrease in temperature more plagioclase crystallizes while the amount of melt decreases. Since the amount of melt decreases less water is required for saturation, and the melt will move along the S_1–S_2 curve.

The phase relations for a system with several solid phases is shown in Fig. VII-8 for a granitic composition. Plagioclase has the highest melting temperature, and alkali feldspar and quartz melt at lower temperatures. The crystallization sequences are the same within the vapor saturated and under-saturated fields. This is a quite common feature for rock compositions without hydrous phases, but the crystallization sequence may change and is dependent on composition (Whitney 1975).

The phase relations of hydrous minerals like micas and amphiboles are quite different from those of the nominal anhydrous minerals. The phlogopite–H_2O system was investigated by Yoder and Kushiro (1969) at 10 kbar and is shown in Fig. VII-9. At water contents less than 2.5% H_2O phlogopite is not in equilibrium with melt, the water content of the melt is too small. Above 2.5% H_2O there is sufficient water present to stabilize phlogopite at melting temperatures, and the stability of phlogopite increases with increasing water content until a maximum is reached at (c). Here the melting temperature begins to decrease because water now decrease the melting temperature in the same manner as for anhydrous minerals. The melting temperature decreases until the melt becomes saturated with water. The phlogopite–H_2O system shows that hydrous minerals may display a melting curve with a temperature maximum within the water undersaturated field. The temperature maximum is dependent on total pressure as evident from Fig. VII-10 which shows the melting curve for pargasite at different pressures.

The phase relations of gas-bearing systems like phlogopite–H_2O are relatively complicated, and as the melting of phlogopite may have implications for the formation of basalts, it is convenient to determine the P–T topology of phlogopite. The topology can be estimated when the reacting phases are known. These phases were

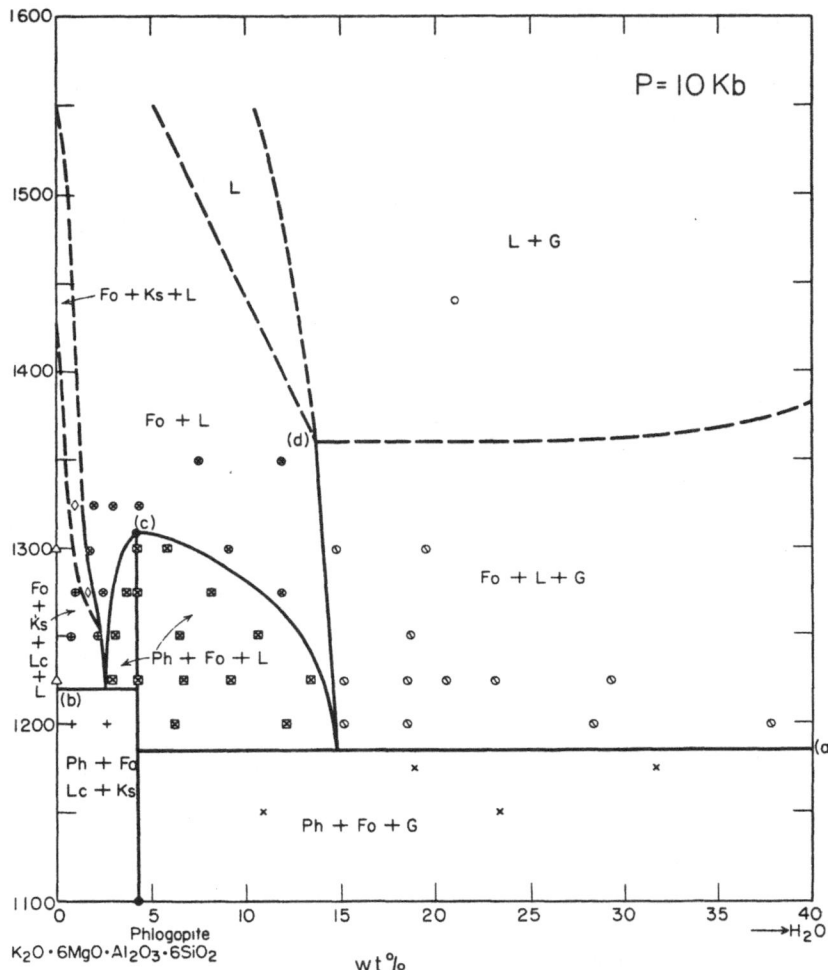

Fig. VII-9. Temperature-composition projection at 10 kbar for the phlogopite-H₂O system (Yoder and Kushiro 1969). (*a*) is the melting temperature in the presence of excess H₂O; (*b*) the beginning of melting in the absence of a gas phase; and (*c*) is the maximal thermal stability of phlogopite

estimated by Yoder and Kushiro (1969) and are the following: phlogopite (Ph), forsterite (Fo), orthorhombic kalsilite (Ok), leucite (Lc), liquid (L), and a separate gas phase (G). The reactions between these phases are:

(V)   Ph = Fo + Ok + Lc + L

(Ph) Fo + Ok + Lc + G = L

(L)   Ph = Fo + Ok + Lc + G

(Fo) Ph + Ok + Lc + G = L

(Ok) Ph + Lc + G = Fo + L

(Lc) Ph + G = Fo + Ok + L

The P–T topology defined by these reactions is shown in Fig. VII-11, and the experimentally estimated curves for (L), (Lc), and (Ph) are shown in Fig. VII-12. In addition, the latter diagram also shows the melting curve for phlogopite in the absence of a vapor phase.

The molar volumes of leucite and kalsilite are so high that the melting curve for phlogopite will not display an inversion in slope, at least beneath 30 kbar. According to the Clausius-Clapeyron equation, the slope of the melting curve may become

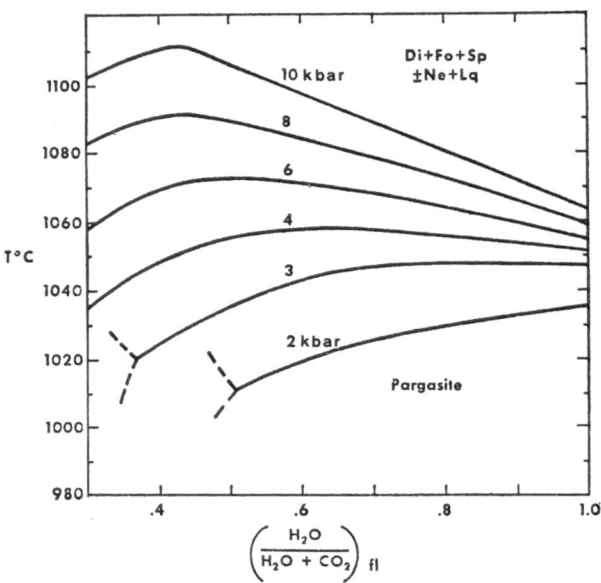

Fig. VII-10. The stability of the amphibole pargasite at different $H_2O/(H_2O + CO_2)$ ratios and pressures between 2 and 10 kbar (Holloway 1973)

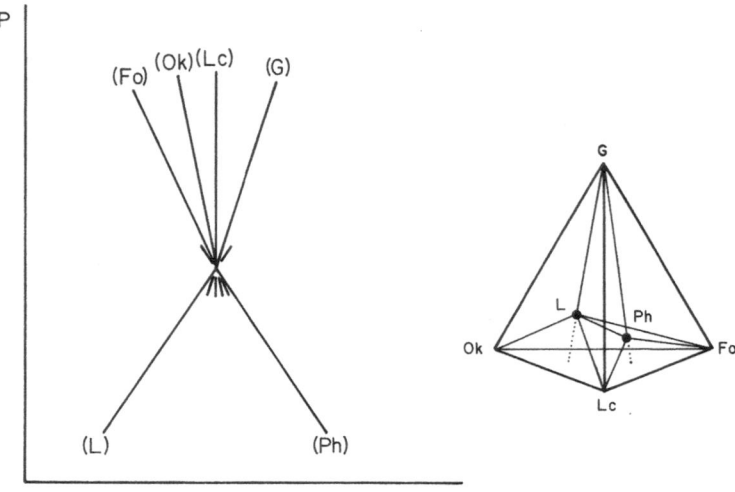

Fig. VII-11. The P-T topology for phlogopite (*left*) and the composition of the initial melt formed by melting of phlogopite (*right*)

Fig. VII-12. The estimated P-T phase relations for phlogopite (Yoder and Kushiro 1969). *A* The maximum stability of phlogopite in the presence of H_2O gas; *B* the beginning of melting of the gas absent phase assemblage $Ph + Fo + Ks + L$; *C* the maximum stability of phlogopite in the absence of a gas phase; and *D* this curve is close to the minimum liquidus in the presence of an H_2O gas phase

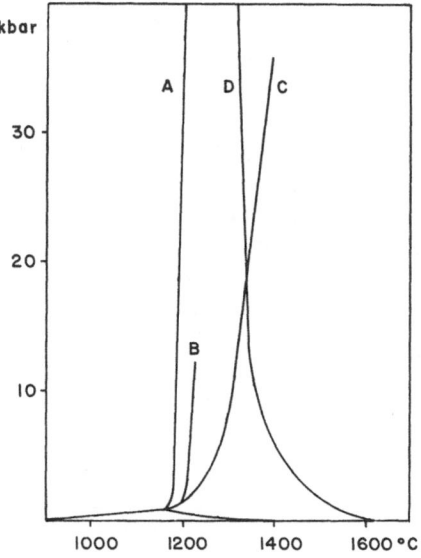

Fig. VII-13. Experimentally based on deduced curves for the solidus of amphibole peridotite, with less than 0.4% H_2O (Wyllie 1978). Three different estimates are shown. *CRS* The shield geotherm; *CRO* the oceanic geotherm (Clark and Ringwood 1964); *RO* the oceanic geotherm, according to Ringwood (1966)

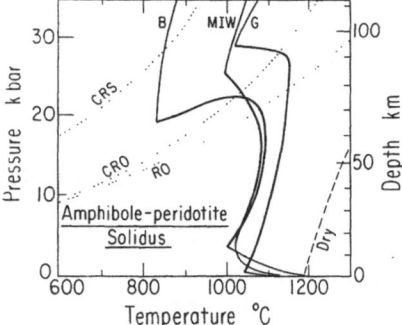

negative if the volume of the melting products is less than that of phlogopite. This has not been observed for phlogopite, but such an inversion is present for amphibole (Fig. VII-13).

The difference in melting behavior between the hydrous and anhydrous minerals may imply changes in the crystallization sequences in the vapor undersaturated and saturated fields. The melting relations of a granite with biotite are shown in Fig. VII-14. It is evident that the crystallization sequence varies with the water content. By comparing the crystallization sequences at different water contents with the crystallization sequence of the granite, estimated from textural relationships,.it becomes evident that the granite magma must have contained less than 1.2% H_2O. The amount of biotite in the granite shows that the water content must have been higher than 0.4% H_2O, and the water content of the granite magma must have been between 0.4% and 1.2% H_2O.

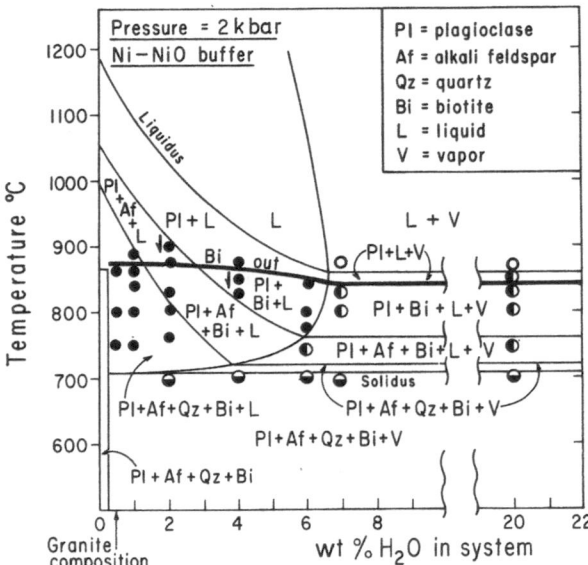

Fig. VII-14. The phase relations for a biotite granite at 2 kbar and different water contents (Maaløe and Wyllie 1975). The melting curve for biotite intersects the melting curves for quartz and alkali feldspar, and the crystallization sequence consequently changes with varying water contents

3 P–T Diagrams of Rock–H₂O Systems

The water saturated P–T diagrams for rock compositions are similar in their general appearance, and do not differ that much from the albite–H_2O system.

The P–T diagram for a rhyolite is shown in Fig. VII-15. The liquidus curves at different water contents are shown and the liquidus phases are indicated. Quartz and plagioclase are on the liquidus at low pressures, while quartz alone is on the liquidus at high pressures.

The P–T diagram for an andesitic composition is shown in Fig. VII-16. At pressures above 15 kbar the subsolidus assemblage has eclogitic mineralogy, and hornblende is stable within the pressure range from 5 to 15 kbar.

The P–T diagram for an olivine tholeiite is shown at pressures up to 30 kbar in Fig. VII-17. The subsolidus assemblage is eclogitic at pressures above 20 kbar. The phase relations of a tholeiite composition at pressures up to 10 kbar is shown in more detail in Fig. VII-18. Amphibole is not stable at water absent conditions, but is stable above the solidus at water excess conditions, at pressures above 1 kbar. A basaltic composition will, thus, form an amphibolite if sufficient water is present. The compositions of melts formed by melting of a basaltic composition have been estimated by Helz (1976) at 5 kbar, and at water undersaturated conditions by Holloway and Burnham (1972) at 2 and 3 kbar at oxygen fugacity controlled conditions.

These three P–T diagrams of the phase relations at water excess conditions, as well as other similar diagrams, show that the water content of magmas generated at vapor excess conditions must be fairly high, at least 5% H_2O, the exact percentage being dependent on the pressure. If it is considered that an andesite has been generated by partial melting of a hydrous tholeiite at, say 30 kbar, then the water

Fig. VII-15. The liquidus curves for a rhyolitic composition at different water contents up to 30 kbar (Wyllie 1977)

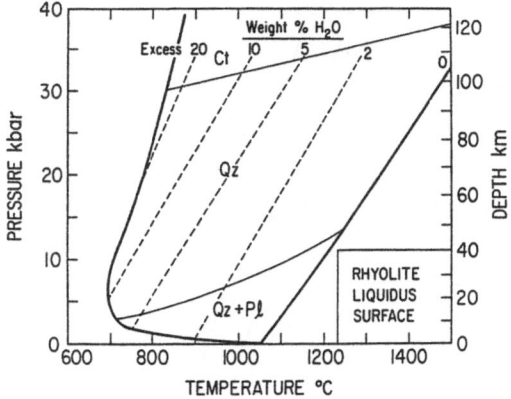

Fig. VII-16. The liquidus curves for an andesitic composition at different water contents. Note that the liquidus phase assemblage changes with both pressure and water contents (Wyllie 1977)

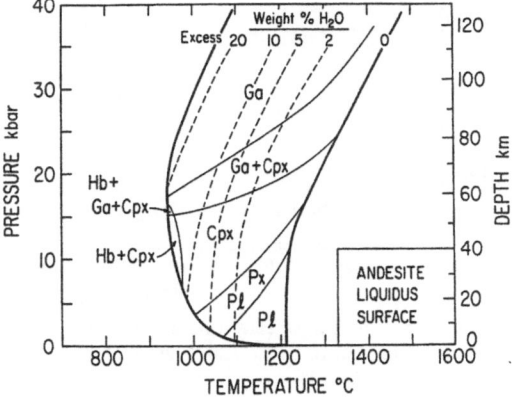

Fig. VII-17. The liquidus curves for an olivine tholeiite at different water contents (Wyllie 1977)

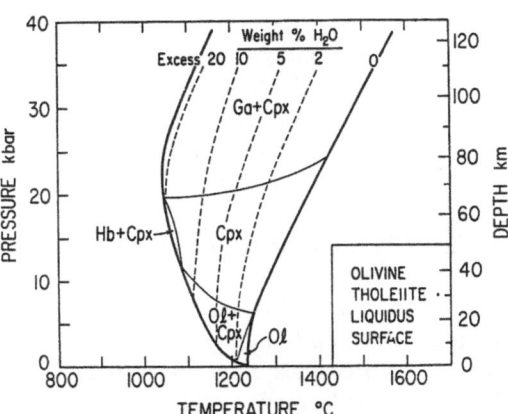

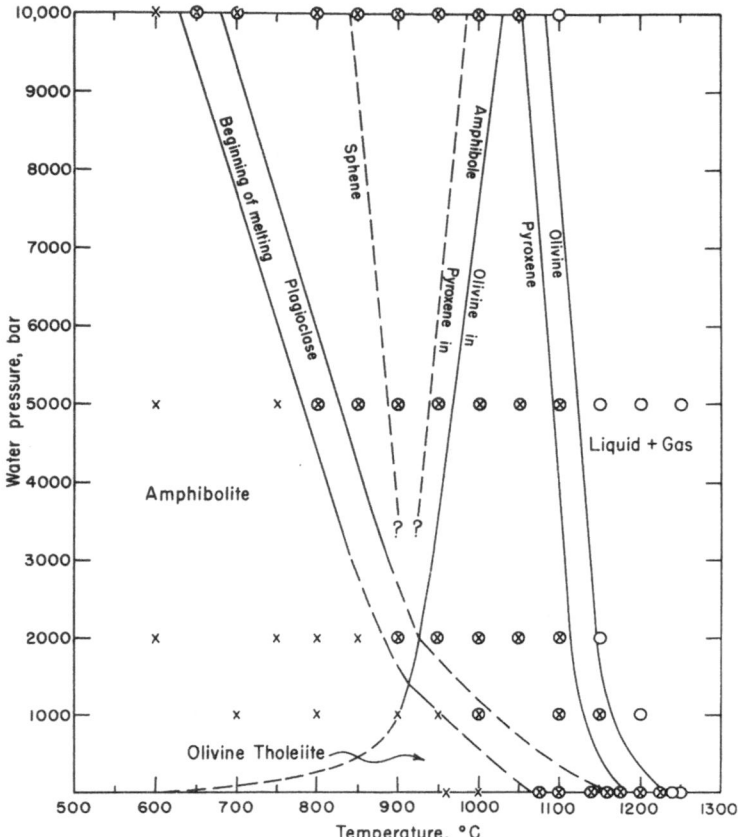

Fig. VII-18. The phase relations at water saturated conditions for an olivine tholeiite, Kilauea, Hawaii (Yoder and Tilley 1962). At water pressures above ca. 500 bar the basaltic liquid will form an amphibolite

content of the andesite must be at least 10% H_2O (Fig. VII-17). The actual water content of magmas will, therefore, indicate if they were formed at water excess conditions. As shown below the water contents of magmas are generally low, and there is no reason to suggest that magmas, like andesites and basalts, are generated from a source containing large amounts of water.

4 Phase Relations in the Presence of Carbon Dioxide

The addition of CO_2 to a phase assemblage will decrease the solidus and liquidus temperatures, and change the compositions of the melts generated. The melts will become more silica undersaturated, and at low pressures, a separate carbonate liquid may be formed in alkalic magmas, like syenitic and nephelinitic magmas.

Within the mantle, at pressures lower than 30 kbar, the CO_2 component will only act as a gas phase that is dissolved in the melts generated. At higher pressures the silicate minerals will react with the CO_2 gas, and one or more of the carbonate minerals, like magnesite, dolomite, and calcite, may be formed. At high pressures forsterite will react with CO_2, according to the equation: $Mg_2SiO_4 + CO_2 = MgSiO_3 + MgCO_3$ (Newton and Sharp 1975). This means that forsterite will coexist with magnesite, and it is said that forsterite has been carbonated. By partial melting the first liquid is always formed in equilibrium with all the minerals present, and the appearance of a carbonate mineral will, thus, ensure that the first formed melts have a high content of carbonate. It is, thus, considered that the presence of CO_2 in the mantle may explain the compositions of kimberlites and nephelinites, which are silica undersaturated magmas with a relatively high CO_2 content. The carbonation of silicate minerals above 30 kbar also result in a substantial decrease of the solidus temperature of lherzolitic rocks, and it is possible that the low velocity layer is caused by the presence of CO_2 in the mantle, however, the low velocity layer could also be due to the presence of water (Lambert and Wyllie 1968; Eggler 1978; Wyllie 1979). As the vapor composition of the mantle is still unknown, we cannot yet specify the cause of the low velocity layer, but it is likely that it is caused by H_2O or CO_2 or a mixture of these two gases.

We will subsequently consider the effect of CO_2 for partial melting in the mantle, and thereafter deal with the generation of carbonatite magmas, which are caused by liquid immiscibility.

5 CaO–MgO–CO₂

Calcite melts to form liquid at all pressures up to at least 30 kbar, and does not undergo polymorphic transitions near the liquidus within this pressure interval (Fig. VII-19) (Irving and Wyllie 1975). Magnesite decarbonates at pressures lower than 25 kbar according to the reaction $MgCO_3 = MgO + CO_2$. At pressures above 25 kbar magnesite melts in the same manner as calcite. At high pressures there is extensive solid solution between calcite and magnesite, there is continuous solid solution between calcite and dolomite, and magnesite contains some calcite in solid solution (Fig. VII-20) (Irving and Wyllie 1975).

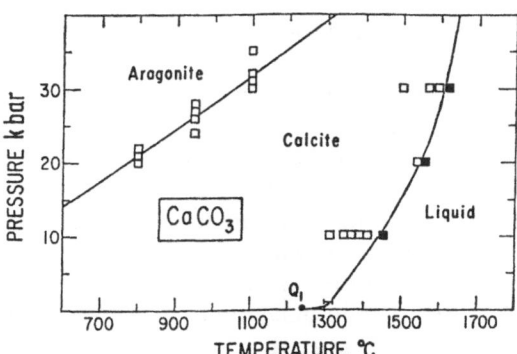

Fig. VII-19. The phase relations of calcite at 30 kbar. Calcite is stable at low pressures while aragonite becomes stable at high pressures. However, the phase transition does not intersect the liquidus at pressures below 50 kbar (Irving and Wyllie 1975)

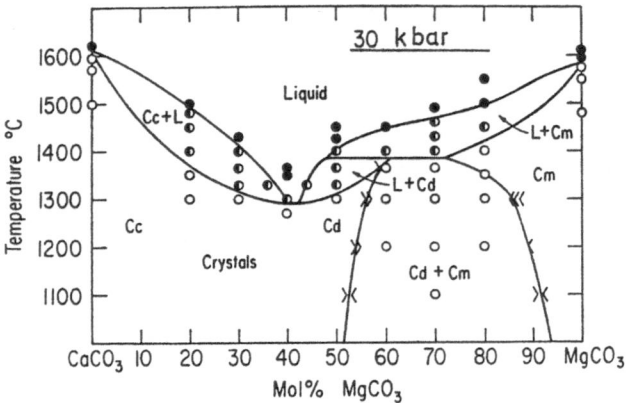

Fig. VII-20. The binary system calcite–magnesite at 30 kbar (Irving and Wyllie 1975)

6 Carbonation Reactions

As mentioned above, the addition of CO_2 may result in the formation of a carbonate mineral, or CO_2 may be present as a gas phase that decreases the liquidus and solidus temperatures. An example of the latter is shown in Fig. VII-21 for the binary system diopside–CO_2 at 20 kbar. The phase relations of this system are similar to the phase relations of the diopside–H_2O system.

Within the mantle the silicate minerals become carbonated at relatively high pressures. Let us consider the carbonation of forsterite, which can be described within the system MgO–SiO_2–CO_2. We can estimate the P–T topology of this system if we know the composition of the liquid. The composition of the first formed liquid has been estimated by Wyllie and Huang (1976), and is situated on the CO_2–rich side of the join enstatite–magnesite (Fig. VII-22). With this composition of the liquid the following reactions occur within the ternary system:

(L)	Mc + En = Fo + V	Mc: magnesite
(Mc)	Fo + V = En + L	En: enstatite
(En)	Mc + Fo + V = L	Fo: forsterite
(V)	En + Mc = Fo + L	V: vapor
(Fo)	Mc + En + V = L	L: liquid

The (V) reaction shows that the liquid is formed at a reaction point where forsterite is reacting with liquid. Using Schreinemakers' theory we obtain the phase topology shown in Fig. VII-22. The P–T parameters for this diagram were estimated by Wyllie and Huang (1976) and are shown in Fig. VII-23. At temperatures lower than 1500 °C and pressures below 25 kbar forsterite is in equilibrium with CO_2 vapor, and no magnesite is formed. By an increase in pressure forsterite and CO_2 react to form magnesite according to reaction (L). This univariant reaction extends to the invariant point Q_2 where liquid appears as the fifth phase. Five univariant curves emanate from the invariant point Q_2, however, the (V) curve is only indicated in Fig. VII-23. The univariant curve (Mc) is divided into two curves at the singular point S at 41 kbar. Below this pressure forsterite + vapor and forsterite + enstatite

+vapor melts congruently. At S forsterite starts to melt incongruently forming enstatite and liquid. This reaction is absent from the vapor-free system and is caused by the presence of CO_2 which increases the stability of enstatite. The increased stability of enstatite is due to an increased degree of polymerization of the melt in the presence of CO_2 (Eggler 1978). Note that the Fo + V = L curve in Fig. VII-23 is a melting curve and not an univariant curve.

The mantle does not contain large amounts of CO_2, and melting within the mantle will be represented by a composition consisting mainly of Fo and En. The melting curve of the mantle is, therefore, represented by the (Mc) curve in the model

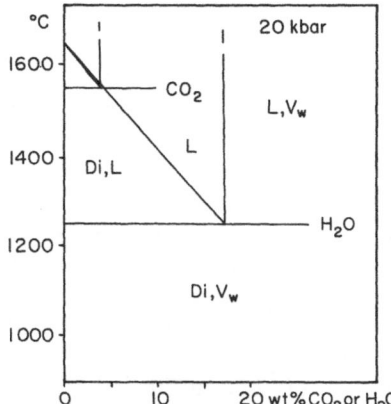

Fig. VII-21. Phase relations for the diopside-H_2O and diopside-CO_2 systems at 20 kbar. The decrease in liquidus temperature is much larger for liquids with H_2O than with CO_2 (Eggler 1973)

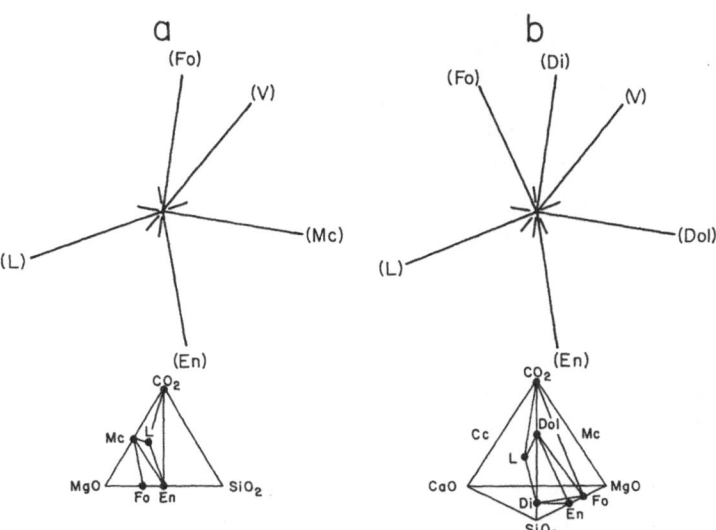

Fig. VII-22a, b. The phase topology for the systems MgO–SiO$_2$–CO$_2$ (a) and MgO–CaO–SiO$_2$–CO$_2$ (b). The composition of the initial liquid formed by melting is shown below the diagrams

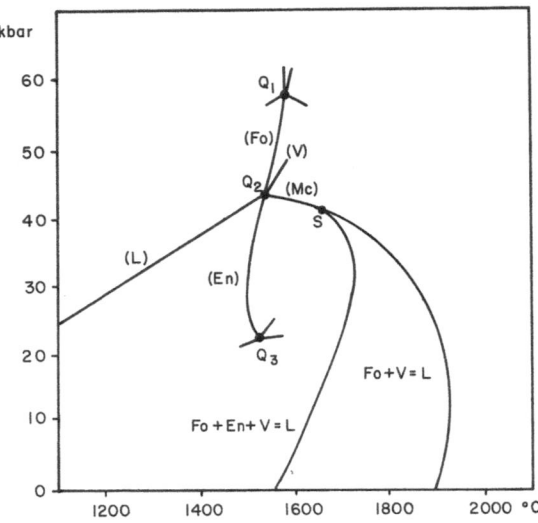

Fig. VII-23. An experimental estimate of some of the univariant curves for the MgO–SiO₂–CO₂ system. The *points Q* are invariant points (Wyllie and Huang 1976)

system just considered. This univariant curve displays an inversion in slope, which has important bearings on melting within the mantle. We will evaluate this feature in more detail on the basis of the system forsterite–diopside–CO₂ which is more similar to mantle compositions.

The phase relations of the join forsterite–diopside with excess CO_2 was estimated by Eggler (1978), and the phase relations of this join at 20 and 30 kbar is shown in Fig. VII-24 and Fig. VII-25. The principal features of these diagrams are that enstatite enters the liquidus at 30 kbar, in agreement with the experimental results of Brey and Green (1977), and that a carbonate mineral is stable at subsolidus conditions at 30 kbar. The diopside–forsterite system is part of the quaternary system MgO–CaO–SiO₂–CO₂ at vapor present conditions, and the P–T topology of this system can be estimated from the following reactions:

(Fo)   Di + Dol + V = En + L
(Di)   En + Dol + V = Fo + L
(V)   En + Dol = Di + Fo + L
(Dol) Di + Fo + V = En + L
(En)   Di + Dol + V = Fo + L
(L)   En + Dol = Di + Fo + V

The P–T topology is shown in Fig. VII-22b, and the experimental determination of the topology is shown in Fig. VII-26. This diagram is somewhat more complicated than the topology diagram of Fig. VII-22b because pigeonite apparently is stable at high temperatures below 25 kbar. There is, therefore, a new invariant point, I_2 on the (Dol) curve. However, the essential point of the diagram is that that the melting curve at CO₂ excess conditions has an inverse slope. Note that the (En) curve only will be the melting curve if enstatite is absent from the phase assemblage.

The CO_2 content of the mantle will not result in CO_2 excess conditions, for which the phase relations of Fig. VII-26 apply. At least it is unlikely that the upper

Fig. VII-24. Vapor-saturated phase relations on the
join diopside–forsterite
–CO₂ at 20 kbar. Volatile
absent phase relations are
shown by stippled curves
(Eggler 1978)

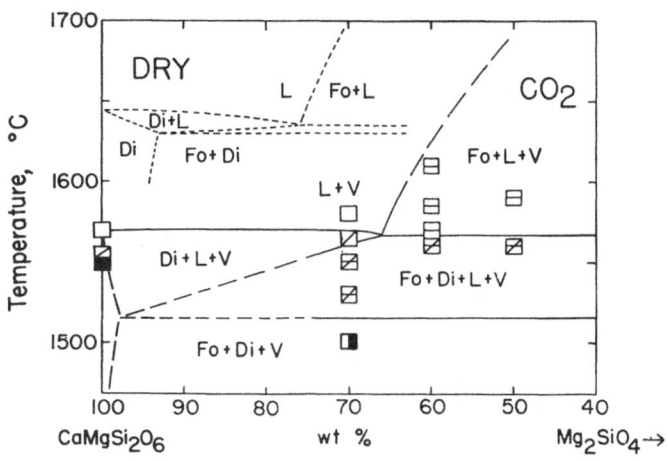

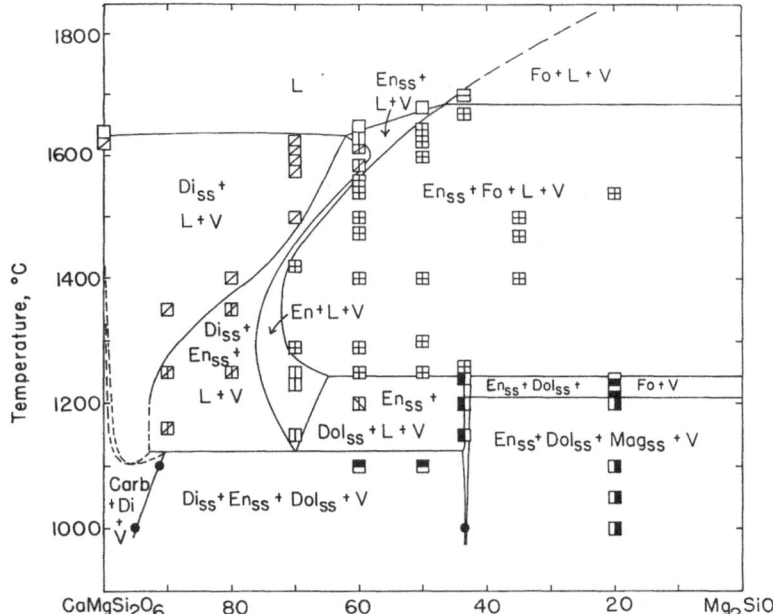

Fig. VII-25. Vapor-saturated phase relations on the join diopside–forsterite–CO₂ at 30 kbar. *Carb* refers to calcite-dolomite solid solutions. Note that enstatite appears on the liquidus at this pressure, while enstatite was absent at 20 kbar (Eggler 1978)

mantle may attain CO_2 excess conditions, while it might be a possibility for the lower mantle. Eggler (1978), therefore, deduced the phase relations for a composition with 0.2% CO_2, which may be compared with the excess phase relations in Fig. VII-27. The solidus curve is here compared with the oceanic geotherm, and it is evident that the geotherm intersects the solidus at a depth of about 90 km. The geotherm is believed to apply for the average oceanic lithosphere, the temperatures would be

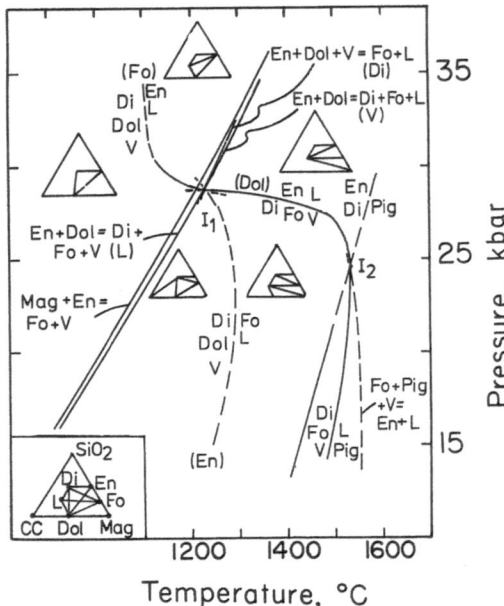

Fig. VII-26. Topology of reactions relevant to peridotite compositions in the system $CaO-MgO-SiO_2-CO_2$ deducted from experiments. The phase relations shown here are vapor-saturated. The *stippled curves* are inferred and are not based on experiments (Eggler 1978)

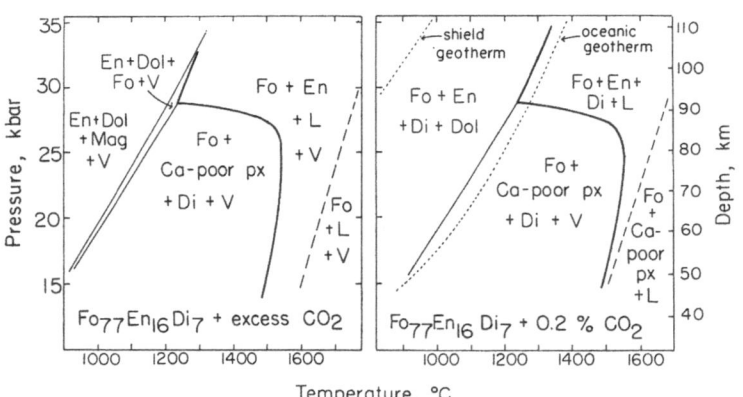

Fig. VII-27. Phase relations of a peridotite composition in the system $CaO-MgO-SiO_2-CO_2$. *Left* The phase relations at CO_2-saturated conditions; *right* the phase relations for a composition containing 0.2% CO_2 (Eggler 1978)

higher near the ridge axes. As the mean depth of the low velocity layer is at about 90 km in the matured lithosphere, the diagram suggests that CO_2 could account for the low velocity layer. It is apparent from the shape of the solidus curve that relatively large variations in temperature would not change the depth of the low velocity layer, and temperatures in excess of 1500 °C are required to decrease the depth of this layer – if the CO_2 model is correct. The temperature gradient of the oceanic lithosphere is not yet known sufficiently well to evaluate the CO_2 model further, but the phase relations shown in Fig. VII-27 constrain the relationship be-

Fig. VII-28. The shape of the solidus surface for peridotite–CO_2–H_2O, illustrated by contours for constant vapor phase compositions in molar fractions (Wyllie 1979). The contour for H_2O implies the absence of amphibole above the solidus

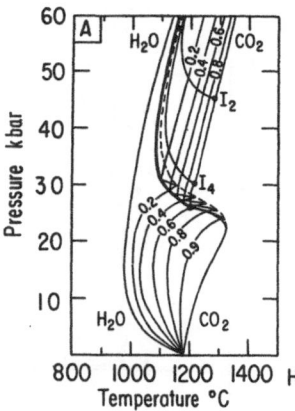

tween the lithosphere thickness and the geotherm if CO_2 is present in the upper mantle.

As the CO_2/H_2O ratio of the mantle is unknown it is relevant to estimate the variation in the solidus temperature for different CO_2/H_2O ratios, and this has been done by Wyllie (1979), and the results are shown in Fig. VII-28. The presence of only H_2O in excess amounts will lower the solidus quite substantially. The H_2O content of abyssal tholeiites is so low that water saturated conditions are excluded. The solidus temperature must be higher than the temperature of the water excess melting curve, and the solidus temperature will depend on the CO_2/H_2O ratio as shown in Fig. VII-28. The presence of amphibole and phlogopite will change the shape of the solidus curve, however, we will not consider these further complications here. The phase relations of partial melting of a lherzolitic composition with both phlogopite and a carbonate has been studied by Holloway and Eggler (1976), and further aspects of magma generation in the mantle with CO_2 and H_2O has been reported by Eggler and Baker (1982).

7 Liquid Immiscibility

Intrusions of syenitic and foyaitic rocks sometimes contain central plugs or cone sheets of carbonatites, which appear to have formed during the late stage of magmatic activity. An excellent account of these intrusions has been given by Tuttle and Gittins (1966). These carbonatites have formed from the latest most fractionated magmas, and their presence suggests that a carbonatite magma somehow may be formed from a silicate magma. Carbonatite has even formed lava flows, the Oldinoy Lenga volcano of the East African rift zone has repeatedly extruded carbonatite lavas (Dawson 1966). The early intrusions of these intrusive complexes consist of pyroxenites, syenites, and ijolites, and the carbonatites must have formed from these magmas.

It was shown by Wyllie and Tuttle (1960) that a carbonatitic liquid could exist at relatively low temperatures of about 600–700 °C in the presence of water, which decreases the solidus temperature of carbonates. It was later shown by Koster van

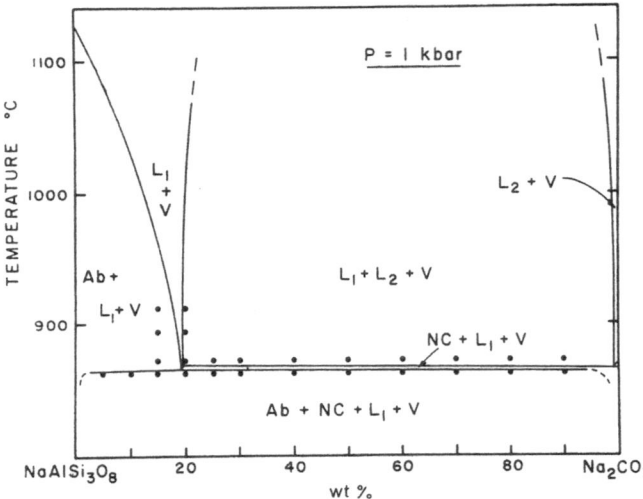

Fig. VII-29. The system albite–Na₂CO₃ at 1 kbar (Koster van Groos and Wyllie 1966). Two different liquids are formed by the liquid immiscibility, a silicate-rich liquid with some Na₂CO₃ in solution, and a carbonate liquid with a very small amount of albite in solution

Groos and Wyllie (1966) that a carbonate liquid may be formed by liquid immiscibility from an alkalic silicate liquid. An example of liquid immiscibility is shown in Fig. VII-29, which shows the pseudobinary phase relations for the albite–Na₂CO₃ system at 1 kbar. Two different liquids are formed, a silicate liquid with about 20% Na₂CO₃, and a carbonatitic liquid with almost no albite in solution. These and similar phase diagrams show that carbonatites may be generated by liquid immiscibility. At first the carbonate liquid will form small bubbles in the silicate liquid. The carbonatitic liquid has a smaller density than the silicate liquid, and the bubbles of carbonatitic liquid may, therefore, become accumulated in the topmost part of magma chambers, whereby a large body of carbonatitic magma may be formed. This magma may then intrude the surrounding rocks or extrude on the surface.

8 Gas Contents of Magmas

The gas contents of magmas have proven particularly difficult to estimate, and it is only during recent years that fairly reliable estimates have been done using rather different methods.

One of the more obvious methods for an estimation of the gas contents of magmas is the direct measurement of volcanic emanations. However, this method has been proven to be without much value, as an essential part of the water vapor stems from groundwater. Therefore, other methods have to be employed.

The solubility of gases in silicate magmas decreases with decreasing pressure (Figs. VII-30 and VII-31). An ascending magma will begin to vesiculate when the total pressure becomes less than the vapor pressure, since bubbles are formed as soon as saturation occurs (Murase and McBirney 1973). The vesicularity of dykes may, therefore, provide some evidence about the gas contents of magmas. It was noted from extensive field work by MacDonald and Abbot (1970) that: "The rock in most dykes is very dense, with almost no vesicles. Only in a zone that must have been

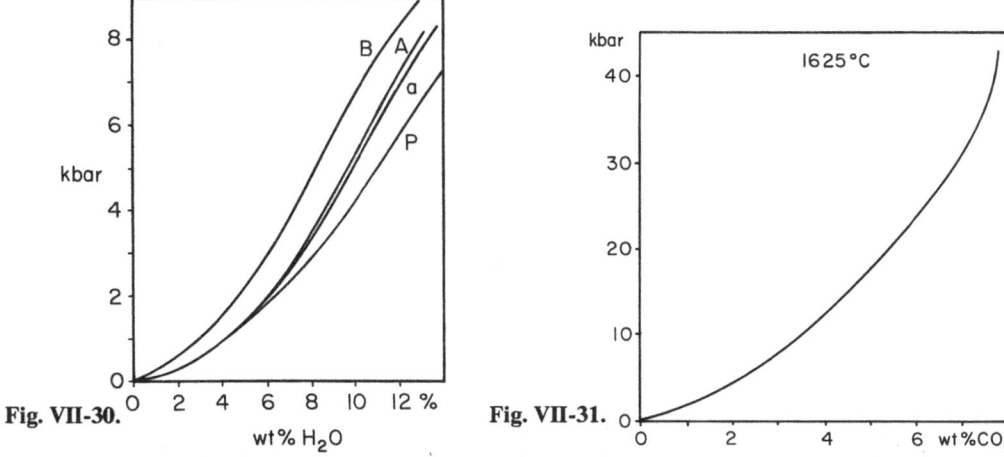

Fig. VII-30.

Fig. VII-31.

Fig. VII-30. Weight percent solubility of H_2O in melts of *A* andesite; *B* basalt; *a* albite; and *P* pegmatite (Burnham 1975)

Fig. VII-31. Weight percent solubility of CO_2 in a melt of olivine melilite nephelinite at 1625 °C (Eggler and Mysen 1976)

within a few hundred feet of the surface at the time of intrusion does a significant number of vesicles appear. It seems probable that below that level the pressure due to the weight of the overlying magma in the dyke fissure was too great to permit rapid release of gas from solution to form bubbles in the magma." This description was considered mainly to apply for tholeiitic dykes.

The pressure within the magma at a depth of a few hundred feet would not be more than about 30 bar, and the percentage of gas must be about 0.1 wt% if the gas is water vapor, and about 0.01 wt% if the gas is CO_2. These values should only be considered approximate as the amount of phenocrysts and the temperature also influence the vesiculation, but it is apparent that we are dealing with low gas contents in the tholeiitic magmas.

The first systematic study of the variation in vesicles was made by Moore (1965) who made an excellent investigation of samples of lava dredged from the submarine part of the eastern rift zone of the Kilauea volcano, Hawaii. The samples of lava were dredged from depths between 400 m and 5.5 km (Fig. VII-32). The bulk density of the lava decreases abruptly at a depth of about 800 m, and the diameter of the vesicles increase with decreasing depth. The volume of gases increases with decreasing pressure, and the amount of gas exsolved is, therefore, not identical to the volume of the vesicles. The amount of gas exsolved was calculated by Killingley and Muenow (1975) and their result is shown in Fig. VII-32. The bend in the curve at a depth of 800 m suggests that the magma began to undergo strong vesiculation at this depth. If the gas forming the vesicles is mainly water vapor then the gas content of the magma should be about 0.45% H_2O (Moore 1965). With this gas content there should be no vesicles at depths larger than 800 m, but there is nevertheless some vesicle formation down to a depth of 5 km. These vesicles may have formed during the cooling of the glass as the solubility of gas decreases with decreasing temperature.

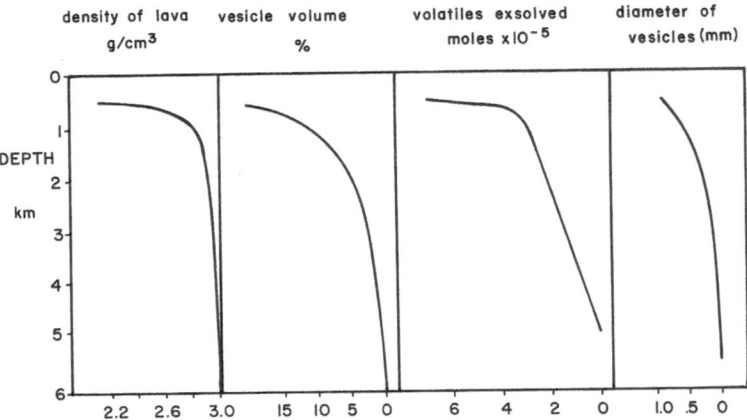

Fig. VII-32. The vesicularity relationships of submarine samples of tholeiite from the submarine part of the Kilauea rift zone, Hawaii (Moore 1965; Killingley and Muenov 1975). The curves suggest the exsolution of volatiles at a depth of about 800 m

The water content of the glassy rim of the pillow lavas was estimated by Moore (1965) and the water contents varied between 0.3% and 0.55% H_2O. The same samples were analyzed by Killingley and Muenow (1975), and they obtained similar water contents using mass spectrometry, while Moore (1965) used the Penfield method. These results suggest that the water content of the plume derived Hawaiian tholeiites is within the range of 0.3% to 0.5% H_2O.

This result has been cast in doubt by later investigations by Muenow et al. (1979). They investigated the H_2O and CO_2 contents of glass inclusions in olivine phenocrysts and found, using mass spectrometry, that the water content of these inclusions is virtually nil, the inclusions only contained CO_2 and SO_2. In contrast Harris (1979) estimated that such inclusions from a subaerial extrusion contain 0.3% H_2O and 0.05% CO_2. It was considered by Muenow et al. (1979) that the water of the glass of the pillow rims was introduced by seawater, but the hydrogen isotope work by Craig and Lupton (1976) does not suggest the presence of seawater in the glassy parts of pillow margins. If the water in these glassy rims stem from seawater, one would not expect vesiculation, as the glassy hot lava would only become saturated with water. However, the water added at high temperature could be exsolved at low temperature. In conclusion, it is considered here that the absence of H_2O in the glassy inclusions in olivine phenocrysts are not well understood. The absence could be due to the diffusion of water away from the crystallizing olivine phenocrysts, but that is only one possibility.

The H_2O and CO_2 contents of andesitic magmas were determined by Garcia et al. (1979), using mass spectrometry, and the average contents were estimated as 1.1 wt% H_2O and 0.245% CO_2 (Table VII-1). They also investigated the gas contents of glassy inclusions in plagioclase phenocrysts and found less water than in the glassy rims of pillow lavas.

The first reliable estimate of H_2O in andesitic magmas was made by Eggler (1972), who compared experimental crystallization sequences with phenocryst as-

Table VII-1. The water contents of rocks estimated by the Penfield method (H_2O^+%) (Nocholds 1954; MacDonald and Katsura 1964), and by mass spectrometric methods (Delaney et al. 1978; Garcia et al. 1979)

Rock type	Penfield method Average H_2O%	Mass spectrometry Average H_2O%
Dunite	0.44 (9)[a]	
Lherzolite	0.50 (11)	
Pyroxenite	0.42 (20)	
Gabbro	0.57 (91)	
Tholeiite	0.91 (157)	
Abyssal tholeiite		0.21 (28)
Hawaiian tholeiite		0.62 (18)
Andesite	0.86 (49)	
Andesite		1.07 (4)
Diorite	0.80 (50)	
Granodiorite	0.65 (137)	
Rhyodacite	0.68 (115)	
Granite	0.53 (72)	

[a] The numbers in parentheses show the number of analyses

semblages. The experimental results showed that the Paracutin andesite must have contained less than 2.2% H_2O, in agreement with the results obtained by Garcia et al. (1979). A similar experimental method was used for a determination of the water content in a granite by Maaløe and Wyllie (1975), and their results suggested that the water content of the granitic magma must have been within the range of 0.4% to 1.2% H_2O.

Mass spectrometric analysis of the glassy rims of abyssal pillow lavas of tholeiite suggest gas contents of 0.21% H_2O and 0.13% CO_2 according to Delaney et al. (1978).

All these investigations suggest that the H_2O and CO_2 contents of andesitic, granitic, and tholeiitic magmas must be quite small, less than 1%. What remains to be evaluated is the significance of the apparent absence of water in the glass inclusions in olivine phenocrysts. If the gas contents of these inclusions are correct, then the mantle forming the Hawaiian lavas must be almost completely devoid of water containing only CO_2 and SO_2 in minute amounts. However, it is clear from noble gas isotopes that even the mantle forming abyssal tholeiites is not completely degassed (Chap. XIII). The compositions of magmas generated at small degrees of partial melting is dependent on the CO_2/H_2O ratio of the source, and the value of this ratio is, therefore, of great significance, but is as yet not known with certainty.

9 Nominal Anhydrous Minerals

In dealing with the partial melting of water-bearing systems, it is generally considered that the nominal anhydrous minerals like plagioclase, quartz, clinopyro-

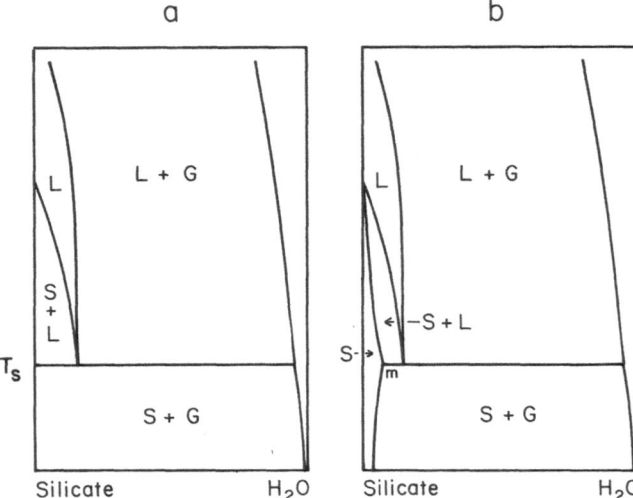

Fig. VII-33 a, b. a The typical silicate-water system for a silicate that does not contain water in solid solution; b The hypothetical phase relations for a silicate-water system where the solid silicate has some water in solid solution. This diagram is hypothetical since the solubility of water in solid silicates has not been estimated experimentally

xene, olivine, and garnet contain no water at all. If the anhydrous minerals are completely devoid of water, then the partial melting of a rock will always occur at the water saturated melting temperature (T_s in Fig. VII-4). Consider the binary phase diagram in Fig. VII-33 a, which is similar to the albite–H_2O diagram of Fig. VII-4. Evidently even the smallest amount of water will cause melting at T_s. If some water is dissolved in the crystalline lattice of the anhydrous mineral, then the melting behavior will be dependent on the water content of the phase assemblage. The phase relations for a binary system where the solid phase dissolves some water is shown in Fig. VII-33 b. If the water content is less than m, then partial melting will occur at a higher temperature than T_s, and the melt will contain less water than the saturated melt. We should, therefore, consider the water content of the minerals considered anhydrous in order to evaluate the significance of water for the partial melting of the mantle and the crust.

Most determinations of the water content of minerals have been made using the Penfield method, which estimates the amounts of water expelled between 105 °C and 900 °C. The water content estimated using the Penfield method for some minerals are listed in Table VII-2. The water content of the lavas are probably too high, as lavas absorb water after their extrusion. Hawaiian lavas contain about 0.05% H_2O just after their extrusion, while prehistoric lavas may contain up to about 0.5% H_2O (Wright 1971). The water content of minerals was compiled from the analyses collected by Deer et al. (1965). It is obvious that the nominal anhydrous minerals contain some water, about 0.3% H_2O. This water may be present in the crystalline lattice as OH^-, H^+, or H_3O^+ (Martin and Donnay 1972), or as minute inclusions. The amount of water that can be present in the crystalline lattice at room temperature was estimated by Wilkins and Sabine (1973) and is also shown in Table VII-2. These analyses were made using infrared absorption methods and an electrolytic technique. Their results suggested that OH^- occupies oxygen sites associated with vacancies or charge imbalance in cation sites. The number of analyses is too few to

Table VII-2. The water contents of minerals (H_2O^+), as estimated by the Penfield method (Deer et al. 1965), and infrared absorption (Wilkins and Sabine 1973)

Mineral	Deer et al. (1965)			Wilkins and Sabine (1973)
	Average	Range	Number of analyses	Average or range
Biotite	2.42	0.60–3.64	16	
Hornblende	1.72	0.22–2.68	38	
Orthopyroxene	0.33	0.09–0.69	12	
Clinopyroxene	0.36	0.02–1.70	22	0.02
Garnet	0.24	0.08–0.28	4	0.009–2.6
Olivine	0.22	0.05–0.34	10	0.008
Plagioclase	0.29	0.00–1.30	52	0.55
Alkali feldspar	0.31	0.04–0.64	17	
Quartz	0.30	0.20–0.39	2	

make a meaningful comparison with the analyses obtained by the Penfield method, but the water content in the lattice appears to be smaller than that obtained by the Penfield method, which includes water both from the lattice and inclusions.

The water present as inclusions in minerals has probably been present in the lattice at high temperatures, and was exsolved during cooling of the rocks. The liquid inclusions within microcline is frequently oriented in relation to the pericline and albite twins, which suggests that the inclusions were formed below 600 °C. The plagioclase phenocrysts of the lavas of the Heimaey eruption, Iceland, were transparent just after the eruption, while they became cloudy some time afterwards (Jakobsson pers. com.), which suggests the formation of minute inclusions.

The available evidence suggests that the nominal anhydrous minerals may dissolve some water at high temperatures, but at present no experimental determinations of the water contents of minerals at high pressures and temperatures have been made. Apparently, the water content of minerals may not be much smaller than the water content of basaltic and andesitic magmas. While this comparison might suggest that the generation of these magma types occurs at water deficient conditions, the compelling evidence for water deficiency is the low water content of the magmas themselves.

VIII Oxygen Fugacity

1 Introduction

The FeO/Fe_2O_3 ratio of magmas will depend on the oxygen fugacity, and the oxygen fugacity will itself be dependent on the composition of the magma and its gas content. The stability of the iron–bearing minerals depended on the FeO/Fe_2O_3 ratio, and the oxygen fugacity will, therefore, exert control on the fractionation trends of magmas. The oxygen fugacity is also significant for experimental work, as the fugacity of the charges should be the same as the fugacity of the magma (cf. Ulmer 1971).

The relationship between the oxygen fugacity and the crystallization sequence is shown for two basaltic compositions in Fig. VIII-1 (Biggar 1974). The crystallization sequence of the chilled margin of the Muskox intrusion varies with the fugacity, while the sequence of the Skaergaard intrusion is independent of the fugacity. The relationship between the oxygen fugacity of the FeO/F_2O_3 ratio for basaltic and andesitic melts was estimated by Fudali (1965), and the variation for a hawaiite is shown in Fig. VIII-2.

The significance of the oxygen fugacity for the fractional crystallization of basalts was considered by Osborn (1959), and Roeder and Osborn (1966) have given a comprehensive account of the system forsterite–anorthite–magnetite–SiO_2 at different

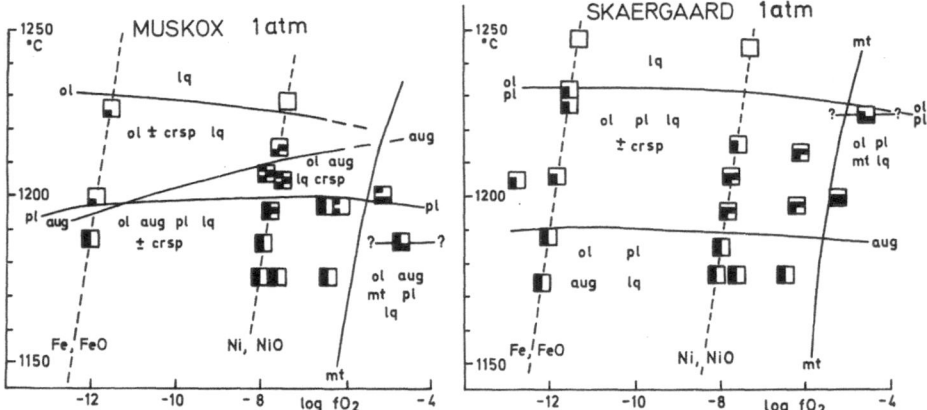

Fig. VIII-1. The variation in crystallization sequences at different oxygen fugacities for two basaltic compositions. The basaltic compositions are the chilled margins of the Muskox and Skaergaard intrusions (Biggar 1974)

Fig. VIII-2. The variation in the 2 FeO/Fe$_2$O$_3$ ratio for a hawaiite at 1200 °C (Fudali 1965)

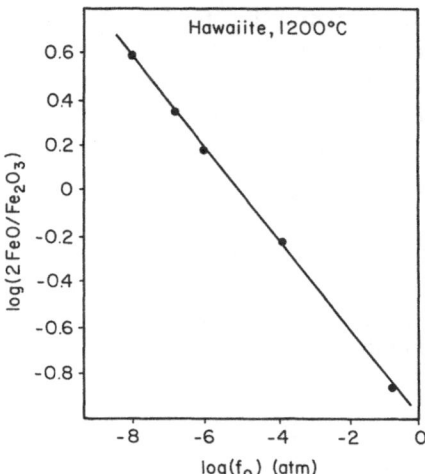

oxygen fugacities. We will consider this system in Chap. X, as it has important bearings for the fractional crystallization of basalts.

2 Equations of State for Gases

The thermodynamic properties of a gas may be calculated if the equation of state for the gas is known. The ideal gas equation is such an equation:

$$PV = nRT \tag{1}$$

where:

P: pressure
V: molar volume
n: mol
R: the gas constant
T: absolute temperature

This equation has been one of the most important equations for the development of thermodynamics, but it has only satisfactory accuracy near or below 1 bar. Another fundamental equation of state, and somewhat more accurate is Van der Waals' equation (Van der Waals 1873):

$$(P + a/V^2)(V - b) = nRT \tag{2}$$

where a and b are constants. For an accurate representation of the physical parameters of a gas the so-called virial equations of state are used. They are polynomials of higher degree:

$$PV = RT + aP + bP^2 + cP^3 \ldots \tag{3}$$

The virial equations of state do not involve any physical model of the gases, as does Van der Waals' equation, their advantage is that they allow a very accurate functional relationship between P, T, and V. The constants are estimated from experimental work. A disadvantage of this type of equation is that one of the parameters, like P in Eq. (3), is of higher degree, and might, therefore, be cumbersome to estimate P as a function of T and V.

3 Ideal Gases

The total differential of Gibbs' free energy, G, is given by (Denbigh 1971):

$$dG = VdP - SdT \tag{4}$$

and the variation in Gibbs' free energy with pressure and temperature is, thus, given by:

$$\frac{\delta G}{\delta P_T} = \bar{V} \tag{5}$$

$$\frac{\delta G}{\delta T_P} = - \bar{S} \tag{6}$$

The variation in Gibbs' free energy with pressure and temperature, thus, depends on the variation of the partial molar volume $\bar{V}$ and entropy $\bar{S}$. The variation of $\bar{V}$ and $\bar{S}$ may be estimated if the equation of state for the gas is known. Using the ideal gas equation the variation in G with pressure is given by:

$$\Delta G = \int_{P_0}^{P} \frac{RT}{P} = RT \ln \left(\frac{P}{P_0} \right) \tag{7}$$

This equation shows the relationship between G and pressure, the actual value of G will depend on the standard state chosen, i.e., on the value of P_0.

The temperature dependence of G may be estimated from:

$$\Delta G = G - G_0 = \int_{T_0}^{T} - S \, dT \tag{8}$$

The variation in S with temperature is given by:

$$dS = \frac{c_p \, dT}{T} \tag{9}$$

and as c_p is constant for an ideal gas:

$$c_p = R + c_v = 4R \tag{10}$$

then at constant pressure:

$$\Delta S = \int_{T_0}^{T} \frac{4R \, dT}{T} = 4R \ln \left(\frac{T}{T_0} \right) \tag{11}$$

Using this result in Eq. (8) we get:

$$\Delta G = - S_0 (T - T_0) + 3R \ln T_0 (T - T_0) - 3R (T \ln (T) - T)$$
$$+ 3R (T_0 \ln (T_0) - T_0) \tag{12}$$

4 Mixtures of Ideal Gases

The intensive properties of an ideal gas is independent of other gases, and the ideal gas with a given pressure P_i has the same properties whether or not the gas is isolated or forms part of a mixture. In both cases the chemical potential is given by:

$$\mu_i = \mu_i^0 + RT\ln(P_i) \tag{13}$$

If the total pressure of a mixture is P, and the partial pressures of a single gas species is P_i, then:

$$\sum_1^n P_i = P \tag{14}$$

and if the total volume is V, then:

$$P_i = n_i \frac{RT}{V} \quad (n_i: \text{mole fraction}) \tag{15}$$

The partial molar enthalpy of an ideal gas is independent of pressure, and the enthalpy of mixing of ideal gases is zero.

The entropy of an ideal gas depends on pressure, and is given by:

$$\Delta S = -R\ln\left(\frac{P}{P_0}\right) \tag{16}$$

Let the total pressure be P, and the mole fraction of a gas component be y_i, then:

$$\Delta S = -R\ln\left(\frac{y_i P}{P}\right) = -R\ln(y_i) \tag{17}$$

Gibbs' free energy will vary with pressure as the entropy is pressure dependent, thus, from:

$$\Delta \bar{G} = \Delta \bar{H} - T\Delta \bar{S} \tag{18}$$

and Eq. (17) we obtain:

$$\Delta G = RT\ln(y_i) \tag{19}$$

For the total mixture the difference in Gibbs' free energy will be given by:

$$\Delta G_{mix} = RT\sum_1^n n_i \ln(y_i) \tag{20}$$

5 Van der Waals' Equation of State

Van der Waals' equation of state for gases takes into account two of the physical properties of gaseous molecules which are ignored by the ideal gas concept. Firstly, the molecules have a certain volume, they are not indefinitely small spheres as assumed by the ideal gas equation. The volume available for the molecules is smaller than V, and the volume is represented by the term $(V-b)$ in Van der Waals' equation. Secondly, there are intermolecular forces between the molecules, they are not

Table VIII-1. Van der Waals' constants, and critical values for some gases (T_b: boiling point at 1 bar)

	a l²atm/mol²	b l/mol	T_b °K	T_c °K	P_c bar
He	0.034	0.0234	4.2	5.2	2.26
H_2	0.244	0.0266	20.0	33.2	12.8
O_2	1.36	0.0318	90.0	190.6	45.8
H_2O	5.72	0.0319	373.0	647.0	217.7
CO_2	3.61	0.0429	184.0	304.1	73.0

just elastic spheres as assumed by the ideal gas law. The pressure is, therefore, corrected by an additive term, $(P + a/V^2)$. With these improvements the Van der Waal's equation becomes:

$$(P + a/V^2)(V - b) = RT \tag{21}$$

The constants a and b are shown for some gases in Table VIII-1. This equation does not allow an exact account of the physical properties of a gas at high pressures, but it is more accurate than the ideal gas equation, and gives an account for the phase transition from gas to liquid.

The Van der Waal's equation is a third degree polynomial with respect to V. The three roots will coincide at the critical point, and the constants a and b may be estimated from the critical temperature T_c, and pressure P_c:

$$a = \frac{27\,R^2T_c^2}{64\,P_c} \tag{22}$$

$$b = \frac{RT_c}{8\,P_c} \tag{23}$$

6 Reduced Equation of State

The oxygen fugacity may be estimated with good approximation within a wide range of temperatures and pressures using the reduced values for pressure and temperature, and we will, therefore, consider the reduced parameters here.

The critical pressure and temperature of a gas is related to the molecular properties as evident from Eqs. (22) and (23). The larger the molecular forces, the larger is a, and the larger the radii of the molecules, the larger is b. Two gases at the same reduced temperature, say at half the critical pressure, may have similar properties, because their temperature and pressure are the same in relation to the critical temperature and pressure. It might, thus, be reasonable to introduce the reduced temperature, T_r, pressure P_r, and volume V_r. These reduced parameters are defined as follows:

$$T_r = T/T_c, \quad P_r = P/P_c, \quad V_r = V/V_c \tag{24}$$

where T, P, and V are the actual temperature, pressure, and volume, respectively. By introducing the reduced parameters in Van der Waals' equation we obtain:

$$(P_r + 3/V_r^2)(V_r - 1/3) = (8/3)T_r \tag{25}$$

This equation is completely devoid of natural constants and is the same for all gases. If Van der Waals' state of equation was strictly correct, Eq. (25) could be used as an equation of state for all gases. This is not the case, but experimental results show that the reduced equation represents the physical properties af gases with an accuracy of 20–30% or better.

7 Fugacity

The behavior of a real gas does not deviate substantially from the ideal gas, and it is, therefore, convenient to retain the thermodynamic functions that describe the ideal gas by introducing a correction factor (γ), the coefficient of fugacity, which really is an activity coefficient.

At constant temperature we have:

$$dG = VdP \tag{26}$$

Using the ideal gas equation we obtain:

$$dG = \frac{RT}{P}dP = RTd\ln(P) \tag{27}$$

The variation in the chemical potential for an ideal gas with varying pressure is given by:

$$\mu = \mu_0(T) + RT\ln\left(\frac{P}{P_0}\right) \tag{28}$$

where P is the actual physical pressure of the gas. For a real gas the chemical potential does not vary linearly with $\ln(P/P_0)$, and the concept of a corrected pressure was, therefore, introduced by Lewis (1908). The fugacity of a gas is defined from the chemical potential (Darken and Gurry 1953):

$$\mu = \mu_0(T) + RT\ln\left(\frac{f}{f_0}\right) \tag{29}$$

or using the equation similar to Eq. (27):

$$dG = RTd\ln(f) \tag{30}$$

The value of the fugacity f is, thus, estimated from the free energy or the chemical potential. The fugacity is the pressure that the gas would have if it behaved as an ideal gas. Equation (29) defines the ratio f/f_0, and in order to complete the definition of the fugacity, so that its absolute value can be estimated, then it is necessary to define a standard state in the same manner as a standard state is defined for the chemical potential. Since all gases approximate ideal behavior at low pressures, the standard state of the fugacity is defined from:

$$\lim(f/P) \Rightarrow 1 \quad (P \Rightarrow 0) \tag{31}$$

Thus, the fugacity is defined so that fugacity and pressure become equal at low pressures. For practical purposes of calculation and tabulation, the fugacity and pressure are sometimes considered equal at, say 0.05 atm, or even 1 atm. This procedure is acceptable as long as the error introduced is small.

The fugacity may be estimated from the actual physical pressure, using the fugacity coefficient defined as follows:

$$\gamma_i = \frac{f_i}{P_i} \tag{32}$$

where P_i is the partial pressure of the gas. Using Eqs. (28), (29), and (32) we get:

$$\mu = \mu_0(T) + RT \ln \left(\frac{\gamma P}{f_0} \right) \tag{33}$$

In this equation the functional relationship for an ideal gas has been retained, and the deviation from ideal behavior has been corrected for by the coefficient of fugacity. Since the deviation from ideal behavior is small γ will generally have values near 1.0. As an example, the value of γ is 1.826 for water at 1000 °C and 10 kbar.

The fugacity coefficient may be estimated from virial equations of state (Burnham et al. 1969), which have been estimated on the basis of experimental work. The experimental approach is the most accurate one, however, it is possible to obtain good approximations for γ using the reduced parameters for P_r values up to 100 (Newton 1935). Thus, the fugacity can be estimated over a large pressure range which is very convenient for calculating gaseous equilibria.

8 Fugacity of Van der Waals' Gas

Before we consider the reduced parameters for an estimate of the fugacity, it may be convenient to consider the fugacity of a Van der Waals' gas.

The variation in chemical potential with pressure is given by:

$$\left(\frac{\delta \mu}{\delta P} \right) = \bar{V} \tag{34}$$

so that:

$$d\mu = \bar{V} dP \tag{35}$$

Let V_0 and f_0 denote the volume and fugacity of the gas at low pressure, respectively, then:

$$\mu - \mu_0 = RT \ln \left(\frac{f}{f_0} \right) \sim \int_{P_0}^{P} V \, dP \tag{36}$$

Using Van der Waals' equation it follows:

$$P = \frac{RT}{V - b} - \frac{a}{V^2} \tag{37}$$

Upon integration it follows:

$$RT \ln \left(\frac{f}{f_0}\right) = PV - P_0 V_0 - RT \ln \left(\frac{V-b}{V_0-b}\right) - \frac{a}{V} + \frac{a}{V_0} \qquad (38)$$

Hence, we derive:

$$RT \ln (f) = RT \ln \left(\frac{f}{f_0}\right) + RT \ln P_0 (V_0 - b) - P_0 V_0 + \frac{a}{V_0} + PV$$

$$- RT \ln (V-b) - \frac{a}{V} \qquad (39)$$

Now letting $P_0 \Rightarrow 0$, so that $V_0 \Rightarrow \infty$ and $P_0 V_0 \Rightarrow RT$, then Eq. (39) becomes:

$$\ln (f) = \ln \left(\frac{RT}{V-b}\right) - \frac{2a}{VRT} + \frac{b}{V-b} \qquad (40)$$

This equation shows the variation in fugacity for a gas that follows Van der Waals' equation. Using the reduced parameters, the equation becomes:

$$\ln (f) = \ln \left(\frac{\frac{8}{3} T_r}{V_r - \frac{1}{3}}\right) - \frac{2.25}{V_r T_r} + \frac{1}{3 V_r - 1} \qquad (41)$$

9 Coefficient of Fugacity from Reduced Parameters

The variation in γ with temperature and pressure was estimated for different gases by Newton (1935), and the results showed that γ could be estimated with good accuracy using reduced parameters, at pressures up to $P_r = 100$, and T_r values of 35. The results were reported in two diagrams, one for T_r values less than 3.5 (Fig. VIII-3),

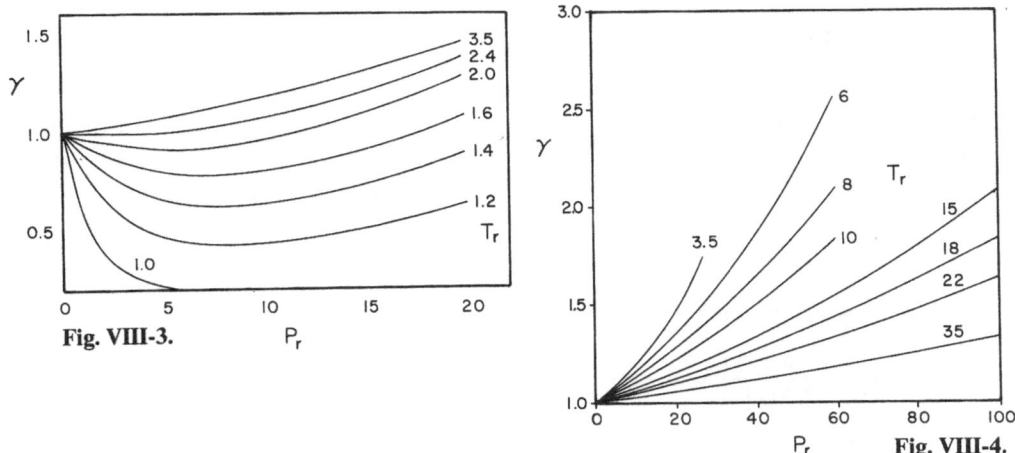

Fig. VIII-3.

Fig. VIII-4.

Fig. VIII-3. The coefficient of fugacity as function of reduced pressure (P_r) and reduced temperature (T_r) for P_r less than 20 (Newton 1935)

Fig. VIII-4. The fugacity coefficient for P_r values above 20 and T_r values above 3.5. The fugacity coefficient displays a maximum for $T_r = 3.5$

and another one for T_r values larger than 3.5 (Fig. VIII-4). These diagrams are used by first estimating the reduced values, T_r and P_r, and the critical values required are shown for some gases in Table VIII-1. Thereafter the value for γ is estimated by interpolation from Fig. VIII-3 or Fig. VIII-4. If more accurate results are required, the reader may use the original diagrams by Newton (1935). Still more accurate data for the oxygen fugacity of gas mixtures in the system C–H–O can be obtained from the tables of Deines et al. (1974), which provide the most accurate data available for the oxygen fugacity of gas mixtures containing carbon, hydrogen, and oxygen.

10 Solid Buffers

The temperature and pressure dependence of the oxygen fugacity of solid systems may be calculated using equations which represent the variation in chemical potential with good approximation.

Let us first consider the temperature dependence using the NNO buffer as an example. This buffer has the equilibrium:

$$2\,Ni + O_2 = 2\,NiO \tag{42}$$

At constant pressure the chemical potentials are given by:

$$
\begin{aligned}
\mu_{Ni} &= \mu_{Ni}^0\,(T) \\
\mu_{NiO} &= \mu_{NiO}^0\,(T) \\
\mu_{O_2} &= \mu_{O_2}^0\,(T) + RT \ln\,(f_{O_2})
\end{aligned} \tag{43}
$$

The chemical potential of oxygen will depend on the partial pressure which generally will be much smaller than the total pressure. At constant pressure ΔG for the reaction is given by:

$$\Delta G = 2\mu_{NiO}^0 - (2\mu_{Ni}^0 + \mu_{O_2}^0) - RT \ln\,(f_{O_2}) \tag{44}$$

so that at equilibrium we have:

$$RT \ln\,(f) = -\Delta G^0\,(T) \tag{45}$$

This equation shows that the oxygen fugacity will be constant as long as both Ni and NiO are present, and the fugacity is, therefore, buffered. The variation in the free energy $\Delta G^0\,(T)$ is represented with good approximation with the following equation:

$$\Delta G^0\,(T) = a - bT \tag{46}$$

so that, using Briggs' logarithm:

$$\log\,(f_{O_2}) = \frac{-a + bT}{2.303\,RT} = B - \frac{A}{T} \tag{47}$$

Eugster and Wones (1962) listed the constants A and B for various buffers (Table 2), and for the NNO buffer we have $A = 24{,}709$ and $B = 8.94$, so that:

$$\log\,(f_{O_2}) = 8.94 - \frac{24{,}709}{T} \tag{48}$$

Table VIII-2. Buffer constants for the equation $\log(f_{O_2}) = -A/T + B + C(P-1)/T$ (T: degrees Kelvin) (Eugster and Wones 1962)

Buffer	Abbr.	A	B	C
Fe_2O_3–Fe_3O_4	HM	24,912	14.41	0.019
Fe_3O_4–FeO	MW	32,730	13.12	0.083
FeO–Fe	WI	27,215	6.57	0.055
Fe_3O_4–Fe	MI	29,260	8.99	0.061
NiO–Ni	NNO	24,709	8.94	0.046
SiO_2–Fe_2SiO_4–Fe_3O_4	QFM	27,300	10.30	0.092

Note here that T is in degrees Kelvin.

The pressure dependence of the oxygen fugacity is estimated in the following manner (Eugster and Wones 1965). It was shown above that for an ideal gas:

$$dG = RTd \ln (P) \tag{49}$$

and for a real gas we similarly have:

$$dG = RTd \ln (f) \tag{50}$$

By differentiation of Eq. (50) with respect to pressure we obtain:

$$\left(\frac{\delta G}{\delta P}\right)_T = RT \left(\frac{\delta \ln (f)}{\delta P}\right)_T \tag{51}$$

This equation defines the variation in G with respect to P and f. The variation in ΔG for a reaction with pressure is given by:

$$\Delta G = \int V dP \tag{52}$$

By the NNO reaction (42), there will be a change in volume due to the change in volume of the solid reactants and oxygen, so that ($\Delta V_s = \Delta V_{solid}$):

$$\Delta G = \Delta V_s dP + RT \int \frac{\delta \ln (f)}{\delta P} dP \tag{53}$$

By equilibrium $\Delta G = 0$, Eq. (54) is obtained:

$$\Delta V_s (P_2 - P_1) = RT \ln \left(\frac{f_2}{f_1}\right) \tag{54}$$

Using the standard state $P_1 = 1$ atm, and Briggs' logarithm we have:

$$\log (f_{O_2}) = \frac{\Delta V_s (P-1)}{2.303\,RT} \tag{55}$$

The molar volumes of the solids are: $V_{NiO} = 10.88$ cc/mol, $V_{Ni} = 6.50$ cc/mol, so that $\Delta V_s = 8.76$ cc/mol. As $R = 82.058$ cc · atm/mol · °K, we obtain:

$$\log (f_{O_2}) = 0.0464 \left(\frac{P-1)}{T}\right) \tag{56}$$

Let the constant on the right side be C, then:

$$\log (f_{O_2}) = C \left(\frac{P-1}{T} \right) \tag{57}$$

so that $\log (f_{O_2})$ is given by (Eugster and Wones 1965):

$$\log (f_{O_2}) = -\frac{A}{T} + B + C \left(\frac{P-1}{T} \right) \tag{58}$$

Constants for some solid buffers are listed in Table VIII-2.

The oxygen fugacity in the presence of graphite has important experimental applications, since graphite is the ideal container material at high pressures by gas absent experiments. The oxygen fugacity is controlled by the following two reactions:

$$C + \frac{1}{2}O_2 = CO$$
$$CO + \frac{1}{2}O_2 = CO_2$$

The oxygen fugacity of this buffer (GC0) was estimated both accurately and by an approximate equation by French and Eugster (1965), and their approximate equation is given here:

$$\log (f_{O_2}) = -20586/T - 0.044 + \log (P_{gas}) \\ - 0.028 (P_{gas} - 1)/T \tag{59}$$

where P_{gas} is the pressure ($P_{CO_2} + P_{CO}$), and T is in degrees Kelvin. This equation is intended by use above 100 bar.

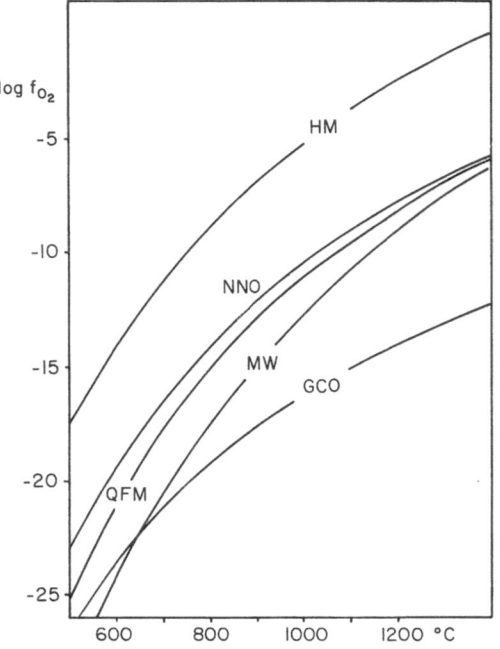

Fig. VIII-5. The variation in oxygen fugacity for some solid buffers between 600 °C and 1200 °C

Fig. VIII-6. The variation in oxygen fugacity for two buffers, *QFM* and *GCO,* at different pressures

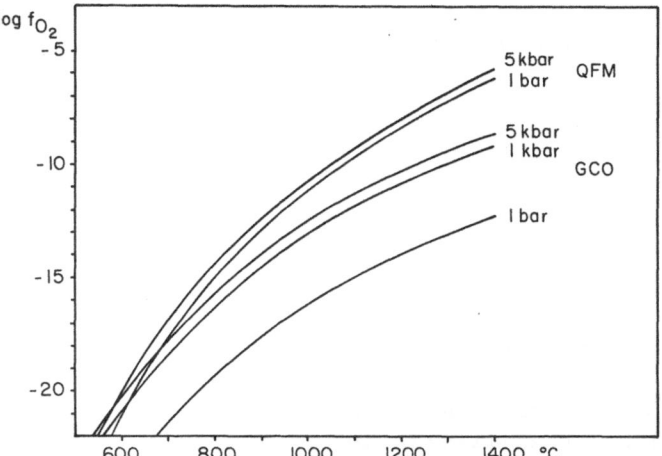

The variation in oxygen fugacity for the GCO buffer and some of the solid buffers are shown in Fig. VIII-5. As Eq. (48) suggests, there is an increase in the fugacity with temperature. The variation in oxygen fugacity at selected pressures is shown for the QFM and GCO buffer in Fig. VIII-6. It is evident from this diagram and Eq. (55) that the fugacity increases with pressure. This is due to the fact that ΔV_s is positive, increased pressure will, therefore, move reaction (42) to the left. By the GCO buffer there are two gas species involved in the reaction and this buffer is more pressure dependent than the solid buffers (Fig. VIII-6).

IX Partial Melting

1 Introduction

The magmas extruded on the surface of the Earth originated by partial melting of a solid source that was transformed into a primary magma and a residuum. It appears that magmas were formed by partial melting during the main part of the Earth's history, however, it is possible that the very first continental crust was formed by fractionation of magmas stemming from the oceans of magma that initially covered the Earth.

The partial melting may occur by two processes, the decompression of ascending material, and the heating of descending material. Partial melting occurs by decompression within ascending convection currents and mantle plumes. Partial melting by heating is considered to take place in the subducted oceanic crusts, whereby andesites are generated. Some continental andesites have so high $^{87}Sr/^{86}Sr$ ratios that they probably were formed by partial melting of continental crust in the deep root zones of the continents that form during orogenesis.

It has proven particularly difficult to estimate the composition of the primary magmas, and the compositions of the primary magmas for the various magma suites are still unknown, at least, there is no general consensus on their compositions. Considering the primary magmas for basalts, then we could estimate their compositions if we knew the detailed composition of the mantle, however, the composition of the lower mantle that is believed to generate mantle plumes is still unknown. The composition of the upper mantle is also unknown, although isotopic evidence might indicate that it consists of a mixture of lherzolite and eclogite, the eclogite being former subducted oceanic crust. The composition of the lherzolitic part is probably that of average lherzolite (Maaløe and Aoki 1977). On the other hand, we could estimate the composition of the mantle sources if we knew the compositions of the primary magmas, but there is presently also no general agreement on their compositions.

The situation for the determination of the primary magmas for andesites appears somewhat better, as the oceanic crust must be the main source, especially for the intraoceanic island andesites, but the source for the continental andesites is less well defined, as isotopic evidence shows that continental material forms part of these magmas.

2 Parameters Controlling Primary Magmas

The composition of a primary magma will depend on the following parameters.

1. *The Composition of the Source Material Being Partially Melted.* The different compositional models for the mantle that have been proposed are not that different, their MgO contents vary from 34% to 42% MgO. The compositions of the partial melts obtained from these different models will be similar for intermediate degrees of partial melting, but quite different for small degrees of partial melting.

2. *The Relative Concentrations of Volatiles.* Experimental data show that the CO_2/H_2O ratio may have a profound influence on the compositions of partial melts.

3. *The Confining Pressure.* The invariant point at which the initial melt is formed will vary with pressure. Mantle-derived primary magmas become enriched in olivine by increasing pressure.

4. *The Degree of Partial Melting.* By increasing degree of partial melting, the MgO content will increase, and the content of incompatible elements will decrease.

3 Estimation of Primary Magma Compositions

The compositions of primary magmas may be determined using one of the three following methods which, however, all involve some assumptions.

A composition believed to represent the composition of the upper or lower mantle may be melted at different pressures and temperatures. Subsequent microprobe analyses of the glasses in the experimental charges will yield the compositions of the melt (Ringwood 1975; Jacques and Green 1980). This method requires that the composition of the source is known with some accuracy. The compositions of the melts will be very sensitive on the Na_2O and K_2O contents of the source as well as the CO_2/H_2O ratios. Some melting experiments have been carried out using platinum capsules (Mysen and Boettcher 1975; Mysen and Kushiro 1976), but these experiments do not provide reliable results as iron is absorbed by platinum. The loss of iron will not only change the MgO contents of the minerals, but will also distort the phase relations. Experiments carried out using graphite capsules yield good results, and the glasses obtained from these experiments may indicate at which pressures and temperatures given magma types have formed.

The second method starts with the assumption that a given composition is primary. If the composition investigated really is primary then the liquidus phase assemblages should fulfill some restrictions. If the magma is considered generated from a garnet lherzolite by small degrees of partial melting, then olivine, diopside, and garnet should be in equilibrium with the liquid near the liquidus at a pressure above about 20 kbar. Enstatite should not be present as enstatite reacts with the liquid at these pressures. An abyssal tholeiitic composition with 17.7% MgO has olivine, garnet, and diopside on the liquidus at 25 kbar and 1580 °C, which suggests that this composition might be primary and has been generated at these conditions (Maaløe and Jakobsson 1980). If an andesitic composition is primary then it should have garnet or clinopyroxene on the liquidus as abyssal tholeiites form eclogite at high pressures. This method is problematic as the primary nature of the chosen composition may be incorrect. However, if that is the case, then this will probably be evident from the phase relations, although coincidence may produce the expected phase relations. Considering basalts there will be an additional control. The forsterite content of lherzolites is about 90% Fo, and does not vary that much. The olivine phenocrysts of primary basalts should, therefore, contain at least about 90% Fo if they are primary.

The third method is based on a geochemical analysis. The trace element contents of primary magmas is controlled by the mineralogical proportions of the source, the degree of partial melting, and the distribution coefficients. The most significant trace elements in this connection are the highly refractory elements like Ni and Cr. Fractionation of the magma will result in a rapid decrease of Ni and Cr. Thus, by comparing the Ni content of a magma composition with the Ni content calculated for the primary magma, it may be possible to estimate whether the magma composition is primary. The practical problem with this method is that accurate distribution coefficients are required, and the available values of these presently show more variation than is warranted.

In general, it is not possible to estimate how much material has been fractionated from a given magma. There is one exception, however, and that is the abyssal tholeiites. The ophiolite complexes provide cross-sections of the abyssal crust, and the initial composition of the abyssal tholeiite that forms the oceanic crust must be identical to the average composition of the oceanic crust. Elthon (1979) estimated the detailed composition of the primary abyssal tholeiite on this basis and obtained a composition with 17.78% MgO, in agreement with an estimate based on a different ophiolite complex (Malpas 1978). This estimate is probably one of the best estimates of a primary composition that has been obtained.

4 Partial Melting in a Ternary Eutectic System

Crustal rocks and the upper mantle consist of at least four solid phases, and anatexis or partial melting will, therefore, occur in a multiphased system. Some of the fundamental relationships may be demonstrated considering a ternary eutectic system where the solid phases have the end member compositions A, B, and C (Fig. IX-1b). The change in composition of a liquid generated by partial melting of the composition 20% A, 15% B, and 65% C will now be considered. The melt initially formed will remain at the eutectic point e until all solid B has been dissolved in the melt, whereafter the liquid moves along e–e_3. At f all solid A has entered the melt, and the liquid moves directly from f towards C and M. The variation in amount of liquid, and its compositional variation is shown in Fig. IX-1a. At the eutectic temperature all heat delivered to the mixture M will be used for melting, and the composition of the liquid and its temperature will remain constant until B has disappeared. Thereafter the temperature will increase by further addition of heat to the system. The increase in amount of melt with temperature will depend both on the phase relations of the system and the composition of the mixture. No general melting curves can be estimated. The melting curves in Fig. IX-1a show the characteristic features for the eutectic system. There will be a large increase in the amount of melt at the eutetic temperature. When the liquid leaves this temperature, there will be an abrupt change in the increase of amount of melt as function of temperature.

Fig. IX-1 a, b. The variation in amount and composition of a melt formed by partial melting in a ternary system. The *heavy curve* (a) shows the variation in amount of liquid, the *other curves* show the variation in concentration of components. The position of the composition *M* undergoing melting is shown in **b**

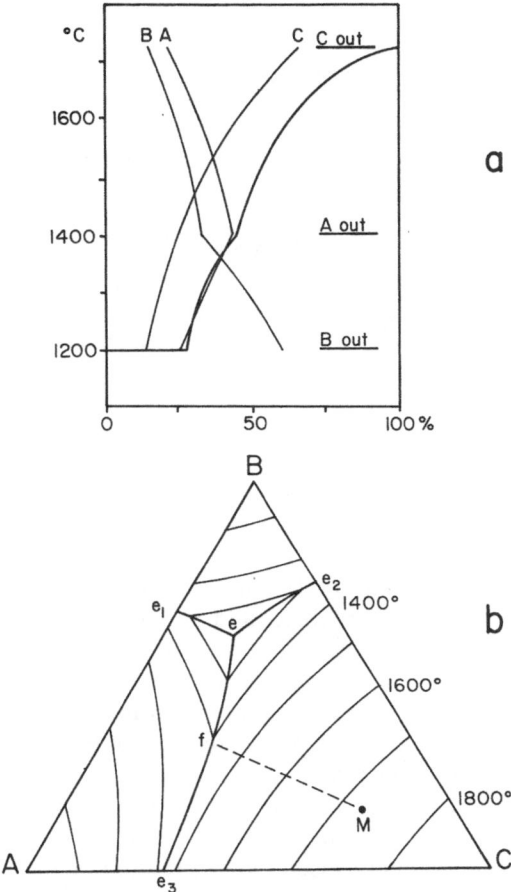

When in the next phase A disappears, there will again be an abrupt change in the melting curve. Each time a phase has been dissolved in the melt there will be an abrupt change in the slope of the melting curve. This relationship also holds for systems where the solid phases display solid solution. The variation in composition of the liquid is also shown in Fig. IX-1 a. The solid C is the solid phase with the highest melting temperature of mixture M, and its percentage in the melt is steadily increasing. All solid B is entering the melt at the eutectic point e, and the percentage of B will be steadily decreasing. The percentage of component A in the liquid displays a maximum at 1400 °C, as solid A has the intermediate melting temperature for a mixture of composition M.

5 Partial Melting of Lherzolite

The lherzolitic phase assemblage consists of olivine, enstatite, diopside, and spinel at low pressures, and the high-pressure assemblage consists of olivine, enstatite, di-

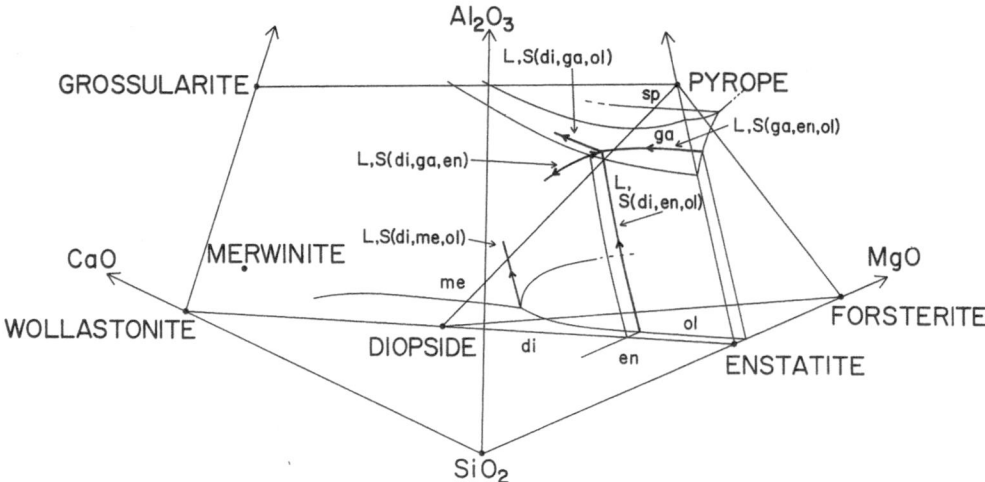

Fig. IX-2. Part of the CMAS system and the phase relations at 30 kbar (Maaløe and Pedersen 1976)

opside, and garnet. In addition there might be accessory, but significant amounts of phlogopite, amphibole, and perhaps carbonate. Carbonate has as not yet been observed in nodules, but may be present at high pressures. Some nodules contain CO_2 inclusions, which suggest the presence of CO_2 in the mantle. Due to the unknown amounts of accessory minerals in mantle lherzolite, it is not possible to give a general account of the partial melting of the mantle, but we may consider partial melting within the CMAS system which is a simplified model of the mantle composition.

The CMAS system consists of the four end members, CaO, MgO, Al_2O_3, and SiO_2 (O'Hara 1968). Since these oxides constitute the main part of lherzolite, the CMAS system is useful for an evaluation of the phase relations of the mantle, as far as the major components are concerned. The detailed phase relations of this system have been considered by O'Hara (1968), and we will only consider here the phase relations at 30 kbar. The known phase relations at 30 kbar are shown in Fig. IX-2. The invariant point L,S(ol,en,di,ga) is situated behind the di–py–fo plane, and near the di–py join. The first melt generated from a garnet lherzolitic composition will be formed at this point. Four univariant curves emanate from this point, L,S(di,en,ol), L,S(di,ga,en), L,S(di,ga,ol), and L,S(di,en,ol). When the melt leaves the univariant point, it may proceed along one of these four curves. It is most likely that either garnet or diopside is the first solid phase to melt completely, which one, will depend on the composition of the lherzolite and the pressure. If we assume that garnet disappears first, then the liquid will proceed along the curve L,S(di,en,ol), thereafter the liquid will leave this curve and move along the L,S(en,ol) surface and finally move into the L,S(ol) space.

6 Melting Curves of Lherzolite

The shape of the melting curves will depend on the composition of the lherzolite and the pressure. The forsterite–diopside–silica system may be considered a sim-

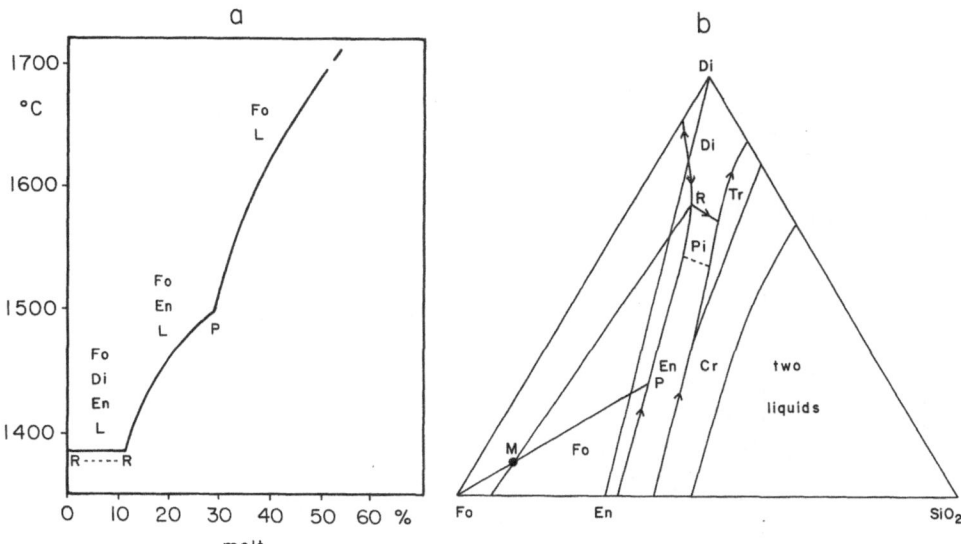

Fig. IX-3a, b. a The variation in amount of partial melt with temperature for a composition (M) in the system forsterite–diopside–silica at 1 bar (Kushiro 1972); **b** The composition M and the phase relations. *R* is a reaction point, and the curve *RP* is a reaction curve. The pigeonite field has been ignored in **a** for simplicity

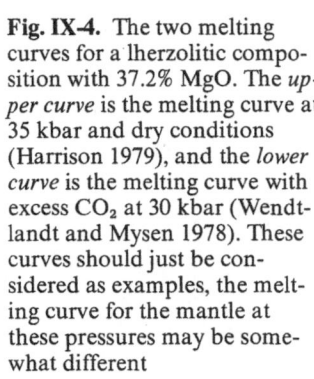

Fig. IX-4. The two melting curves for a lherzolitic composition with 37.2% MgO. The *upper curve* is the melting curve at 35 kbar and dry conditions (Harrison 1979), and the *lower curve* is the melting curve with excess CO₂ at 30 kbar (Wendtlandt and Mysen 1978). These curves should just be considered as examples, the melting curve for the mantle at these pressures may be somewhat different

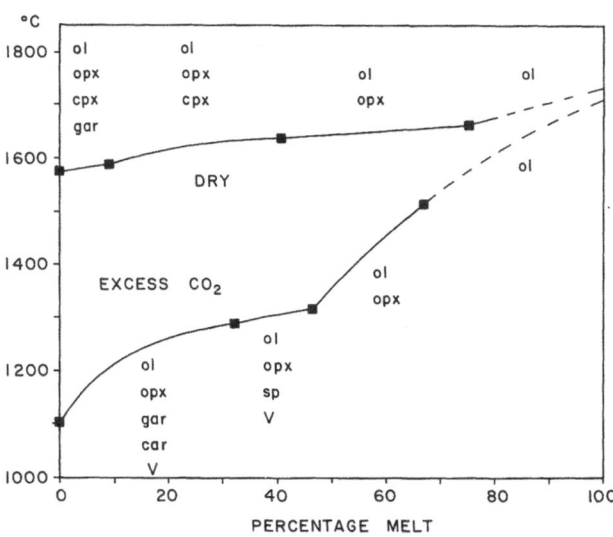

plified lherzolitic system, and a melting curve for this system is shown in Fig. 3. The composition chosen is similar to the composition of a spinel lherzolite (point M). A fairly large amount of melt is formed at the reaction point R, about 12% (Fig. IX-3a). When all diopside has entered the melt, the melt will move along the univariant curve R–P and towards P, and finally the melt will move from P to M. The melting curve gets a kink each time a solid phase is completely melted, and the slope of the

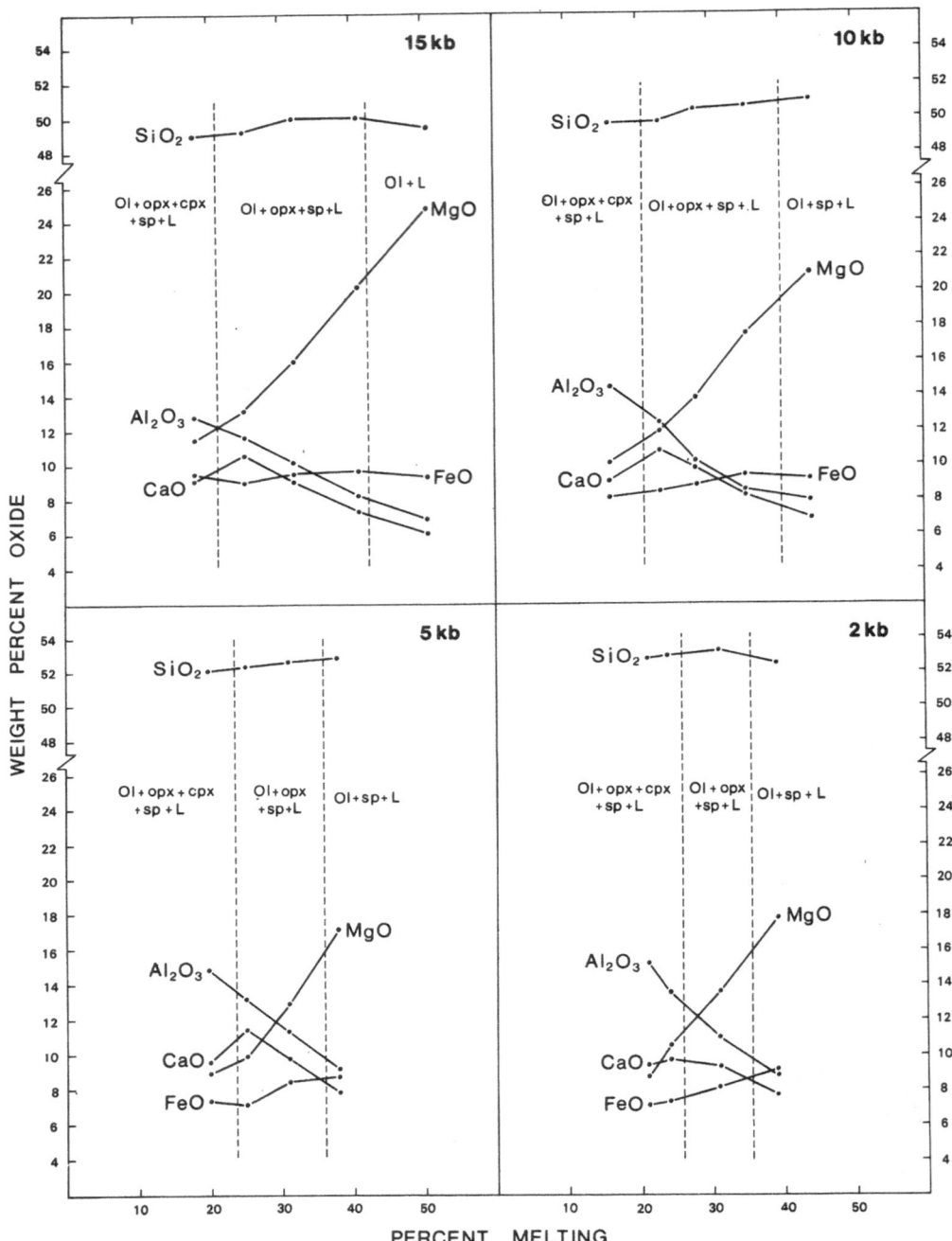

Fig. IX-5. The variation in composition by partial melting of a lherzolitic composition with 37.5% MgO (pyrolite), at pressures between 2 and 15 kbar (Jaques and Green 1980)

curve decreases along each segment of the curve. These two features are commonly observed by the melting curves lherzolites. Also one would expect the slope of the curve segments to increase with increasing temperature. Consider for example a ternary system where the eutectic curves normally have a smaller slope than the liquidus surfaces. The system considered here has limited solid solution, and the initial amounts of melt are formed at constant temperature at the reaction point R. The minerals of rocks display solid solution, and in the minerals of a lherzolite there will mainly be an exchange between the Mg^{++} and Fe^{++} ions. The initial amount of melt is, therefore, not formed at constant temperature; instead, the amount of melt will increase with temperature. The phase relations of partial melting of minerals with solid solution has been analyzed by Presnall (1979).

The variation in melt of a garnet lherzolite with 37.2% MgO is shown in Fig. IX-4. The upper curve is the melting curve at 35 kbar (Harrison 1979), and the lower curve is the melting curve at 30 kbar at CO_2 excess conditions. The presence of CO_2 decreases the melting temperatures, and increases the slope of the first curve segment. The variations in composition with the degree of partial melting has been estimated by Jaques and Green (1980), and some of their results are shown in Fig. IX-5. The SiO_2 content is rather constant, while the MgO content of the melt increases with temperature, while the CaO content displays a maximum.

7 Partial Melting of a Granitic Composition

The continental basement rocks have largely a granodioritic composition, and we may consider the partial melting or anatexis of a dioritic composition. The phase relations of the albite–orthoclase–anorthite–silica system is shown in Fig. IX-6 at 5 kbar H_2O pressure. This water pressure is higher than the water pressure prevalent by the anatexis of natural granitoid rocks, but this pressure has been chosen because the leucite field is absent at this pressure. The melting process will be similar at dif-

Fig. IX-6. The phase relations of the granite system at 5 kbar H_2O pressure compiled from phase relations reported by Yoder et al. (1957), Franco and Schairer (1951), Luth et al. (1964), and Yoder et al. (1957). The phase relations of the interior part of the diagram is based on interpolation

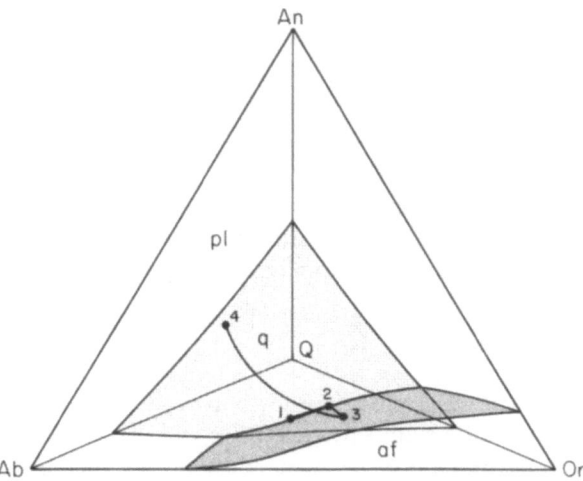

ferent pressures. The composition shown at point 4 is that of an average andesite (Chayes 1969). The phase relations of the interior parts of the system have not been estimated and we can, therefore, only give a qualitative description of the partial melting. By partial melting of composition 4, the first melt will form somewhere along the curve L, S(pl, or, q), and the first melt has been indicated at point 1. By increasing degree of melting, the melt will move along this curve to point 2 where all quartz has become melted. The liquid will thereafter move along the L, S(pl, or) surface to point 3, where orthoclase has melted completely, and the melt will finally move from point 3 to point 4 along a curve, as plagioclase changes its composition during the partial melting.

The detailed melting relationships of the quaternary system with H_2O as an additional component has been considered by Pressnal and Bateman (1973), and Wyllie (1977) has given a review of the phase relations related to crustal anatexis.

8 Influence of Pressure on the Liquid Composition

The melting temperatures of minerals increase with increasing pressure, some of the very few exceptions being ice and graphite. The temperatures of the isobaric invariant points will, therefore, also increase with increasing confining pressure at volatile absent conditions. The variation in composition of the isobaric invariant point will, for a given composition, be dependent on the relative change in the melting temperatures of the different mineralogical constituents. For an ideal binary eutetic system the liquidus curves may be calculated from (Lewis and Randall 1961):

$$\ln(x) \quad = \frac{\Delta H^a}{R}\left(\frac{1}{T_m^a} - \frac{1}{T}\right) \tag{1}$$

$$\ln(1-x) = \frac{\Delta H^b}{R}\left(\frac{1}{T_m^b} - \frac{1}{T}\right) \tag{2}$$

where the variables are: x: the mole fraction of component a, ΔH^a and ΔH^b: the latent heats of melting for a and b, respectively, R: the gas constant, T_m^a and T_m^b: the melting temperatures for pure a and b, respectively. At the eutectic point the two liquidus curves intersect, so from Eqs. 1 and 2:

$$\frac{R}{\Delta H^a}\ln(x) - \frac{1}{T_m^a} = \frac{R}{\Delta H_2}\ln(1-x) - \frac{1}{T_m^b} \tag{3}$$

In order to evaluate the change in eutectic composition with variations in T_m^a and T_m^b, the latent heats of melting will be considered equal for convenience. With this simplication Eq. (3) becomes:

$$\frac{R}{\Delta H}\ln\frac{x}{1-x} = \frac{1}{T_m^a} - \frac{1}{T_m^b} \tag{4}$$

If the melting temperature for b increases more than that of a with increasing pressure, then the right side of Eq. (4) will increase. Thus, the eutectic point will move towards the solid phase which displays the smallest increase in melting temperature with increasing pressure. Some idea about the change in the compositions of partial melts may, therefore, be obtained considering the melting curves for the pure components.

The melting curves for various minerals of interest are shown in Figs. IX-1 and IX-5. Most notably the melting curve for forsterite displays a very small increase in temperature with increasing pressure compared to pyrope, enstatite, and diopside. The isobaric invariant point for the phase assemblages olivine, enstatite, diopside, and pyrope will, therefore, move towards olivine with increasing pressure. Experimental investigations of the CMAS system actually confirms this relationship, the liquidus volume of forsterite contracts by increasing pressure, and the isobaric invariant point where the above mentioned solids are in equilibrium with the liquid will therefore move towards olivine. Note here that the invariant point L, S (fo, en, di, gar) not is strictly invariant, since enstatite, diopside and garnet display solid solution. The initial melting will take place at increasing temperatures and not at constant temperature, and the point is actually a melting interval. With increasing pressure this pseudo-invariant point moves towards forsterite and the liquids generated will contain an increasing amount of the forsterite component. This phase behaviour is evident from the contraction of the liquidus field of forsterite shown in Fig. IX-7.

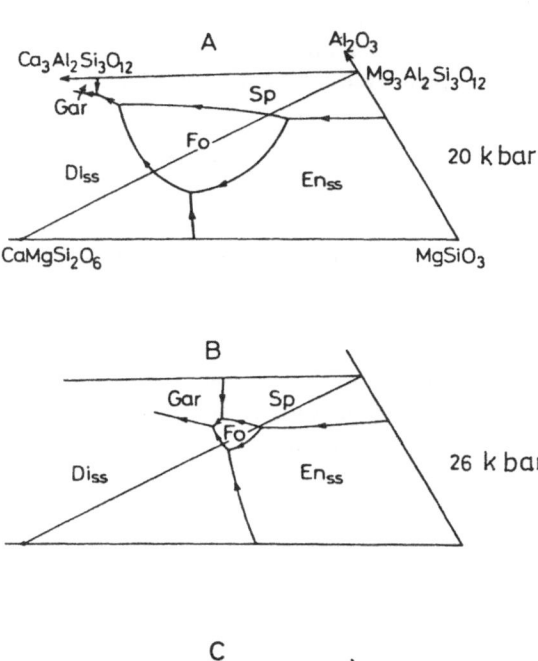

Fig. IX-7 A–C. The contraction of the liquidus region for forsterite in the CMAS system at pressures between 26 and 30 kbar (Kushiro and Yoder 1974)

9 Partial Melting by Adiabatic Ascent

The partial melting of the mantle occurs by the transfer of material from high to low pressures. The heat energy required for the partial melting is, thus, transferred with the material that undergoes partial melting. The alternative possibility that a mantle region becomes hot in a local region because of some heat source can be ruled out because the heat generation caused by radioactive elements is relatively small compared to the heat loss due to thermal conductivity. Mantle material may ascend as part of a convection current or a plume (Holmes 1931; Morgan 1972). In both cases materials are transferred from high pressures towards lower ones. If the ascent occurs at a slow rate, say less than 1 mm/yr, then the cooling exerted by the mantle surrounding the ascending system will be too large for partial melting to occur. Mantle plumes that do not undergo partial melting might exist, at least in theory, but according to the scale model experiments by Ramberg (1967), plumes will not be activated and are, unable to ascend before they have a certain size. At some rate of ascent partial melting becomes possible, whereafter primary melts may accumulate and form magma chambers. The ascending material will be cooled for two reasons. Firstly, as the ascending material is hotter than the surrounding mantle it has to surrender heat. Secondly, the ascent towards lower pressures results in expansion at isobaric conditions, and the amount of cooling caused by this effect will be about 1 °C/kbar. The heat balance relationships during adiabatic pressure release will first be estimated for a monary and ternary system, whereafter the ascent of mantle material is considered taking the geothermal gradient into account.

10 Adiabatic Ascent of a One-Component System

The estimation of the degrees of partial melt formed by adiabatic ascent will be entirely dependent on the phase relations of the system undergoing pressure release. The relationships for a one-component system are the most simple ones, and diopside will be used as an example.

Some material consisting entirely of diopside is transferred from pressures above 30 kbar towards lower pressures. The initial temperature of the material at 30 kbar is 1754 °C, the melting temperature of diopside at this pressure (Williams and Kennedy 1969). The first melt is, thus, formed at a pressure slightly below 30 kbar. The amount of melt formed at subsequent pressures will be dependent on the following parameters c_p^s: the specific heat of solid diopside; c_p^l: the specific heat of diopside glass; ΔH_m: the latent heat of melting for diopside; α: the coefficient of thermal expansion; δ_s: the density of solid diopside; and δ_l: the density of diopside liquid.

The amount of melt formed will mainly be dependent on the temperature difference between the initial temperature and the melting temperature of diopside. However, the initial temperature will be lowered due to the temperature decrease caused by adiabatic expansion. This effect may be calculated from (Adams 1924; Jeffreys 1962):

$$\frac{dT}{dP} = \frac{\alpha T}{\delta_s c_p} \tag{5}$$

The adiabatic decrease in temperature will be significantly smaller than the increase in temperature difference between the melting temperature of diopside and the initial one. An accurate estimate of dT/dP is, therefore, not essential, but will be considered here in order to demonstrate the approximations made.

The specific heats, c_p^s and c_p^l vary with temperature (Kelly 1960):

$$c_p^s = 52.87 + \ 7.84 \cdot 10^{-3} \, T - 15.74 \cdot 10^5 \, T^{-2} \tag{6}$$

$$c_p^l = 51.32 + 10.30 \cdot 10^{-3} \, T - 13.24 \cdot 10^5 \, T^{-2} \tag{7}$$

These are the specific molar heats for 216.56 g diopside. At 1673 °K or 1400 °C one gets, $c_p^s = 0.3021$ cal/g °C, and $c_p^l = 0.3144$ cal/g °C. These two values are very similar, and the difference in c_p between diopside and diopside glass may be neglected. During the initial part of partial melting there will mainly be solid diopside, and during the last part there will mainly be diopside glass. The value for c_p will, therefore, vary between the values for solid and liquid diopside, but the difference is too small to be taken into consideration. Similarly the difference in density between solid and liquid diopside may be neglected, but with less justification; they are 3.275 and 2.846 g/cm³ for diopside and diopside glass, respectively (Yoder 1976). A density of 3.1 g/cm³ will be used below. The thermal expansion in diopside is nearly linear, and a constant value may be assumed as $24 \cdot 10^6$ °C^{-1} (Kozu and Ueda 1933). For $c_p = 0.31$ cal/g °C, dT/dP may be calculated as:

$$\frac{dT}{dP} = \frac{24 \cdot 10^{-6} \cdot 1673}{3.1 \cdot 0.31 \cdot 0.0413} = 1.0117 \cdot 10^{-3} \ °C/atm \tag{8}$$

The constant 0.0413 is the conversion factor from calories to liter atmospheres. As 1.01325 bar equals 1 atm, the value for dT/dP becomes 0.9984 °C/kbar, or 1 °C/kbar. The latent heat of melting of diopside has been estimated at 18.500 cal/mol, or 85.4 cal/g (Robie and Waldbaum, 1968). With these constants the amount of partial

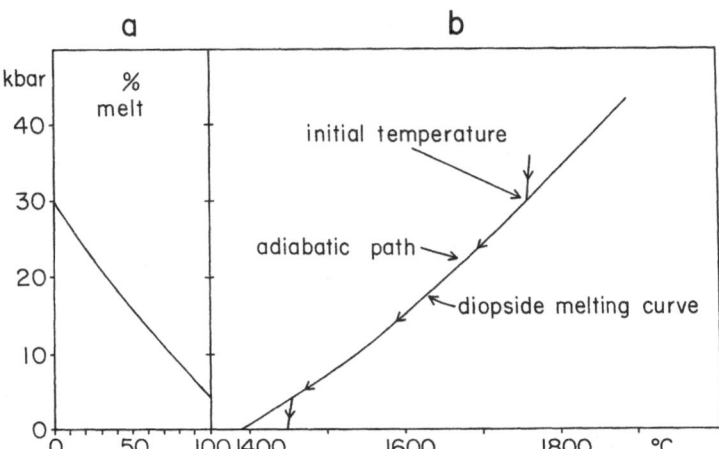

Fig. IX-8a, b. The increase in percentage of melt by adiabatic pressure release on diopside starting at 30 kbar and 1754 °C. **a** The variation in amount of partial melt; **b** the P-T path of the melting system

melting may be calculated. At 20 kbar the melting temperature of diopside is 1650 °C. For each degree the temperature is lowered there is $100 \cdot 0.31$ cal available for melt formation, while 85.4 cal is required for generation of 1% melt (Yoder 1976). If the diopside material is still completely solid at 20 kbar, its temperature would be $1754 - 10° = 1744$ °C. The temperature difference between the melting temperature and this temperature is, thus, 94 °C. There is then $94 \cdot 100 \cdot 0.31$ cal available for melt formation, and the amount of melts is $2914/85.4 \, g = 34.1$ or 34.1%. The amount of melt at other pressures is calculated in a similar manner, the resulting melt percentage curve is shown in Fig. IX-8 a. The amount of melt at 4.2 kbar is 100%. If the cooling by adiabatic expansion was not taken into account, the calculations would have suggested that all diopside was melted at 1479 °C and 5.5 kbar. The temperature decrease caused by adiabatic expansion by a pressure release of 30 kbar is 30 °C, and this difference will be equivalent to 10.9% melt for diopside. Accurate estimates of degrees of partial melting should, therefore, take the adiabatic cooling into account. By calculations of the degrees of partial melting of mantle material this effect might be ignored, as the geothermal gradient of ascending convection currents are hardly known with a accuracy better than ± 100 °C at great depths.

11 Adiabatic Ascent of a Ternary System

The melting interval for a ternary system is generally divided into three sections, as was shown in Fig. IX-1. The melt percentage is a discontinuous function of temperature, the percentage of melt formed by a given temperature increase depends on the number of solid phases. The amount of partial melt formed in a ternary system undergoing a certain pressure release will, therefore, also vary with pressure or the degree of partial melting.

The ternary system shown in Fig. IX-9 consists of three components having the composition of the solid phases A, B, and C, and the partial melting of 100 g of the mixture with composition c will now be treated. The mixture is initially at a pressure above 45 kbar and at a temperature of 1450 °C. In the subsequent calculations the latent heat of melting of all three solids will be assumed as 100 cal/g, and the specific heats of the liquid and three solids are taken as 0.25 cal/g °C. The adiabatic cooling will be ignored for simplicity in the calculations, but it can easily be taken into account if warranted. At 40 kbar the solidus temperature is 1400 °C, and 50 °C below the initial temperature. The amount of heat available for melting is, thus, $100 \cdot 0.25 \cdot 50 \, cal = 1250$ cal, and the amount of melt formed at this pressure is 12.5 g or 12.5%. The composition of the melt will be the compoisition of the eutectic point until one of the solid phases has been completely melted. Solid B will be the first solid to be completely melted, using the lever rule it will be evident that this happens when 65% liquid has been formed. This requires 6500 cal, or a temperature difference of 260 °C, the temperature for complete melting of B being 1190 °C, and the pressure 19 kbar as evident from Fig. IX-9.

The amount of liquid formed at lower pressures than 19 kbar may be estimated as follows. Let the initial temperature of the solids be T^0, and the actual temperature to be estimated T. The percentage of partial melt at any given pressure is then es-

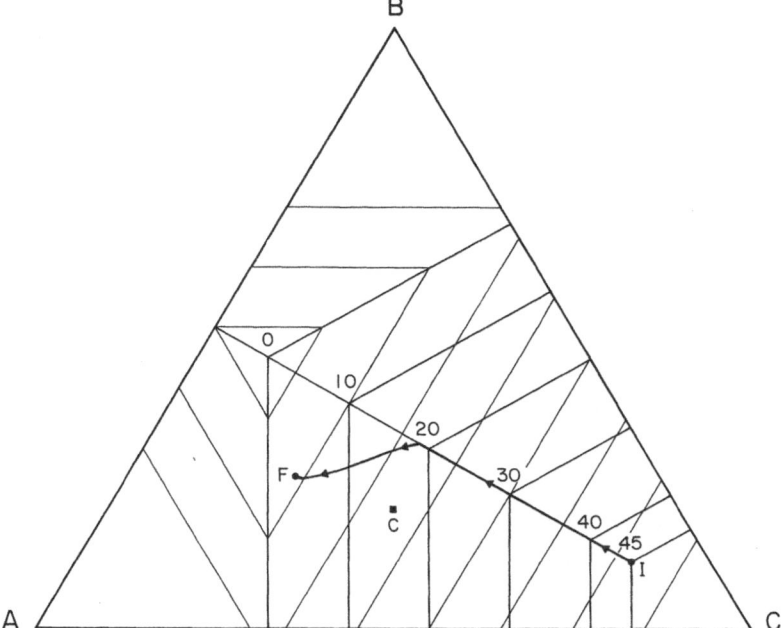

Fig. IX-9. The phase relations at various pressures for the ternary system undergoing partial melting between 45 kbar and 0 kbar. The compositional path for the melt is indicated by the *heavy curve*. The composition undergoing partial melting is *c*

timated from (Ramberg 1972):

$$(T^0 - T) \, 0.25 = 100 \, p \, (T) \qquad\qquad (9)$$

If adiabatic cooling is taken into account, and the cooling is $\varDelta T_a$, then the equation becomes:

$$(T^0 - T - \varDelta T_a) \, 0.25 = 100 \, p \, (T) \qquad\qquad (10)$$

This equation is quite simple, however, the function $p \, (T)$ will not only be a discontinuous function of T, but will also vary with pressure and composition (Fig. IX-10). Knowing the $p \, (T)$ function for a given pressure, the percentage of partial melting may be estimated. Let us consider the situation at 15 kbar. Brought from 45 to 15 kbar, the mixture c has undergone a certain amount of partial melting. The heat for its melting has been stored in the mixture as specific heat, for a decrease in 1 °C there will be 25 cal available for melting, and 0.25% melt may be formed. The temperature of the mixture will decrease until its temperature intersects the melting curve for the mixture at the prevailing pressure, whereafter the melting stops because the mixture has obtained thermal equilibrium. It should be noted here that the solidus temperature, here 1150 °C, has no significance for the melt generation after the eutectic temperature has been left. When the material is superheated due to the pressure release, the melt will increase in temperature until its temperature becomes that of the host material. The temperature of the superheated material, on the other hand, will decrease until its temperature becomes that of the melt,

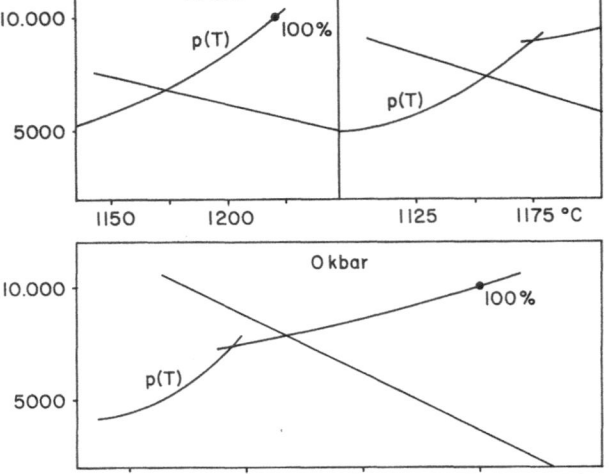

Fig. IX-10. The variation in amount of partial melt at 15, 10, and 0 kbar for the composition c shown in Fig. IX-12. The *straight lines* show the amount of heat required for melting

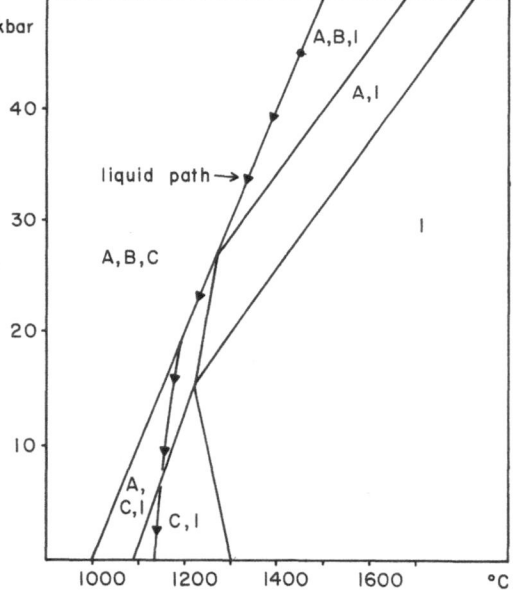

Fig. IX-11. The P-T path for the liquid generated by partial melting of composition c in Fig. IX-14. The total heat content of the ascending system is assumed constant, and the amount of melt increases with decreasing pressure, i.e., the difference in actual temperature and solidus is increasing. The temperature of the melt is, however, steadily decreasing

whereafter it loses its potential for generating melt. The curves p (T) shown in Fig. IX-10 show the increase of temperature of the melt by increasing degree of partial melting, while the straight lines show the amount of heat available for generating melt. When these two curves intersect, the system has obtained internal thermal equilibrium. The P–T path for the melting as function of pressure is shown in Fig. IX-11. By the adiabatic ascent there will be no heat loss from the ascending system to its surroundings, and the amount of melt, therefore, increase steadily with decreasing pressure. The temperature of a natural convecting system will decrease near the topmost part of the system, and its geothermal gradient will display a maximum. Consequently, the partial melting curve will also display a maximum as will be shown in the next section.

12 Partial Melting of Lherzolite

It is not possible to give an accurate account of the variation in amount of melt by the ascent of some mantle material as the required data are not available. Before an approximate calculation is considered it might be reasonable to summarize which data are required. An accurate estimate of the amount of partial melt would involve the following parameters:

1. The composition of the mantle, i.e., the proportions of minerals at different pressures should be known.
2. The geothermal gradient.
3. The isobaric increase in amount of partial melt at various pressures.
4. The latent heats of melting of the solids.
5. The specific heats of solids and the liquid.
6. The adiabatic cooling of the generated melt.

The composition of the mantle is not known in detail, but an lherzolite composition is considered representative (Maaløe and Aoki 1977).

The geothermal gradient for ascending convection currents or plumes cannot be estimated from mineralogical data, like the two-pyroxene geotherm (Body 1973; Mercier 1976). This geotherm applies for matured lithospheres and indicates only subsolidus temperatures. The geotherms may be calculated from theoretical considerations applying convection physics. Several estimates have been made, but they differ to such a degree that no conclusive result can be obtained (Oxburg and Turcotte 1968; Parmentier et al. 1975; Houston and De Bremaecher 1975; Schubert

Fig. IX-12. The variation in amount of partial melt for a garnet lherzolitic composition. ΔT is the temperature difference between the initial temperature of the system and the solidus temperature. The melt curve used here should only be considered an example, because the melting curve for the mantle is not known in detail

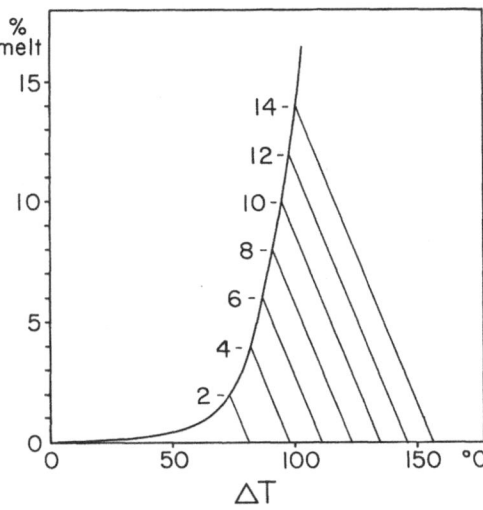

et al. 1976; Forsyth 1977). Most of the geotherms are more than 50 °C below the dry solidus of lherzolitic compositions at any pressure. Further, they ignore the endothermic reaction occurring at the olivine phase transition at 350 km depth in ascending mantle material which will result in a temperature decrease of 100 °C (Ringwood 1975). The geotherm estimated by Schubert et al. (1976) has been adopted in Fig. IX-13, because it appears attractive from a petrological point of view.

The isobaric increase in amount of melt at 20 kbar has been estimated for an lherzolitic composition by Mysen and Kushiro (1976), and has been used by the present estimate (Fig. IX-12).

The latent heats of melting are only known for the pure magnesian end members of lherzolitic minerals, they are as follows:

forsterite: 208.2 cal/g
clinoenstatite: 146.4 cal/g
diopside: 85.4 cal/g
pyrope: 82.4 cal/g

A representative latent heat of melting is considered as 100 cal/g as diopside and pyrope mainly enter the melt.

The specific heat will be assumed as 0.25 cal/g °C, which is a generally accepted value for rocks (Jaeger 1968). The temperature decrease caused by adiabatic expansion will be ignored, as the effect is small compared to the uncertainty of the geothermal gradient.

With these approximate data it is possible to calculate the variation in degree of partial melting with pressure. The relationship between temperature and degree of partial melting is shown in Fig. IX-12 and the variation in degree of partial melting with pressure is shown in Fig. IX-13. The applied data suggest that the partial melting starts at 40 kbar, and that a maximum is reached at 18 kbar where the degree of partial melting is 14.5%. The amount of melt increases slowly towards this maxi-

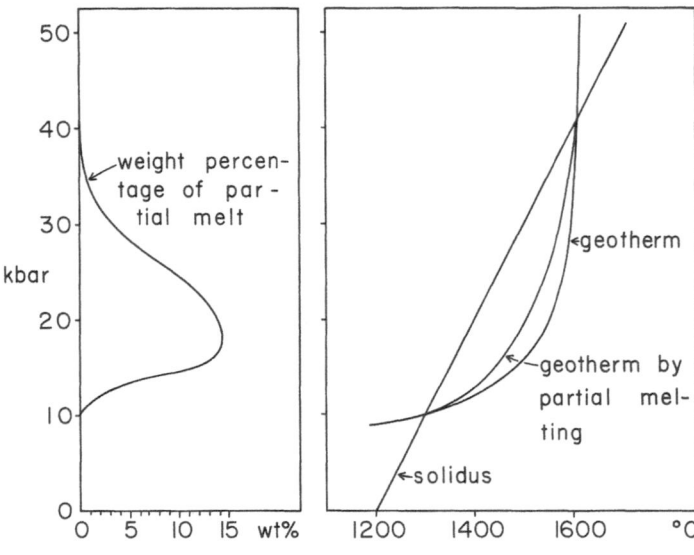

Fig. IX-13. The hypothetical variation in amount of partial melt for the ascending mantle for the geotherm (*right diagram*). This geotherm is perturbed because of the formation of partial melt, resulting in lower temperatures in the pressure interval of melting

mum, whereafter it drops rather fast at lower pressures. This variation is related to the curvature of the geotherm, and as most geotherms based on convection display this type of variation, the obtained type of variation in amount of melt is considered realistic. The absolute amounts of melt are, however, very approximate as evident from the above discussion. The curve displaying the degree of partial melting only expounds one aspect of magma generation. The other process to be considered is the accumulative processes of magma which lead to the formation of magma chambers.

13 Geochemical Relationships of Partial Melting

Interest in the geochemical relationships of partial melting began after the more accurate analytical methods for trace element determination became available. These methods, including a variety of techniques like X-ray fluorescence, optical spectrometry, mass spectrometry, and neutron activation were developed between 1950 and 1970. The first major contributions were made by Schilling and Winchester (1967) and Gast (1968). Gast (1968) also derived some equations for the determination of trace element concentrations by partial melting, but some of the equations were not entirely correct, and Shaw (1970) derived the correct equations.

One of the major problems within theoretical geochemistry is to distinguish between primary and fractionated magma types. This problem has proven difficult to solve, but Treuil (1973) has proposed a calculation method which might prove useful. His method as well as another similar method is considered here.

With the equations for fractional crystallization (Rayleigh 1902) and the equations for trace element distributions by partial melting (Shaw 1970; Hertogen and Gijbels 1976), the fundamental formal apparatus for trace element calculations is available. However, the quantitative application of the theoretical equations have been, and still are, limited by the lack of precise distribution coefficients. The distribution coefficients may be estimated from analyses of phenocrysts and lavas (Philpotts and Schnetzler 1970), and from experimental work (Leeman and Lindstrom 1978).

14 Types of Partial Melting

The dynamic features of the melting processes within the mantle and the continental crust are still partly unknown, although some models have been proposed (Turcotte and Ahern 1978; Weertman 1972; Frank 1968; Maaløe and Scheie 1982). It is, therefore, necessary to consider different melting models as the trace element concentrations in primary magmas depend on the dynamic processes occurring during partial melting.

There are two extreme models presently considered, batch melting and fractional melting. By batch melting the partial melt remains interstitially between the residual crystals until the source is tapped for magma. By fractional melting the melt is extracted as soon as it is formed. An intermediate type of melting is critical melting, where a critical melting threshold should be attained before the interstitial melt can be extracted from the source (Maaløe 1982). The evaluation of the melting processes within the mantle is further complicated by the evidence for metasomatic processes within the mantle (Menzies and Murthy 1980). The REE pattern of basalts may show an enrichment in LREE although their ε_{Nd} values are positive and show that the source must have been depleted at some time before the last generation of magma took place (Clague and Frey 1982).

The equations for determination of trace element concentrations in the presence of one solid phase will first be dealt with, whereafter the equations for a multiphase assemblage will be considered. Only batch melting and fractional melting are dealt with here, a summary of trace element equations has been given by Allégre and Minster (1978) and Hanson (1980).

15 Batch Melting, One Solid Phase

The equation derived here applies for batch melting, and implies a stationary liquid that remains interstitially between the solid grains of the residuum. This type of melting has also been called equilibrium melting, but this terminology is unfortunate as thermodynamic equilibrium also prevails by fractional melting and other types of melting.

Let the fraction of liquid generated be f, the amount of liquid L (g), and the total amount of solid S^0, we then have:

$$f = \frac{L}{S^0} \tag{11}$$

If the total amount of residual solid is given by S, then:

$$S + L = S^0 \tag{12}$$

Let the distribution coefficient D be constant and defined by Nernst's distribution law:

$$D = \frac{c_s}{c_l} \tag{13}$$

where c_s is the concentration of the element considered in the solid, and c_l is the concentration of the element in the liquid. If c_s^0 is the initial concentration of the trace element in the solid, then the concentration in the residual solid is given by:

$$c_s = \frac{c_s^0 W^0 - c_l L}{W^0 - L} \tag{14}$$

Using Eq. (13) we obtain:

$$\frac{c_l}{c_l^0} = \frac{W^0 - \frac{c_l}{c_s^0} L}{D (W^0 - L)} \tag{15}$$

By division with W^0 and by rearrangement we have:

$$c_l = \frac{c_s^0}{D + f(1 - D)} \tag{16}$$

The concentration in the liquid is, thus, estimated from the initial concentration c_s^0 and the degree of partial melting, f. As $c_s = Dc_l$ the concentration in the residual solid is given by:

$$c_s = \frac{Dc_s^0}{D + f(1-D)} \tag{17}$$

Equation (16) implies a constant distribution coefficient, if the distribution coefficient varies linearly with the degree of partial melting, we obtain:

$$D = a + bf \tag{18}$$

and Eq. (16) becomes:

$$c_l = \frac{c_s^0}{a + (b - a + 1)f - bf^2} \tag{19}$$

Equations (16) and (19) define the trace element concentration by batch melting of one solid. When several solids are present, basically the same equation is used, but the distribution coefficient D has to be calculated from the distribution coefficients of each solid. It should be noted here that the application of a constant D value involves diffusive homogenization of the solid phases. The variation in concentration of a liquid for different D values is shown in Fig. IX-14. By small degrees of partial melting there is a large variation in the concentration of the incompatible elements, while the concentration of the refractory elements is rather constant.

Apparently the trace element concentration for a residuum might also be calculated using Eq. (17). However, the values calculated for c_s refer only to the concentration of the solids in the residuum. The rock left as a residuum may contain some interstitial melt, and the trace element concentration of such a residuum should be calculated in a different manner. Consider that a rock has been partially

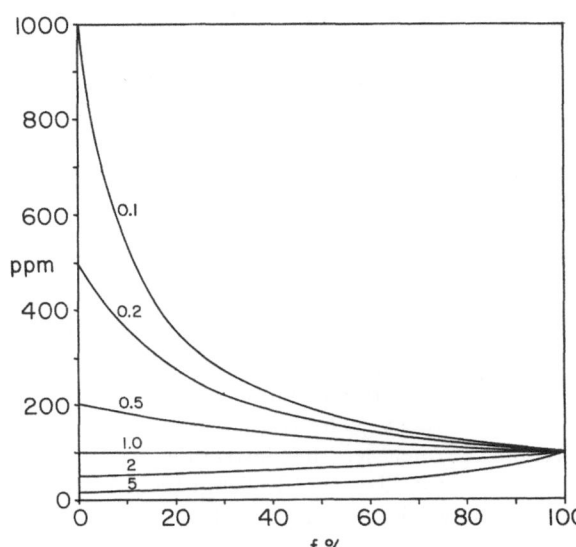

Fig. IX-14. Trace element concentrations for different D values as function of degree of partial melting, f. The calculated concentrations for some D values are shown in Table IX-1

Table IX-1. Trace element concentrations by equilibrium partial melting

					F				
D	0	0.01	0.05	0.1	0.2	0.3	0.5	0.8	1.0
0.1	1000	917	690	526	357	270	182	122	100
0.2	500	481	417	357	278	227	167	119	100
0.3	333	326	299	270	227	196	153	116	100
0.5	200	198	190	182	167	154	133	111	100
0.8	125	125	123	122	119	116	111	104	100
1.0	100	100	100	100	100	100	100	100	100
2.0	50	50	51	53	56	59	67	83	100
3.0	33	34	35	36	38	42	50	71	100
5.0	20	20	21	22	24	26	33	56	100
8.0	13	13	13	14	15	17	22	42	100
10.0	10	10	10	11	12	14	18	36	100
20.0	5	5	5	6	6	7	10	21	100

melted, and a certain amount of melt, L_m, has been extracted from the rock. It is unlikely that all melt has been extracted, and some melt may remain within the rock. Let the amount of residual melt be L_r, and the total amount of melt generated by L, then:

$$L = L_r + L_m \tag{20}$$

and the total amount of residuum is given by:

$$R = S + L_r \tag{21}$$

The trace element concentration in the total residuum will be proportional to L_r. As an example consider that 50% of L remains as residual melt as L_r. If D equals 0.1 and f is 0.2, then the concentration in the melt will be 357 ppm, when $c_s^0 = 100$ ppm (Table 1). The concentration of the trace element in the residual solid is 35.7 ppm, and the total amount of the trace element in the residuum is given by: $0.8 \times 35.7 + 0.1 \times 357 = 64.26$ ppm. The concentration is, thus, given by: $64.26/0.9 = 71.4$ ppm, and is considerably larger than the concentration estimated for the solid alone, i.e., 35.7 ppm. The effect of residual melt is most significant when D is smaller than 1.0, for D larger than 1.0, the effect is negligible.

16 Fractional Melting, One Solid Phase

The fractional melting model assumes a continuous removal of the melt from the source undergoing partial melting, that is, as soon as the slightest amount of melt has formed, then it is extracted from the source. This model is probably not entirely realistic for rocks, as a certain amount of melt has to be generated before the rock becomes permeable. The derivation of the required equations is the same as for fractional crystallization (Rayleigh 1902; Shaw 1970).

Let the total amount of solid be S^0, and the initial concentration of the solid be c_s^0. The total amount of trace element in the solid is given by $c_s S$. If an infinitesimal amount of solid, dS is removed as melt, and the concentration in the melt is c_l, then the total amount of removed trace element is c_l dS. This amount must equal the change in total amount for the solid:

$$c_l \, dS = d \, (c_s \, S) \tag{22}$$

This differential equation is solved by isolation of the variables:

$$\frac{dS}{S} = \frac{dc_s}{c_l - c_s} \tag{23}$$

Using Nernst's distribution law:

$$\frac{c_s}{c_l} = D \tag{24}$$

the solution of Eq. (23) is given by:

$$\ln \left(\frac{S}{S^0} \right) = \frac{D}{1-D} \ln \left(\frac{c_s}{c_s^0} \right) \tag{25}$$

where c_s^0 is the initial concentration of the solid.

17 Critical Melting

The critical melting model assumes that the source rock does not become permeable before a critical melting threshold has been attained. Experimental work suggests that this threshold is attained at relatively small degrees of partial melting, about 3–5%, but the permeability threshold has not yet been estimated for natural rocks. The accumulation of magma will be dealt with in Chap. XII, here we will briefly consider the trace element variations by critical melting. The derivations of the required equations have been given by Maaløe (1982).

The variations in concentrations for a trace element with a D value of 0.2 are shown in Fig. IX-15. The initial concentration of the solid is 100 ppm. The variation for simple batch melting is shown by the x_{batch}^l curve. It is assumed that the source rock undergoes 10% partial melting before it becomes permeable and that the melt is extracted. After the permeability threshold has been attained the melt is extracted continuously as the degree of partial melting increases. The composition of the melt in the interstices is shown by the x_{int}^l curve, and the concentration in the accumulated melt is given by the x_{acc}^l curve. It is considered that the interstitial melt is extracted and becomes accumulated, and the composition of the accumulated melt will, thus, depend on the composition of the interstitial melt. The variation in concentration of the accumulated melt will initially be quite small, but will thereafter decrease more than the concentration of the melt formed by batch melting. For comparison, the composition of the melt formed by fractional melting, x_f^l, is shown. The variations just considered apply for an incompatible element, the variations for a refractory element are shown in Fig. IX-16. The interesting feature of this diagram is that the concentration of the refractory element is almost constant for small degrees of partial melting both by batch melting and critical melting. The initial variation in

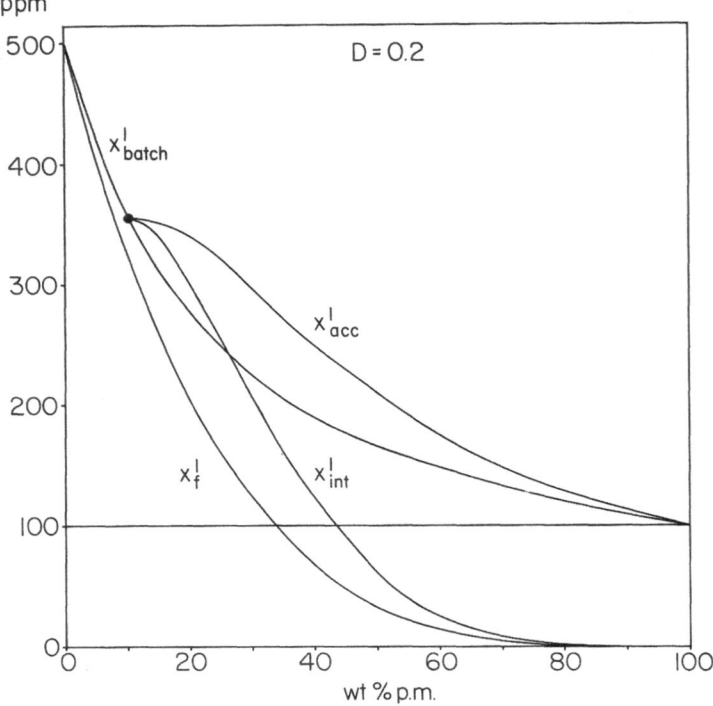

Fig. IX-15. The variations in concentrations of a trace element with $D = 0.2$ by critical melting and fractional melting. x^l_{batch} = the concentration of the liquid by batch melting; x^l_{acc} = the concentration of the accumulated liquid; x^l_{int} = the concentration of the interstitial liquid; and x^l_f = the concentration of the liquids formed by fractional melting. The initial concentration of the solid is 100 ppm

concentration of refractory elements, thus, appears independent on the type of melting, and will nearly depend only on the D value.

18 The Distribution Coefficient for a Multiphased Assemblage

The concentration of a trace element in a melt formed from different solids will be dependent on the relative amounts of these solids and the distribution coefficients for each of the solids. The calculation of the concentration in the melt can be done if the following parameters are known:

1. The distribution coefficient for each solid D_i.
2. The initial amount of each solid S_i^0.
3. The total amount of solid left as residuum, S, or the degree of partial melting, f.
4. The relative proportions of the solids in the residuum. The degree of partial melting is given by:

$$f = \frac{L}{S^0} \tag{26}$$

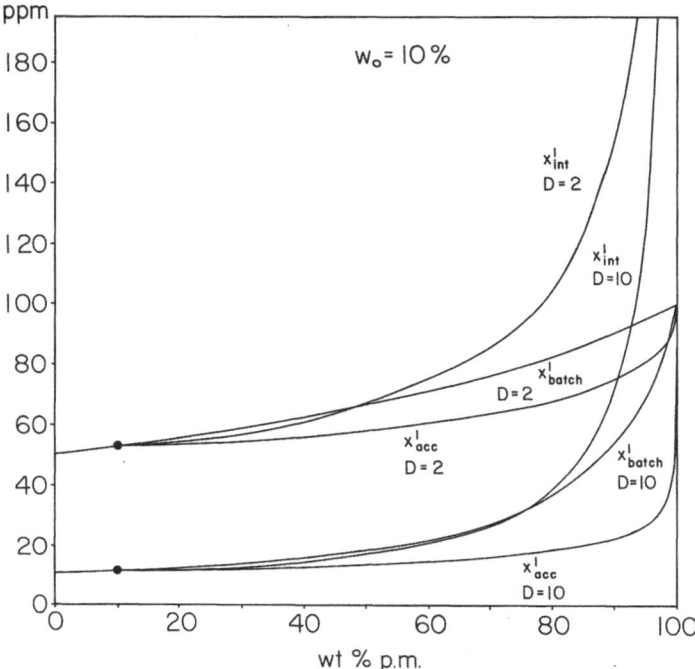

Fig. IX-16. The variations in the concentration of liquids formed by critical melting for two trace elements with $D=2$ and $D=10$. The degree of partial melting required for permeability is $w_o=10\%$. The difference in concentration between batch melting (x^l_{batch}) and accumulated critical melting (x^l_{acc}) is small for small degrees of partial melting

Each of the solids will surrender some material to the melt by eutectic melting, by peritectic melting one of the solid phases will increase in amount. Let x^i_l be the fraction of solid phase "i" in the melt, so that:

$$x^a_l + x^b_l + x^c_l + \ldots x^i_l = 1 \tag{27}$$

If w^i_l is the absolute amount of solid phase i that has melted, and now is part of the liquid, then we have:

$$w^i_l = x^i_l\,L \tag{28}$$

Let m^i_s be the total amount of solid "i" in the solid residuum, and w^i_0 the initial amount of solid "i", so that:

$$m^i_s = w^i_0 - x^i_l\,L \tag{29}$$

The fraction of solid "i" in the residuum is then given by:

$$x^i_s = \frac{m^i_s}{S^o - L} = \frac{w^i_0 - x^i_l\,L}{S^o - L} \tag{30}$$

Note that we have for x^i_s:

$$\sum_1^n x^i_s = 1 \tag{31}$$

The concentration of a trace element in the solid residuum is then given by:

$$c_s = c_s^a x_s^a + c_s^b x_s^b + c_s^c x_s^c + c_s^i x_s^i \tag{32}$$

We then have from Eq. (30) and (32):

$$c_s = \left(\frac{1}{S^0 - L}\right) \sum_1^n c_s^i (w_0^i - x_1^i L) \tag{33}$$

The bulk distribution coefficient D_b may now be obtained by division with c_l:

$$D_b = \frac{c_s}{c_l} = \left(\frac{1}{S^0 - L}\right) \sum_1^n \frac{c_s^i}{c_l} (w_0^i - x_1^i L) \tag{34}$$

Hence, as $D_i = c_s^i / c_l$ we obtain:

$$D_b = \left(\frac{1}{S^0 - L}\right) \sum_1^n D_i (w_0^i - x_1^i L) \tag{35}$$

This equation will enable us to calculate the concentration in the melt when D_i is known for each of the solid phases, however, it was shown by Shaw (1970) that the equation could be rearranged and simplified into a more elegant equation, and we will, therefore, consider that derivation.

Equation (35) can be simplified by introducing the very initial D_b value, i.e., the D_b value for the melting system when the very first melt appears. In that case, L will be almost nil, and for $L = 0$ we obtain from Eq. (35):

$$D_b = D_0 = \sum_1^n D_i \frac{w_0^i}{S^0} \tag{36}$$

and as x_0^i is the fraction of solid phase "i" in the initial solid, we get:

$$D_0 = D_a x_0^a + D_b x_0^b + D_c x_0^c + D_i x_0^i \tag{37}$$

We will now return to Eq. (35). The initial amount of solid phase "i" can be expressed by:

$$w_0^i = x_0^i S^0 \tag{38}$$

Using Eq. (38) in Eq. (35) we get:

$$D_b = \frac{1}{S^0 - L} \sum_1^n D_i (x_0^i S^0 - x_1^i L) \tag{39}$$

This summation may be separated into two summations, where the first one will be related to D_0:

$$D_b = \frac{S^0}{S^0 - L} \sum_1^n D_i x_0^i - \frac{L}{S^0 - L} \sum_1^n D_i x_1^i \tag{40}$$

The second summation may be considered a new parameter for convenience:

$$P = \sum_1^n D_i x_1^i \tag{41}$$

From Eqs. (37), (40), and (41), we obtain:

$$D_b = \frac{S^0}{S^0 - L} D_0 - \frac{L}{S^0 - L} P \qquad (42)$$

and by division with S^0, we finally obtain:

$$D_b = \frac{D_0 - fP}{1 - f} \qquad (43)$$

This value for the bulk distribution coefficient D_b applies for a system that has undergone some partial melting, but still has retained all the initial solids. D_0 depends only on the initial mineral proportions, while P is determined by the amount of material the solids have surrendered to the melt.

We can now use D_b instead of D in Eq. (16) for batch melting, and obtain:

$$c_l = \frac{c_s^0}{\dfrac{D_0 - fP}{1 - f} + f\left(1 - \dfrac{D_0 - fP}{1 - f}\right)} \qquad (44)$$

$$c_l = \frac{c_s^0}{(1 - f)\left(\dfrac{D_0 - fP}{1 - f}\right) + f} \qquad (45)$$

$$c_l = \frac{c_s^0}{D_0 - f(1 - P)} \qquad (46)$$

This equation is valid for the first melting interval where all the solid phases are still present. The equations required for the calculation of concentrations at higher degrees of partial melting where one or more of the solid phases has melted completely have been derived by Hertogen and Gijbels (1976). They also derived equations for partial melting at a reaction point.

19 Treuil's Method

One of the more pertinent problems within igneous petrology is the determination of the magma types that originate as primary magmas. There are several parameters involved in magma genesis and the problem has not been solved from present evidence, even some petrologists disagree with this evaluation (Green 1970). If the incompatible elements like alkalies, light rare earth elements, Zn, Be, Zr, Hf, P, and halogens are considered, then it has to be concluded that these elements can only define relative proportions. The high concentrations of some of these elements in various rock types have suggested to some petrologists and geochemists that these rock types constitute primary compositions formed at low degrees of partial melting. This is not necessarily so. If the LREE content of rock type A is half the amount of that of B, then B might have been derived by 50% fractionation from A, or A might have been generated by twice the degree of partial melting relative to B. In order that the problem may be solved one would have to employ various trace elements with different partition coefficients. An interesting method has been suggested by Treuil (1973) and Ferrara and Treuil (1973), and is described below.

The method will be described considering the melting of one solid, but the results can equally well be applied for a multiphase system.

By batch melting the concentration in the liquid of a trace element is given by:

$$c^l = \frac{c^{0s}}{D + F(1 - D)} \tag{47}$$

By fractional crystallization the variation in concentration is estimated from:

$$c^l = c^{0l} f^{(D-1)} \tag{48}$$

where f is the degree of fractionation. For a strongly incompatible element "i" one gets $D_i = c_i^s / c_i^l \ll 0.01$ and then c_i^l is approximately given by:

$$c_i^l = \frac{c_i^{0s}}{F} \tag{49}$$

and

$$c_i^l = \frac{c_i^{0l}}{f} \tag{50}$$

Thus, the variation in concentration of an incompatible element in the liquid will be inversely proportional to F and f.

For a less incompatible element "j" the partition coefficient D_j will have values between 0.01 and 1.0. For these values of D_j the product FD_j may be ignored, but D_j has a value similar to F, so Eq. (47) now becomes:

$$c_j^l = \frac{c_j^{0s}}{D_j + F} \tag{51}$$

and Eq. (48) remains the same:

$$c_j^l = c_j^{0l} f^{(D_j - 1)} \tag{52}$$

The variation in concentration of the two types of trace elements i and j, will be of different functional types. By fractional crystallization the relationship between c_j and c_i is given by:

$$c_j^l = \left(\frac{c_j^{0l}}{c_i^{0l}} \right) f^{D_j} c_i^l \tag{53}$$

By partial melting, c_j is given by:

$$c_j^l = \left(\frac{c_j^{s0}}{c_i^{s0}} \right) \cdot \frac{F}{D_j + F} c_i^l \tag{54}$$

As an example consider the case where $D_j = 0.1$ and $D_i = 0.01$, and $c_j^{os} = 100$ ppm and $c_i^{os} = 100$ ppm. The variation in c_i^l is first estimated from Eq. (49), whereafter c_j^l may be calculated from Eq. (54). The calculated values for c_i^l and c_j^l are tabulated in Table IX-2, and c_j^l is plotted as a function of c_i^l in Fig. IX-17. The variation in c_i^l and c_j^l by fractional crystallization may be calculated using the same values for D_i and D_j, and the result is also shown in Fig. IX-17. The diagram shows the different behavior

Fig. IX-17. The variation in trace element concentrations c_i and c_j by different degrees of partial melting and fractionation. The initial concentration of c_i and c_j is 100 ppm, and the distribution coefficients are $D_i = 0.01$ and $D_j = 0.1$

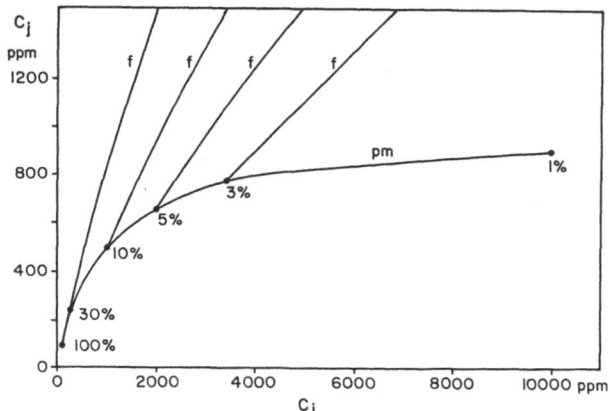

Table IX-2. Concentrations of c_i and c_j by partial melting with $D_i = 0.01$ and $D_j = 0.1$

F	c_i^l	c_j^l	f	c_j^l	c_i^l
1.00	100	100	1.00	100	100
0.80	125	111	0.80	122	125
0.70	143	125	0.70	137	143
0.60	167	143	0.60	158	167
0.50	200	167	0.50	187	200
0.40	250	200	0.40	228	250
0.30	333	250	0.30	296	333
0.20	500	333	0.20	426	500
0.10	1,000	500	0.10	794	1,000
0.05	2,000	667	0.05	1,482	2,000
0.01	10,000	909	0.01	6,310	10,000

of the trace element concentrations by partial melting and fractional crystallization. By partial melting c_i^l displays a much more rapid variation than c_j^l, and the partial melting curve, therefore, becomes nearly horizontal in the diagram. For other values of D_i and D_j, the slope of the curve for partial melting may be larger, but the slope is generally less than for the fractionation curves.

20 R/I Discrimination Method

Treuil's method, considered above, applies the variation in two trace elements, one being slightly incompatible, and the other one being strongly incompatible. Both elements are, thus, incompatible, and similar in behavior. The D values of trace elements may vary with the degree of partial melting, and it should also be taken into account that the D values used frequently are approximate. It might, therefore, be relevant to consider a method that applies trace elements whose D values are substantially different, and it is, thus, obvious to consider the variations for an incompatible element and a refractory element. Let us evaluate the correlation between two such elements by partial melting and fractional crystallization. Let the

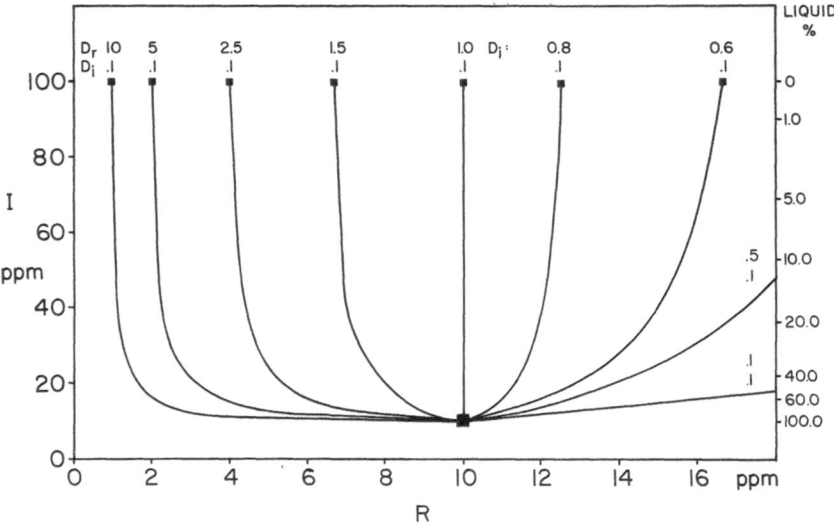

Fig. IX-18. The variations in concentrations by partial melting for different D_r values of a refractory trace element. The D_i value for the incompatible trace element is constant and equal to 0.1. The concentration of the refractory element R is nearly constant at small degrees of partial melting for D_r smaller than 1.0. The variation in the amount of liquid or the degree of partial melting is shown on the *right hand side* of the diagram

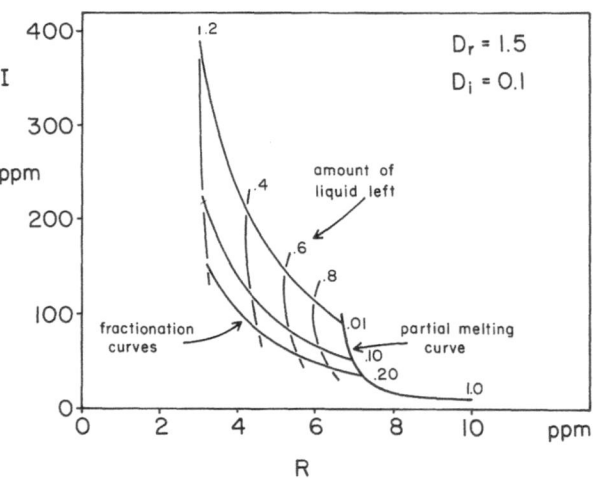

Fig. IX-19. The variation in concentrations by partial melting and fractional crystallization for two trace elements with $D_r = 1.5$ and $D_i = 0.1$. The incompatible element I displays a large variation by partial melting, while the refractory element R displays a large variation by fractional crystallization

source contain 10 ppm of a refractory element R, and 10 ppm of an incompatible element I, and let the value of D_i be 0.1 The variations by partial melting are shown in Fig. IX-18 for different D_r values from 10 to 1. The conspicuous feature of these variations is that the concentration of R in the liquid is almost constant during the initial part of the partial melting. This variation applies for all D_r values larger than 1, and this characteristic variation is, thus, insensitive to variations in D_r.

On the other hand, it is evident from Fig. IX-19 that R displays a marked decrease in concentration during the initial part of fractional crystallization, while I is almost constant. The variations in concentration by partial melting and

Fig. IX-20. The *Th* and *Ni* concentrations in nephelinites from Oahu, Hawaii (Claque und Frey 1982). The *Ni* concentration displays a small variation, while the *Th* concentration displays a relatively large variation. This suggests that the composition of the nephelinites may have been controlled mainly by a variation in the degree of partial melting. The *circles* show analyses that deviate from the general trend

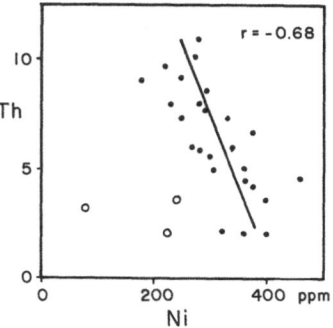

fractional crystallization are compared in Fig. 9 for $D_r = 1.5$ and $D_i = 0.1$. It is obvious that the variations in R and I are very different by the two processes, by partial melting R is almost constant, while I is almost constant by fractional crystallization. This difference suggests that an R/I plot might indicate the process that has controlled a given magma suite. So far this method has not been used extensively, but we might consider two plots in order to see how the method works in practice. Clague and Frey (1982) made a detailed investigation of the nephelinites on Oahu, Hawaii, and their Th/Ni diagram is shown in Fig. IX-20. Thorium displays a relatively large variation, while the variation in Ni is smaller, although the concentration of Ni by no means is constant. The variation in Ni is larger than one would expect from the theoretical diagram in Fig. 20, but the deviation might be due to metasomatic activity in the source or variations in the amount of olivine that enter the partial melt. The variation cannot be related to fractional crystallization, as an increase in Th from 3 ppm to 10 ppm by fractional crystallization would decrease the Ni content below 10 ppm Ni. The scatter displayed by these analyses is what one should expect, as the lavas have formed from different parts of a plume.

A plot which suggests fractional crystallization is shown in Fig. IX-21. This plot shows the variations in Ni and Ta for the trachybasaltic suite on Jan Mayen, and the variations in both major elements and trace elements strongly indicate compositional control by fractional crystallization (Maaløe et al. 1985).

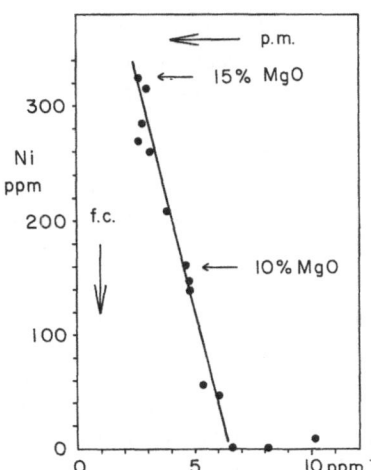

Fig. IX-21. The correlation between *Ni* and *Ta* concentrations for the trachybasaltic lavas on Jan Mayen, North Atlantic. The directions expected by partial melting and fractional crystallization are shown by *p.m.* and *f.c.*, respectively. The correlation suggests that the lavas vary in composition due to fractional crystallization

X Fractional Crystallization

1 Introduction

The chemical composition of igneous rocks from many different igneous provinces displays systematic variation. When the chemical compositions of the rocks are plotted in a diagram, it becomes evident that they apparently define either linear or curved trends. Comparisons of trends for similar rock types have shown that the trends for different provinces are quite similar, neither age nor geographic origin seem to have a major influence on the character of the trends. This consistency suggests that the compositions of igneous rocks are controlled by specific processes that operate under certain circumstances.

The common features of trends may be demonstrated from the alkalic trend of Hawaii (Fig. X-1). The trend might be divided into three major parts, the divisions being defined by the abrupt changes in the slope of the trends for the different oxides. A major break in the trend slopes appear near 5% MgO, and another possible break occurs near 15% MgO, but the abrupt change is only suggested by the variation in CaO content. Within each of the three parts, the weight percentages for the different oxides display a smooth variation that might be represented either by curves or straight lines. Between 5% and 15% MgO, the trend is approximately linear. However, the spread in analyses would also allow a curved trend, but the scatter is too large for a specific curve to be defined. The trend may, therefore, be considered linear as a first approximation.

It is the objective of the igneous petrologist to analyze such trends in order to estimate their origin. The different magma series might either be related to partial melting or fractionation processes, and by an evaluation of the origin of the variations in chemical composition it might be possible to estimate the petrogenesis of the trends. The general procedure required for the definition of a petrogenetic theory may be stated as follows.

First, the trend is estimated from the chemical analyses, preferably both major and trace element concentrations should be estimated. As the chemical analyses display some scatter, the statistical significance of the obtained trends should be calculated. Generally, the trend lines or curves will be calculated using first or second order regression, and the statistical parameters like standard deviations and regression coefficients may be calculated simultaneously using computer methods. It will be shown later that the true standard deviations cannot be calculated from present methods, and the standard deviations obtained using conventional calculation methods are minimum estimates. However, the obtained deviations will afford an indication of the significance of the trend. If an origin of the trend by fractionation is

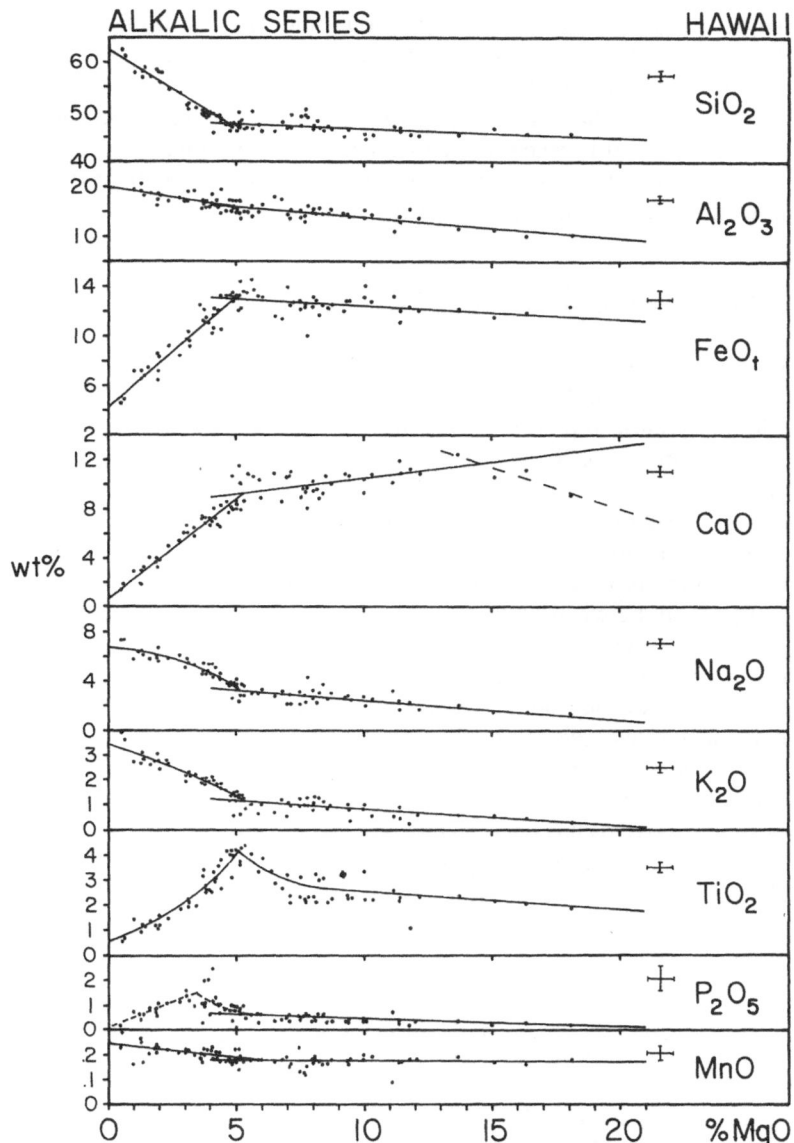

Fig. X-1. The trend for the Hawaiian alkalic series. The diagram is a Bowen diagram, that is, MgO is used as abscissa. The trend lines and curves were estimated by regression. The accuracy of analysis is indicated to the *right*.

considered, the next step is an estimate of the mineral species constituting the fractionate, which may be executed using various methods. After the mineralogical composition of the fractionate has been estimated, the P–T-conditions for the fractionation should be determined. This might be done if the P–T diagrams for the rocks considered are available. If the minerals of the fractionate are in equilibrium in a P–T region of the P–T diagram, then this region might define the pressure and temperature conditions for the fractionation. The last problem will then be to link the obtained P–T conditions together with the geophysical processes that might have resulted in the specific P–T conditions.

The present chapter will deal with the phase relations and geochemical methods that are relevant to fractional crystallization. Some phase relations will first be considered, whereafter the analyses of trends will be dealt with in detail, and finally the statistical analyses of trend lines will be represented.

2 Phase Relations of Fractional Crystallization

We have considered the general relationships between phase relations and trends in Chap. III, and will here deal with some phase relations of relevance for the fractionation of basaltic and granitic magmas.

Tholeiitic basals generally form phenocrysts in the succession olivine –plagioclase–augite, orthopyroxene may be a phenocryst in tholeiitic lavas, but that is rare. Alkalic olivine basalts have the crystallization sequence, olivine or diopside–plagioclase–titanomagnetite. The phase relations of systems containing these minerals are relevant for the study of the fractional crystallization of basalts. A comparison between the P–T phase relations of basalts and the phenocryst assemblages show that the basaltic magma suites form phenocrysts at low pressures, a relationship that is also evident from their major element trends. The mineralogical constitution of the fractionate will vary with pressure, and the trends of the basaltic magma suite will also vary with pressure. Nevertheless, the trends, for say, the alkalic suites are very similar, and of low pressure type. Some systems will here be considered at 1 bar, but we will refrain from a systematic treatment, a detailed analysis of similar systems has been given by Morse (1980).

The phase relations of the forsterite–diopside–silica system is shown in Fig. X-2 (Kushiro 1972). Forsterite reacts with the liquid forming either enstatite or pigeon-

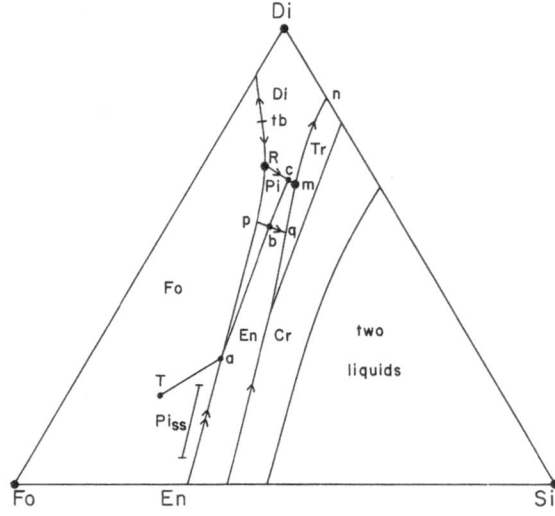

Fig. X-2. The system forsterite–diopside–silica at 1 bar (Kushiro 1972). The reaction curve between forsterite and enstatite and pigeonite is marked with a *double arrow*. The line Pi_{ss} shows the compositions of pigeonite. Both R and m are reaction points. The thermal barrier on the univariant curve between olivine and diopside is indicated with *tb*

Fig. X-3. The system forsterite–anorthite–silica at 1 bar (Anderson 1915). The system is not strictly ternary as the composition of spinel is situated outside the ternary join. However, the crystallization dealt with here is ternary. The point r is a reaction point, while m is the true minimum point of the system, all liquids outside the spinel field will end up at m by fractional crystallization

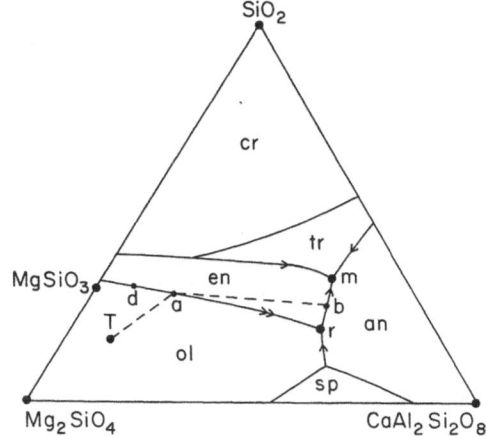

ite. The composition T will first form forsterite. By fractional crystallization it will move to a, where enstatite begins, to form instead of forsterite, and the liquid will then move to b. At b pigeonite is crystallizing instead of enstatite, and the liquid will, therefore, now move away from the composition of pigeonite (Pi_{ss}). The liquid will then reach c, and then move to m, which is a reaction point. Finally, the liquid will move from m to n. The point n is a piercing point, and the crystallization does not cease at n.

Each time a new solid phase appears, the direction of the liquid changes, and the major element trend for composition T will, therefore, display several kinks.

The system forsterite–anorthite–silica is shown in Fig. X-3. The crystallization path of composition T is T–a–b–m. Here m is a eutectic minimum point and the crystallization of all liquids end at m by fractional crystallization. All liquids that reach the reaction curve between d and r will crystallize anorthite as the third solid phase, while liquids that reach the reaction curve to the left of d will crystallize cristobalite as the third solid.

The quaternary phase relations at 1 bar of the forsterite–anorthite–diopside–silica system have been compiled by Presnall et al. (1978) and are shown in Fig. X-4. Two of the ternary systems of the quaternary system have just been considered. This quaternary system displays the major phase relations of basaltic and andesitic magmas at low pressure. The point p is a reaction point, while m is the quaternary eutectic minimum point. The diagram is suitable for an estimate of the possible crystallization sequences of various basaltic and andesitic compositions.

The solid phases that react with the liquid will cease to crystallize by fractional crystallization, while the eutectic solids continue to crystallize. In some cases a solid phase may stop to crystallize and then start again. This behavior is observed for olivine in some of the layered intrusions (Fig. III-10). The phase relations of the forsterite–fayalite–silica system (Fig. X-5) explain this behavior. Consider composition T, which first forms olivine. When the liquid reaches a, olivine ceases to crystallize and enstatite forms from a to b. From b to r enstatite and tridymite crystallize together. Point r is a reaction point and enstatite stops to crystallize at r, while olivine starts to crystallize together with tridymite. The crystallization sequence for

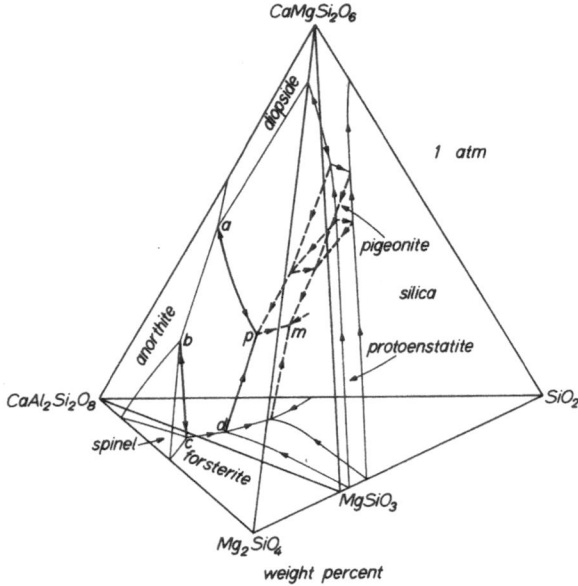

$CaMgSi_2O_6$

diopside

1 atm

pigeonite

silica

anorthite

protoenstatite

$CaAl_2Si_2O_8$

spinel

forsterite

SiO_2

$MgSiO_3$

Mg_2SiO_4

weight percent

Fig. X-4. The forsterite–diopside–anorthite–silica system at 1 bar compiled by Pressnal et al. (1978)

composition T is, thus: ol, en, en + tr, ol + tr. Hence, olivine crystallizes twice interrupted by enstatite.

The trends of the tholeiitic and alkalic suites display a marked kink at about 5% MgO (Fig. X-1). The trends for TiO_2 and FeO suggest that the kink is caused by the appearance of titano magnetite, which apparently begins the fractionate when the magmas attain 5% MgO. The pertinent phase relations are displayed by the olivine–anorthite–silica–magnetite system (Fig. X-6), which was investigated at different oxygen fugacities at 1 atm by Roeder and Osborn (1966; cf. Osborn 1979).

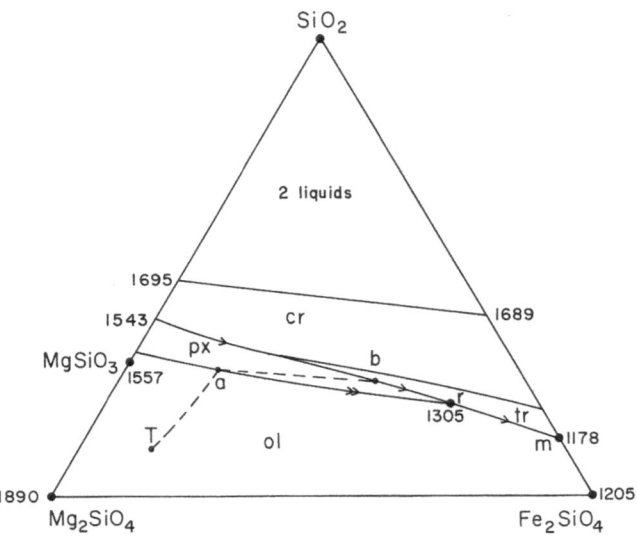

SiO_2

2 liquids

1695

1543

cr

1689

$MgSiO_3$

1557

px

b

a

1305

r

tr

T

ol

m

1178

1890

Mg_2SiO_4

Fe_2SiO_4

1205

Fig. X-5. The forsterite –fayalite–silica system redrawn from the phase diagram estimated by Bowen and Schairer (1935)

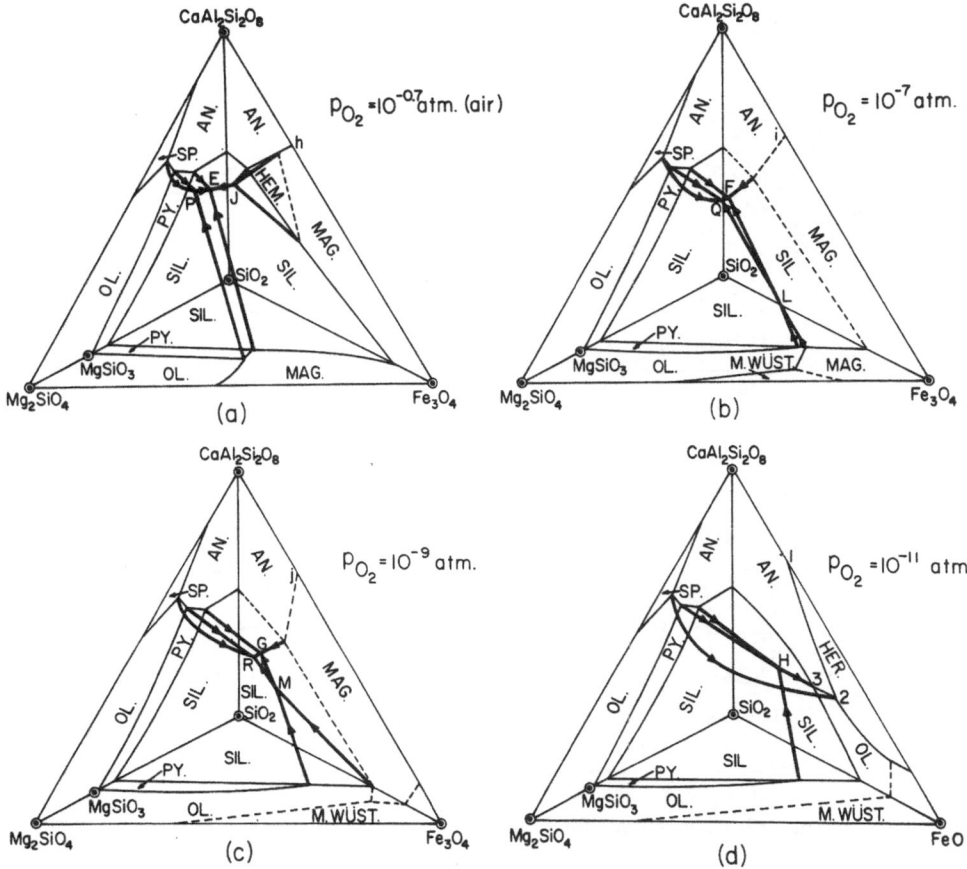

Fig. X-6 a–d. The phase relations of the system forsterite–anorthite–magnetite–silica at different oxygen fugacities (Roeder and Osborn 1966). The shape of the univariant curves will be evident from Fig. X-7

The expiration of the univariant curves becomes clear from the projections shown in Fig. X-7. The oxygen fugacity of most relevance is 10^{-9} atm, and we will consider the phase relations at this pressure in particular. We may choose a liquid which is similar to ankaramite, and assume olivine on the liquidus followed by pyroxene. This liquid will at first be situated in the olivine volume, and thereafter move to the L, S (ol, py) surface (Fig. X-6). Thereafter the liquid will hit the L, S (ol, py, an) curve and move along this curve to the reaction point R [Figs. X-6c and X-7c (p′–R′)]. This later part of the fractionation is equivalent to the fractionation of the alkaline olivine basalts with 10% to 5% MgO. At R olivine reacts with the liquid, and magnetite begins to crystallize instead of olivine along R′–G′. The phase assemblage is L, S (ol, py, an) above the reaction point, and L, S (py, an, mg) below the reaction point. This change causes the predominant kink of the alkalic suites at 5% MgO. The quaternary minimum is at point G (Fig. X-6), however, since basalts contain Na_2O and K_2O, this phase diagram is not representative for the phase relations at this stage of the fractionation.

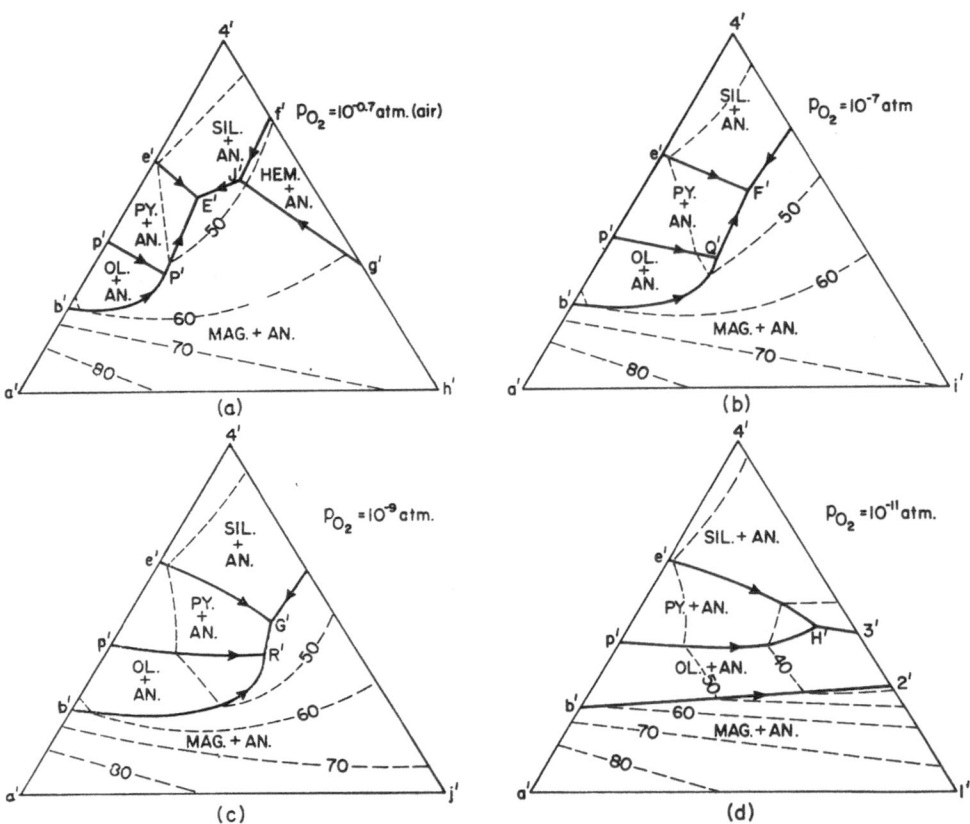

Fig. X-7 a–d. The univariant curves for liquids in equilibrium with anorthite in the system shown in Fig. X-6

By evaluation of the mineralogical constituents of the fractionates of the alkalic suites from 5% MgO to smaller MgO contents, one would have to use the least square method. However, there are four to six minerals involved, which display solid solution, and the least square method does not yield a definite estimate of the mineral constituents. The phase relations of the quaternary system considered here show that the fractionate ceases to contain olivine, and that magnetite begins to crystallize instead.

3 Thermal Barriers

Magmas have been generated by different degrees of partial melting and fraction-ation, and a wide variety of magma types appears possible. Admittedly, there is in-deed a large variety of magma compositions, but by far the majority of magma types may be classified into one of the four major magma series, the tholeiitic, alkalic, nephelinic, and andesitic series. There is, therefore, a high degree of systematic vari-ation in the composition of magmas. Why is this so? Probably part of the answer is that the possible compositional range of magma types is controlled by thermal barriers and eutectic-like minimum points. The phase diagrams of silicate systems can be divided into closed sectors because of the presence of thermal barriers, the existence of which are due to congruently melting compounds. The existence and position of thermal barriers vary with pressure, but at a given pressure a magma is forced to remain within a given closed sector. Within the sector the magma will fractionate towards the minimum point, quite independent of its original composition. A magma with a composition far away from the minimum point will most likely first crystallize one solid phase, then two and finally three or four phases. However, the rate of crystallization will be largest in the beginning as only one phase surrenders latent heat. Near the minimum point several solid phases crystallize, with the result that the fractionation occurs slowly in this region with respect to time. Thus, there will be a tendency for the magmas to have compositions near minimum points. Statistically, there will be a larger chance that such magmas erupt, rather than the erruption of a magma with a composition being far off the minimum point.

The significance of thermal barriers became widely accepted after Yoder and Tilley (1962) considered the fractionation from tholeiitic to alkalic basalts, and showed that this transition was impossible at low pressures. Alkalic magmas are nepheline normative, while tholeiitic magmas are hyperstene or quartz normative. The phase relations for these basaltic compositions might be demonstrated by the quartz–nepheline–diopside system shown in Fig. X-8. Plagioclase forms a congruent compound between nepheline and quartz, consequently the join plagio-

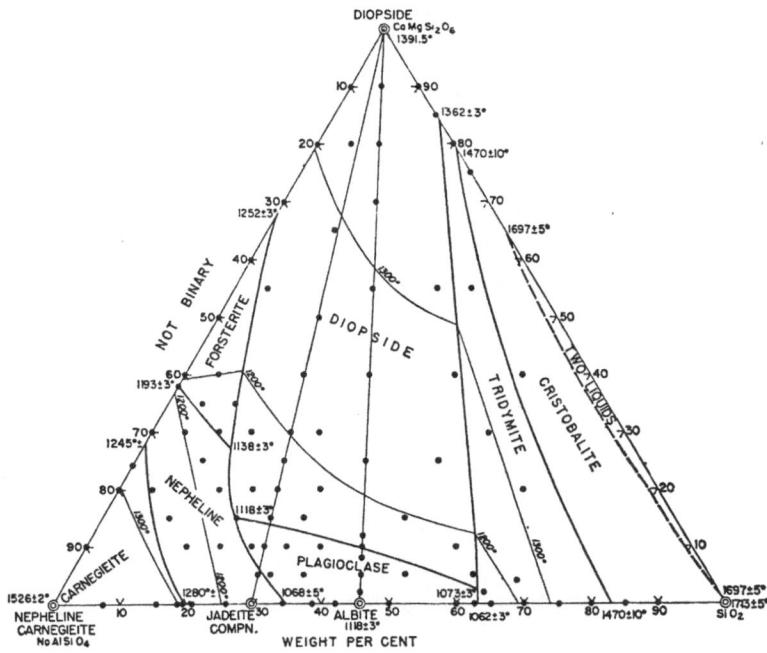

Fig. X-8. The ternary join nepheline–diopside–silica at 1 bar. The join diopside–albite forms a thermal barrier in the system

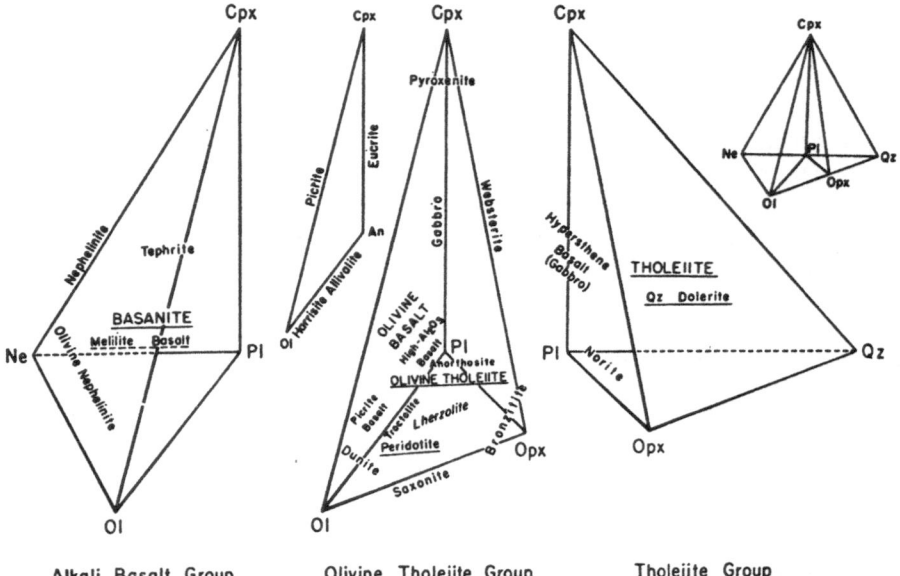

Fig. X-9. The positions of magmatic compositions within the quaternary system nepheline–olivine–quartz–pyroxene. The plane Ol–Cpx–Pl is a thermal barrier at low pressures and basanite cannot have formed by fractionation from tholeiite

clase–diopside is a thermal barrier in the ternary system. Liquids with compositions near the join will end up in widely different minimum points, either L, S(ne, di, pl) or L, S(tr, di, pl), the final liquids being undersaturated or saturated.

The ternary join will be a thermal barrier in the system quartz–nepheline–olivine–diopside (Fig. X-9). Basaltic magmas that crystallize at pressures at which this barrier exists, i.e., at pressures up to 8 kbar (O'Hara 1968),

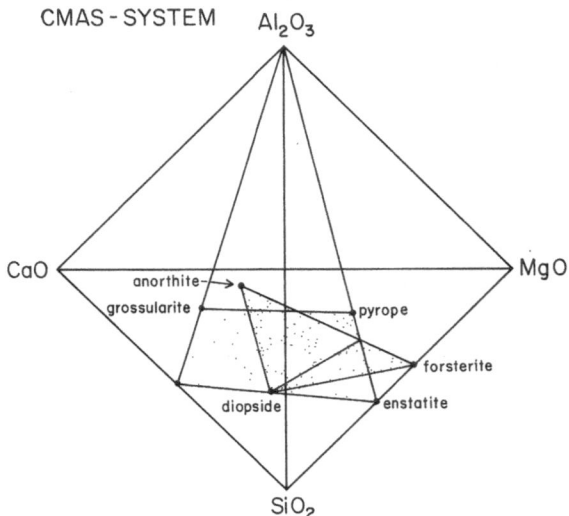

Fig. X-10. The pyroxene-garnet barrier within the system CaO–Al₂O₃–MgO–SiO₂ at high pressures and the low pressure thermal barrier anorthite –forsterite–diopside

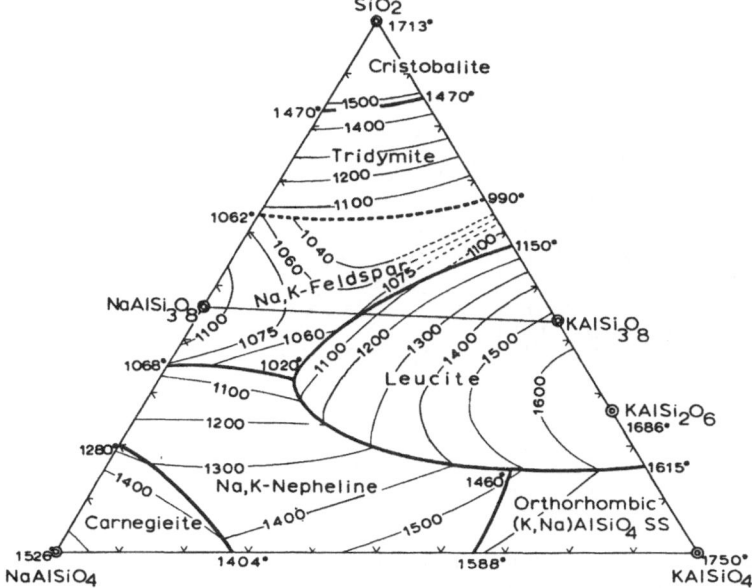

Fig. X-11. Petrogeny's residua system, NaAlSiO₄–KAlSiO₄–SiO₂. The alkali feldspar join forms a thermal barrier within the system, dividing saturated from undersatured compositions (Schairer 1957)

are forced either into the undersaturated or saturated region. Another thermal barrier of importance for basaltic compositions is the garnet–pyroxene barrier which exists at pressures above ca. 30 kbar (Fig. X-10). Magmas that have formed by partial melting above this pressure must remain on the olivine-rich side of the plane grossularite–pyrope–enstatite–wollastonite, as the reaction point is situated on this side of the plane (Maaløe and Wyllie 1979).

Another important barrier is the feldspar barrier in the system quartz–nepheline–kaliophilite (Fig. X-11), also called petrogeny's residua system by Bowen (1928). Magmas that are rich in the feldspar component, such as trachytes, will either fractionate towards the minimum point in the granite system, or towards the minimum point where feldspar, nepheline, and leucite are in equilibrium. The direction of fractionation will be dependent on the original composition of the magma, and small differences might result in fractionation in either direction.

4 The System Albite–Orthoclase–Quartz

The albite–orthoclase–quartz system is the fundamental system for granitic rocks. The phase relations of the system were estimated at different water pressures by Tuttle and Bowen (1958), and their work is now a classic contribution to petrology. They also provided an account of fractional crystallization within the system, and considered fractional crystallization at 1 kbar water pressure (Fig. X-12). The curve DE is the eutectic curve that separates the liquidus fields of quartz and alkali feld-

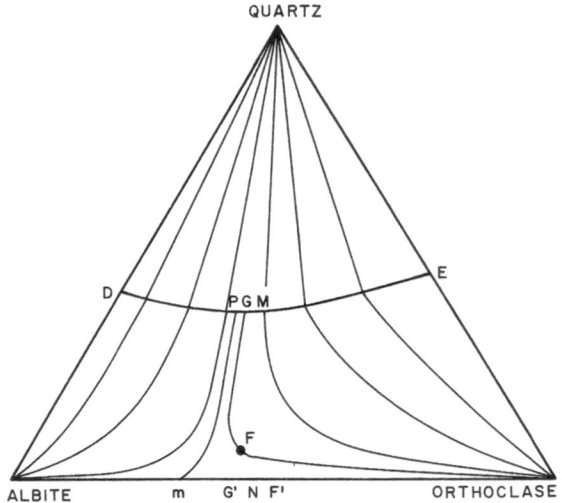

Fig. X-12. The fractionation curves for the system albite–orthoclase–quartz at 1 kbar water pressure (Tuttle and Bowen 1958)

spar. The binary minimum for the albite–orthoclase system is at m, and the ternary minimum is at M. Albite and orthoclase display continuous solid solution with a temperature minimum like the system shown in Figs. X-2 and X-13.

By fractional crystallization all liquids within the area OrME will end up at the minimum point M. The liquids will follow the fractionation curves shown in Fig. X-12. The feldspar crystals formed will all display normal zoning, and the composition of the feldspar formed can be estimated by drawing tangents to the fractionation curves.

The crystallization within the area OrMPm will be different. Reverse zoning is generally related to changes in crystallization conditions, but reverse zoning will be developed here by fractional crystallization. A liquid at F will begin to crystallize feldspar with the composition F', and the liquid will move to G, where the feldspar composition is G'. By further fractionation the liquid will move from G to M, while the composition of feldspar moves from G' to N, and a reverse zone is formed.

Fractional crystallization within the area AbmPD is similar to the crystallization within the area OrME.

5 Rayleigh Fractionation

The compositional trends of liquids fractionating minerals with a constant composition are easily estimated if the phase relations of the liquids are known. If the position of univariant curves and bivariant surfaces within a cotectic quaternary system is known, then the variation of liquid compositions may be determined directly from the phase diagram, and the amount of residual liquid may be calculated using the lever rule and matrix methods. However, nearly all minerals of petrological importance – olivine, pyroxenes, garnets, amphiboles, and feldspars – display a wide range of solid solution, and the fractionation trends of magmas cannot be determined without taking into consideration the variation in chemical composition of the fractionating minerals (cf. Osborn and Schairer 1941).

The compositional variation of liquids fractionating solid solution minerals may be estimated by solving a differential equation for the particular system. Differential equations for perfect fractionation were worked out by Rayleigh (1896, 1902). These equations involve the integration of fractional expressions and may be difficult to solve, and solutions are often based on simplified fractional models (Robinson and Gilliland 1950).

Solutions of fractionation equations assuming a constant distribution coefficient have been presented by Rayleigh (1896, 1902), Neuman et al. (1954), and McIntire (1963). A solution based on a linear variation in the distribution coefficient with respect to the composition of the fractionating liquid was obtained by Greenland (1970), who also considered the compositional variation of liquids fractionating several solid phases simultaneously.

Estimated phase relations of mineralogical systems show clearly that distribution coefficients vary with temperature or degree of fractionation, and assuming constant distribution coefficients for major elements leads to inaccurate or erroneous results.

6 Types of Fractional Crystallization

It has been generally considered that fractional crystallization is the same as gravitative settling of phenocrysts. Many sills show ubiquitous evidence for gravitative settling, but the rhythmic layering of the layered intrusions indicate that the crystallization of these large intrusions mainly occurred near the margins of the magma chamber, i.e., just above the surface of the layered series, and at the walls of the magma chamber (McBirney and Noyes 1979). All the details of this type of crystallization are not yet fully understood, but the crystallization is governed by super saturated crystallization and diffusion in the magma. This type of crystallization is also fractional, but the crystallization process is more complicated than gravitative settling. The structural and petrological features of the layered intrusions and the gabbro units of the ophiolite complexes are quite similar, and it is likely that the features of the layered intrusions are quite general for large magma chambers. These intrusions may reveal the crystallization processes not only in crustal magma chambers, but also those of large magma chambers in the mantle.

The relative volume of the intercumulus liquid has been estimated at 30% (Hess 1960; Jackson 1961), 40–50% (Wager and Brown 1967), 15–24% (Henderson 1970, 1975), and 65% (Maaløe 1975). The estimates display some variation, but the intercumulus liquid clearly forms a significant part of the cumulate.

The monomineralic rocks of some layered intrusions (Hess 1960; Wagner and Brown 1967), and the evidence of adcumulus growth in some cumulates indicate that the accumulated crystals and the intercumulate liquid may exchange material with their surroundings (Wagner et al. 1960). The diffusion rate of magmatic liquids is too small to allow diffusion over a distance of more than a few centimeters according to Hess (1972), and estimates of the convection velocity of intercumulus liquid suggest that it is too small for the development of monomineralic rocks. The nature of the exchange of material between the intercumulus liquid and the fractionating magma is, thus, uncertain (Hess 1972).

Some intercumulus liquid must originally have been present in the cumulate and at least a part of it may have remained between the cumulated crystals. Therefore, two types of fractionation are in operation during fractional crystallization; a part of the fractionating magma is removed as crystals and another part as liquid. The composition of the removed liquid will be that of the fractionating magma, if the exchange of material is negligible, otherwise it will differ.

Two simplified types of fractional crystallization will be considered here, both representing ideal limiting cases; (1) perfect fractional crystallization, involving a complete separation of the fractionating liquid and the cumulating crystals, and (2) partial fractional crystallization, where both crystals and liquid are removed from the fractionating liquid. The composition of the intercumulus liquid is assumed identical to that of the fractionating liquid in the calculations which follow.

7 Perfect Fractional Crystallization

The differential equation for the compositional relations of a fractionating system consisting of several components was first estimated by Rayleigh (1902). Let the system contain the amount W mol liquid and let the composition of the liquid be x^l, where x^l is the mole fraction of a given reference component. If a certain amount dW is removed from the system with fraction x^s of the reference component, then the total amount of removed reference component is given by $x^s dW$. From material balance this amount must be equal to the change in the amount of the component in the fractionating liquid which is $d(x^l W)$:

$$x^s dW = d(x^l W) \tag{1}$$

The fractionation trend of a magma depends, among other factors, on the relative amount of the different fractionating minerals. This factor is included in Eq. (1) as x^s will vary with varying mineral proportions. For a particular relation between x^s and x^l, that is, for a particular phase diagram, the relative proportions of minerals being fractionated will be unambiguously determined by the phase relations, as x^s is a single-valued function of x^l. By differentiation the equation becomes:

$$x^s dW = x^l dW + W dx^l \tag{2}$$

and by definitive integration:

$$\ln \frac{W}{W^0} = \int_{x^{01}}^{x^l} \frac{dx^l}{x^s - x^l} \tag{3}$$

where W^0 is the initial amount of mol liquid, and x^{01} is the mole fraction of the reference component in the initial liquid. Equation (3) is often solved assuming a constant x^l/x^s, in which case the solution is given by:

$$\ln \frac{W}{W^0} = \frac{k}{1-k} \ln \left(\frac{x^l}{x^{01}} \right) \tag{4}$$

where $k = x^l/x^s$ (Rayleigh 1902). The ratio x^l/x^s is not constant for major elements of mineralogical systems, but may vary between 0 and 1. This ratio is shown in Fig. X-13 for the albite–anorthite system at 1 bar, using data by Bowen (1913).

As evident from this figure x^s is not a linear function of x^l for this system.

The solution of the equation requires that x^s be expressed as a function of x^l, or that x^l and x^s are expressed as a function of a third variable. The phase relations of magmas depend on the thermodynamic properties of the constituents of the magma. It would, therefore, be most correct and elegant to express the composition of a fractionating magma as a function of temperature and pressure. A solution based on the thermodynamic relations of a binary system with solid solution, and with temperature as the variable has been made by Maaløe (1976). The solution is a series solution, and may be evaluated as accurately as warranted, but the series is slowly convergent and the thermodynamic solution is not suitable for general use. An approximate but analytical solution may, however, be obtained expressing x^s as a function

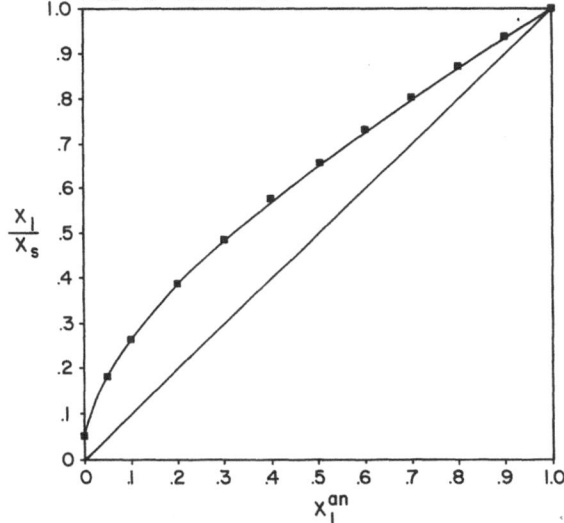

ALBITE - ANORTHITE

Fig. X-13. The variation in the ratio between the anorthite content of liquid and solid for the binary albite–anorthite system estimated from the phase relations of the system (Bowen 1913). The *squares* show the experimental data, while the *curve* has been estimated from Eq. (5)

of x^l. The particular function chosen will depend on the shape of the solidus and liquidus curves of the system under consideration and the accuracy required. Unfortunately, there is no systematic manner by which the most accurate functional form may be estimated when approximating a given curve by a function (Dahlquist and Bjørck 1974). The coefficients of a polynomial may be estimated by regression or least square curve fitting, but it is by no means granted that a polynomial represents the best functional form (Draper and Smith 1966).

The calculation of fractional crystallization for the binary albite–anorthite system will now be considered, the solutions obtained can be generally applied to similar systems, like the fayalite–forsterite system.

The ratio x^l/x^s is nearly constant for trace elements. There is no reason to expect that this ratio is also constant for major elements, however, it is likely that the ratio displays a small variation, or a variation with a slight curvature. Since we have to estimate x^s as a function of x^l, using an approximate equation, it is an advantage to obtain a function with a small curvature, because such functions can be estimated more accurately.

The following equation for the albite–anorthite system was obtained by trial and error, because polynomial regression not did yield satisfactory results (Maaløe 1984):

$$\frac{x^l}{x^s} = a x^l + b \sqrt{x^l} + c \tag{5}$$

where:

$$a + b + c = 1 \tag{6}$$

For the albite–anorthite system the constants were estimated as:

$$a = 0.55, \quad b = 0.40, \quad c = 0.05 \tag{7}$$

For the fayalite–forsterite system the constants are:

$$a = 0.61, \quad b = 0.25, \quad c = 0.14 \tag{8}$$

The x^l/x^s curve estimated from Eqs. (5) and (7) is shown in Fig. X-13, where it may be compared with the experimental data. The Rayleigh integral Eq. (3) may now be estimated using Eq. (5). Let $x = x^l$, we then have:

$$\ln\left(\frac{W}{W_0}\right) = \int_{x^0}^{x} \frac{dx}{\dfrac{x}{ax + b\sqrt{x} + c} - x} \tag{9}$$

By substituting $y = \sqrt{x}$, and by rearrangement we get:

$$\ln\left(\frac{W}{W_0}\right) = \int_{y_0}^{y} \frac{2\,(a y^3 + b y^2 + c y)\,dy}{(1 - c)\,y^2 - a y^4 - b y^3} \tag{10}$$

The determination of this integral is elaborate, the solution is given by:

$$\ln\left(\frac{W}{W_0}\right) = \left[-\ln(a + b - by - a y^2) + \frac{b}{2a+b}\ln\left(\frac{ay + a + b}{-ay + a}\right) \right.$$
$$\left. -\frac{2ca}{(2a+b)(a+b)}\ln\left(y - \frac{c-1}{a}\right) - \frac{2c}{2a+b}\ln(y-1) + \frac{2c}{a+b}\ln(y) \right]\Bigg|_{y_0}^{y} \tag{11}$$

Using the constants for the albite–anorthite system we obtain:

$$\ln\left(\frac{W}{W_0}\right) = \left[-\ln(0.95 - 0.4y - 0.55 y^2) + 0.2667\ln\left(\frac{0.55y + 0.95}{-0.55y + 0.55}\right) \right.$$
$$\left. -0.0386\ln(y + 1.7273) - 0.0667\ln(y-1) + 0.10526\ln(y) \right]\Bigg|_{y_0}^{y} \tag{12}$$

Using Eq. (12) the variation in the composition of the plagioclase liquids may be estimated, and a series of fractionation curves are shown in Fig. II-6. The variation in the anorthite content will be dependent on the initial anorthite content of the liquids. Liquids with a higher anorthite content display a small variation in

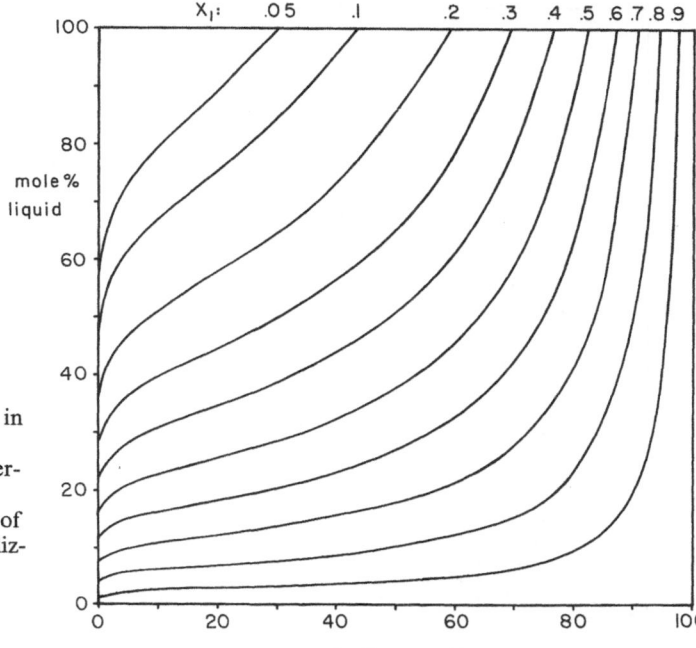

X_l: .05   .1   .2   .3   .4 .5 .6 .7 .8 .9

mole %
liquid

mole % anorthite

Fig. X-14. The variation in the anorthite content of plagioclase formed by perfect fractional crystallization. The composition of the initial liquids crystallizing plagioclase is shown just above the diagram

the anorthite content during the initial fractionation, while the decrease in anorthite content is very rapid during the final stage of the fractionation. Liquids with a small anorthite content display a more constant decrease in anorthite content.

The variation in the composition of the instant fractionate, i.e., the plagioclase that forms at any given temperature is shown in Fig. X-14. The variation is very much the same as for the liquids. The variation in the compositions of solid and

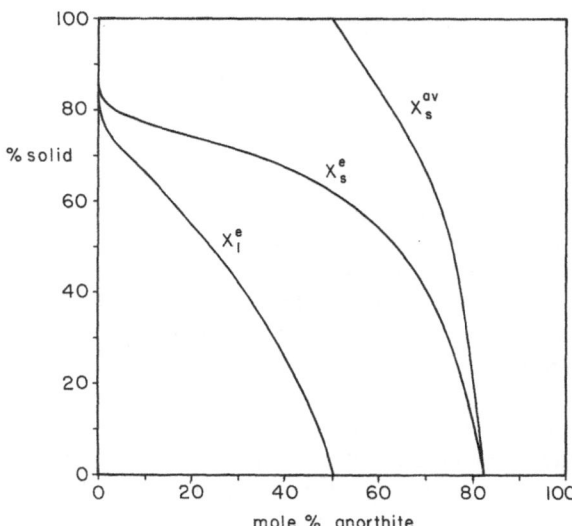

% solid

X_s^{av}

X_s^e

X_l^e

mole % anorthite

Fig. X-15. The variation in the anorthite contents for a liquid that initially contained 50 mol% An. The composition of the liquid is shown by the curve x_l^e, that of the solid that forms at any given temperature is shown by x_s^e, and the average composition of the integral fractionate is shown by the curve x_s^{av}

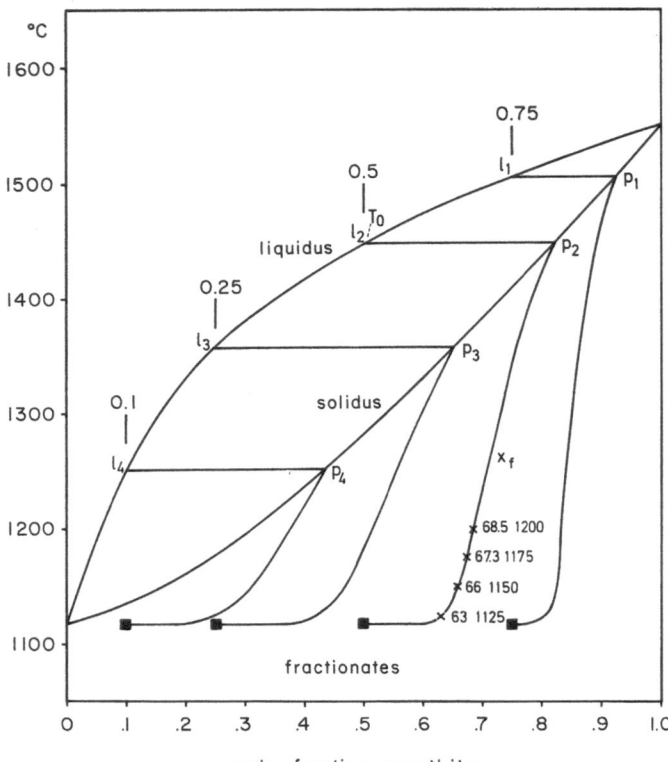

Fig. X-16. The phase relations of the albite–anorthite system and the variation in the composition of the integral fractionate (x_f) for four different liquid compositions

liquid is compared in Fig. X-15 for a liquid with an initial anorthite content of 50 mol% An. This diagram shows the variation in the composition of plagioclase that one would expect for a gabbroic intrusion, x_s^c is the composition of the instant fractionate, being equivalent to the composition of a cumulate. The diagram assumes perfect fractional crystallization, if the plagioclase crystals remain fluentively in the liquid the curves will be different. A comparison between the theoretical variation and the observed variations in anorthite contents show that the crystallization of the layered intrusions is not perfect fractional (Maaløe 1976).

The apparent phase relations by fractional crystallization are summarized in Fig. X-16, which shows the albite–anorthite system at 1 bar, assuming the melting temperature of albite as 1118 °C (Bowen 1913; Greig and Barth 1938). The x_f curves show the variation in average composition of the total fractionate. The fractionating liquids will end up crystallizing almost pure albite, a liquid that starts with 50 mol% An will leave 12.72% of the initial amount of liquid when the anorthite content is as low as 1 ppm. The more viscous magmas like andesitic and granitic magmas may undergo cooling without gravitative fractional crystallization. When that happens the phenocrysts will remain fluentively in the magma, and the fractionation will, therefore, occur on a local scale. The theoretical zoning developed by fluentive crystallization is shown in Fig. II-7 for three different initial anorthite contents.

8 Partial Fractional Crystallization

The presence of significant amounts of intercumulus liquid between the cumulated crystals in fractionating magma chambers results in the fractionation of both crystals and liquid, the composition of the liquid being equal to or differing from that of the fractionating magma. This type of fractionation differs from perfect fractionation, described by Rayleigh's equation (1902), where only crystals are removed from the fractionating system. The fractionation of both crystals and liquid may be taken into account in the Rayleigh (1902) equation by applying a corrected distribution coefficient, but it is more correct to consider the effect of the intercumulus liquid and the distribution coefficient separately as they are completely independent, and a new differential equation is, therefore, required.

Let the amount removed from a liquid system be dW and let the system contain the amount W mol liquid. The mole fraction of a reference component in the removed solid part is x^s, and x^l in the removed liquid part. Further, let the fraction of solid material in the removed part be given by the fractionation factor f. The fractionation factor may vary during the fractionation, as the amount of intercumulus liquids varies, but the fractionation factor has been assumed constant in the calculations below. The fractionation factor is 1 for perfect fractional crystallization and 0 for a nonfractionating system. From material balance we have:

$$(1-f)x^l dW + f x^s dW = d(x^l W) \tag{13}$$

This differential equation is the general equation for partial fractional crystallization. A similar equation assuming a constant distribution coefficient has been represented by Anderson and Greenland (1968). Rearranging by definitive integration, we have:

$$\ln\left(\frac{W}{W^0}\right) = \int_{x_0^l}^{x^l} \frac{dx^l}{f x^s - f x^l} \tag{14}$$

and we, thus, get an integral which is identical to the original Rayleigh integral apart from a factor $(1/f)$:

$$\ln\left(\frac{W}{W^0}\right) = \frac{1}{f} \int_{x_0^l}^{x^l} \frac{dx^l}{x^s - x^l} \tag{15}$$

The usual solutions of the Rayleigh integral can, therefore, be used for partial fractional crystallization, the solution for partial fractionation is obtained just by division by the fractionation factor f.

For a constant distribution coefficient we obtain:

$$\ln\left(\frac{W}{W^0}\right) = \frac{k}{f(1-k)} \ln\left(\frac{x^l}{x_0^l}\right). \tag{16}$$

The significance of partial fractionation for the distribution coefficient may be estimated by calculating an apparent distribution coefficient D:

$$\frac{D}{1-D} = \frac{k}{f(1-k)} \tag{17}$$

and by rearrangement we obtain:

$$D = \frac{k}{f(1-k)-k} \tag{18}$$

Equation (17) which was also derived by Anderson and Greenland (1968), may be used for an estimate of the variation in the concentration of trace elements by partial fractional crystallization, if the distribution coefficient can be assumed constant. The fractionation factor f may theoretically vary between 0 and 1, but probably has a value between 0.2 and 0.7 for basaltic liquids. The actual value may be estimated from the textural relationships of rocks in favorable cases, or from geochemical relationships (Henderson 1970), but in general an approximate value has to be assumed, and a value of 0.5 is considered representative for basaltic liquids.

Layered intrusions provide good examples of the crystallization behavior of large magma chambers, and the cumulated rocks display a continuous record of the crystallization history of the magma. A comparison between theoretical fractionation models estimated from Eq. (17), and the variations observed in basaltic layered intrusions may, therefore, show to what degree basaltic magmas solidify by fractional crystallization. Significant deviations from the theoretical model may indicate variations in the type of crystallization.

The fractionation relationships of basaltic magmas at low pressures are complex because several solid phases displaying solid solution are involved in the fractionation, as evident from layered intrusions (Wager and Brown 1967). Despite extensive investigations of the equilibrium relations of basaltic liquids, the variation of tie lines with varying temperature have not been estimated in detail; and an exact account of the fractionation relations is not possible at present. Basaltic liquids mainly crystallize plagioclase and pyroxene, and some of their fractionation relationships may, therefore, be estimated from the albite–anorthite–diopside system, the system used by Bowen (1915) as a model for the crystallization of basaltic liquids. The phase relations of this system were estimated by Bowen (1915), and the tie lines along the univariant line of the system were estimated by Wyllie (1963) from Bowen's results. Later studies by Hytønen and Schairer (1961) and Kushiro (1973) have shown that the phase relations of the system are more complex than estimated by Bowen (1915). There is no eutectic point on the join albite–diopside, but a piercing point, and the system is, therefore, not truly ternary. These complications will only affect the crystallization of the most albite-rich liquid, and they are ignored in the following account.

The fractionation relations of a haplobasaltic composition modeled from the supposed initial composition of the Skaergaard magma may demonstrate some of the fractionation relationships of basaltic magmas and layered intrusions. A specimen considered representative of the initial composition of the Skaergaard magma contains 57.14% feldspar in the norm, of which 35.56% is normative anorthite and 21.58% is normative albite and orthoclase (Wager and Brown 1967, Table 7). From these figures and assuming 42.86% diopside in the norm, the initial composition of the haplobasaltic Skaergaard magma is estimated to be 31.32% anorthite, 20.17% albite, and 48.51% diopside, in mole percentages.

It follows from Bowen's (1915) phase diagram that this composition falls on the univariant line, and both plagioclase and diopside will, therefore, be on the liqui-

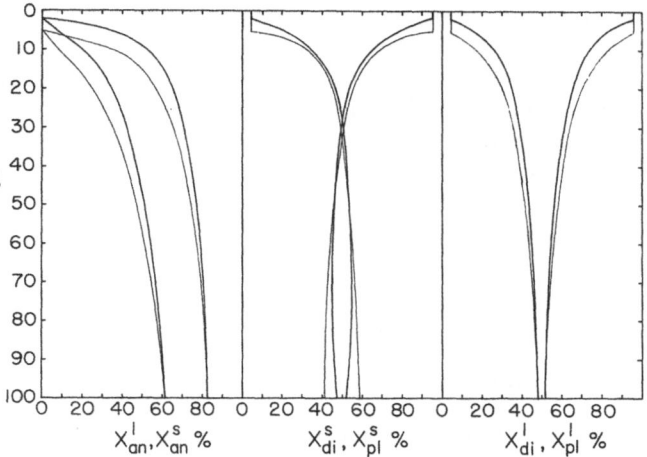

Fig. X-17. The variation in anorthite content of plagioclase (*left*), the molar percentages of the diopside and plagioclase in the cumulate (*middle*), and the molar percentages of diopside and plagioclase in the liquid (*right*). The *ordinate* is the amount of residual liquid. The *heavy curves* show the compositions for partial fractional crystallization, while the *light curves* show the variation in compositions for perfect fractional crystallization

dus. The variations in liquid and solid compositions during perfect and partial fractional crystallization are shown by light and heavy curves, respectively, in Fig. X-17. The diagrams of this figure may be regarded as depicting the compositional variations of a layered intrusion with constant cross-section of haplobasaltic composition, assuming no exchange between the fractionating liquid and the intercumulus liquid. The bottom and top of the diagrams are then equivalent to the bottom and top parts, respectively, of the layered intrusion. The general variation in compositions by the two types of fractionation are the same, but the composition of liquids and solids differ at the same degree of fractionation, the rate of variation in compositions being larger for perfect fractional crystallization than for partial fractional crystallization. For perfect fractional crystallization the anorthite content of the material subtracted from the fractionating liquid is higher, and the amount of residual albitic liquid is, therefore, larger for this type of crystallization. The left diagram shows the variation of the anorthite content of liquid and plagioclase. The anorthite content of both liquid and plagioclase varies only slightly during the first 50% of the fractionation, while there is a marked decrease in the anorthite contents during the last 25% of the fractionation. The middle diagram shows the variation in modes of the cumulated haplogabbro. The modal variation is fairly constant throughout the first 50% of the fractionation, and varies sharply during the final 25% of the fractionation. The right diagram shows the variation of the liquid composition with respect to diopside and plagioclase components. Again the composition displays a slight variation during the initial part of fractionation and varies markedly during the last part of fractionation.

9 The Calculation Procedure for a General Fractionation Formula of Binary Systems

The integration of the differential equations of perfect and partial fractionation requires an equation between x^s and x^l. A simple and quite accurate expression was found for the albite–anorthite system, but in general a polynomial expression may be required. This is, for example, the case for the albite–orthoclase system where the function $x^s = f(x^l)$ has an inflection point. The integration of a polynomial representation of x^s by x^l is, therefore, described below.

We have from Eq. (15):

$$\ln\left(\frac{W}{W^0}\right) = \frac{1}{f} \int_{x^{0l}}^{x^l} \frac{dx^l}{x^s - x^l} \tag{19}$$

The composition of the solid phase may be expressed as a function of the liquid composition by a polynomial. Let $x^l = x$, then:

$$x^s = a + b + cx^2 + dx^3 + ex^4 \tag{20}$$

The coefficient of the polynomial may be estimated by Newton's divided difference interpolation (Dahlquist and Bjørck 1974), or by regression analysis (Draper and Smith 1966; Kreyszig 1970). Newton's divided difference interpolation is the most suitable as least square curve fitting tends to smooth the curves too much. Also, by Newton's divided difference interpolation, the function $x^s = f(x)$ can be specified to attain accurate values at critical values of x. The degree of the polynomial chosen for a representation of x^s depends on the shape of the curve approximated. If the curve is monotonous, then a second or third-degree polynomial may be sufficient. If the curve has an inflection point, as in systems with a binary minimum, then a fourth-degree polynomial should be used. Polynomials with a higher degree than four may of course be used, but the integration of Eq. (19) becomes more difficult. The reason is that the roots of the denominator in Eq. (19) below, have to be estimated before the integration can be performed. The roots of polynomials with a higher degree than three cannot be estimated by analytical methods, but a denominator of the fourth degree may in the present case be used, because x^s has to be zero for x equal to zero. One of the roots of the denominator is consequently known, and only the roots of a third-degree polynomial have to be estimated.

If we substitute Eq. (20) into Eq. (19), then

$$\ln\frac{W}{W^0} = \frac{1}{f} \int_{x^0}^{x} \frac{dx}{a + bx + cx^2 + dx^3 + ex^4 - x} \tag{21}$$

The requirement that $x^s = 0$ for $x = 0$ results in $a = 0$, as evident from Eq. (20), and Eq. (21) may, thus, be rearranged to:

$$\ln\frac{W}{W^0} = \frac{1}{f} \int_{x^0}^{x} \frac{dx}{x[(b-1) + cx + dx^2 + ex^3]} \tag{22}$$

The integration of Eq. (22) may be performed by estimating the roots of the denominator following the procedure described in the *Handbook of Chemistry and Physics* (Weast 1968, p. A 245). Some of the roots may be complex, but the imaginary parts of the roots will cancel as the complex roots are conjugate. After the roots of the denominator have been estimated, the integration may be performed following the procedure given by Gradshteyn and Ryzhik (1965, p. 56).

10 Olivine Fractionation in Basalts

The analysis of the trends of magma series demands a rather detailed knowledge about the variation in the composition of minerals as a function of liquid composition. All the minerals that may form part of the fractionate of basaltic and andesitic compositions display solid solution, i.e., olivine, augite, garnet, amphibole, and feldspar. The binary solid solution series of these minerals are known in some detail, but the phase relations of the ternary solid solution series are not known in sufficient detail. However, the binary or quaternary phase relations do not have much relevance for exact calculations of fractionation trends, as the phase relations of the minerals are different in the magmas, which are multicomponent systems. The variation in composition of minerals in equilibrium with magmatic liquids have been investigated in a few cases, most notably, olivine and plagioclase (Roeder and Emslie 1970; Kudo and Weill 1970; Mathez 1973; Drake 1976). Olivine is a typical phenocryst phase in basaltic magmas, and the fractionation of olivine from basaltic liquid will, therefore, be considered.

The compositional relationship between the basaltic magma and the composition of olivine is expressed by the distribution coefficient K_D, which is a function of temperature:

$$\log K_D = \log \frac{x^s_{FeO}\, x^l_{MgO}}{x^l_{FeO}\, x^s_{MgO}} = \frac{171}{T\,^\circ C} - 0.64 \tag{23}$$

where x^s_{FeO} and x^s_{MgO} are the molar percentages of the oxides in olivine, and x^l_{FeO} and x^l_{MgO} are the molar percentages of FeO and MgO in the magma. As an example, the tholeiitic trend from Hawaii will be considered. Using the regression equations shown in Table X-1, the compositions of tholeiite with 10, 15, and 20% MgO may be estimated. The amount of mol for the different oxides are then calculated for 100 g of rock, the result is shown in Table X-2. The compositions of olivine in equilibrium with these three compositions may then be estimated from the diagram shown in Fig. X-18. The forsterite contents obtained by interpolation from this figure are also shown in Table X-2. The equilibrium compositions have now been estimated. In order to calculate the fractionation trend for a tholeiite fractionating olivine, the

Table X-1. Linear regression equations for the tholeiitic series of Hawaii based on 88 analyses

x = MgO wt%

$SiO_2 = -0.28119\,x + 52.244$
$Al_2O_3 = -0.40882\,x + 17.295$
$FeO = 0.05323\,x + 10.834$
$CaO = -0.21872\,x + 12.461$
$Na_2O = -0.07093\,x + 2.894$
$K_2O = -0.01539\,x + 0.616$
$MnO = -0.00102\,x + 0.184$
$TiO_2 = -0.06072\,x + 3.039$
$P_2O_5 = -0.00527\,x + 0.314$

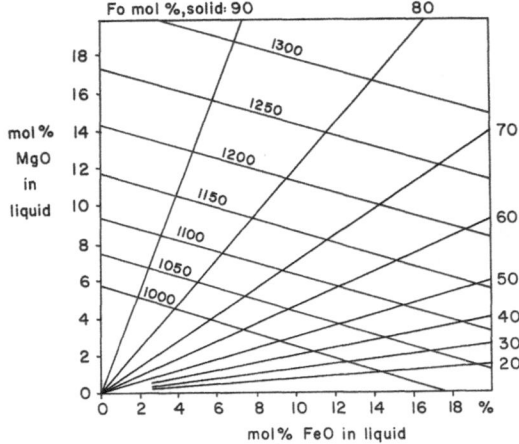

Fig. X-18. The forsterite content of olivine crystallizing from basaltic liquids may be estimated from this diagram. The mole percentages of FeO and MgO are estimated from the major element analysis of the basalt, and the forsterite content and the liquidus temperature may then be estimated by interpolation between the lines shown in the diagram (Roeder and Emslie 1970)

Table X-2. The molar compositions of tholeiites from Hawaii, and their nominal forsterite contents, and the forsterite contents of olivines in equilibrium with the compositions

	20% MgO	15% MgO	10% MgO
SiO_2	0.7759	0.7994	0.8227
Al_2O_3	0.0894	0.1095	0.1296
FeO_t	0.1656	0.1636	0.1583
FeO^a	0.1408	0.1377	0.1345
MgO	0.4961	0.3721	0.2481
CaO	0.1443	0.1637	0.1831
Na_2O	0.0239	0.0295	0.0352
K_2O	0.0033	0.0041	0.0049
MnO	0.0023	0.0024	0.0024
TiO_2	0.0228	0.0267	0.0304
P_2O_5	0.0007	0.0008	0.0009
Sum	1.8651	1.7404	1.6156
Lava Fo (mol%)	77.89	72.99	64.85
Olivine Fo (mol%)	92.35	90.25	86.34

[a] $FeO = 0.85\ FeO_t$

Rayleigh equation should be used. This requires a functional relationship between the composition of the liquid and olivine. The molar contents of MgO and FeO in the liquid have just been calculated. These two oxides will be considered as representing the olivine composition of the liquid, thus, the oceanitic composition contains 26.60 mol% MgO and 7.55 mol% FeO which equals an olivine composition with 92.35% Fo. It is implied by this calculation that all MgO and FeO is available for the formation of olivine. Evidently, this is not the case, but as long as olivine crystallizes alone, the forsterite content of olivine will be determined by the MgO/FeO ratio of the liquid. Let the forsterite contents of liquid and solid be x_l and x_s. The functional relationship between x_l and x_s will be estimated using the ratio x_l/x_s, because this ratio may be expressed as a linear function of x_l using Eq. (23). By exponentiation of Eq. (23) we have:

$$\frac{x_l(1-x_s)}{x_s(1-x_l)} = K \tag{24}$$

where K is constant for a given temperature. The variation in K with temperature is small. The liquidus temperatures for Hawaiian tholeiites with 20% MgO and 10% MgO is 1413 °C and 1227 °C, respectively (Tilley et al. 1965). Between these two temperatures K varies from 0.289 to 0.278, and an average value of 0.290 is adopted here. For $a = 1 - K$ and $b = K$, we obtain from Eq. (24):

$$\frac{x_l}{x_s} = a x_l + b \tag{25}$$

so that:

$$x_s = \frac{x_l}{a x^l + b} \tag{26}$$

By substituting Eq. (26) into Eq. (3), one obtains by definitive integration:

$$\ln\left(\frac{W}{W^0}\right) = -\ln(1-b-ax_l) - \frac{1}{1-b}\ln\left(\frac{1-b-ax_l}{x_l}\right) \tag{27}$$

The trend for oceanite and olivine tholeiite fractionating olivine may now be calculated using Eq. (27), the results are shown in Table X-3 and Fig. X-19. The calculated trends differ from the observed ones, but the scatter in analyses for the observed trend is too large for any evaluation. The observed trend might be controlled by olivine fractionation according to this result, however, other evidence suggests

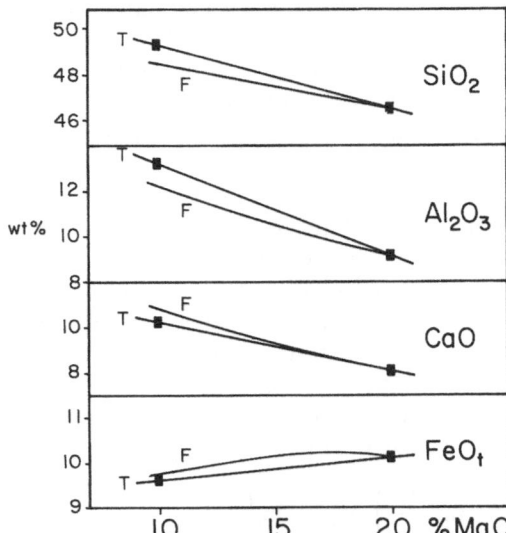

Fig. X-19. Trends obtained by fractionation of olivine from oceanite calculated using the geothermometer by Roeder and Emslie (1970). The calculated trends are denoted *F*, and the observed trends *T*. The trend for FeO displays a small maximum because the FeO content of olivine increases with increasing degree of fractionation

Table X-3. Compositions for the tholeiitic series, observed and calculated values, and φ-values for oceanite as initial composition

SiO$_2$	46.62	48.03	49.43	47.64	48.55
Al$_2$O$_3$	9.12	11.16	13.21	10.53	12.22
FeO$_t$	11.90	11.63	11.37	12.08	11.58
FeO	10.12	9.89	9.66	10.22	9.85
MgO	20.00	15.00	10.00	15.57	10.20
CaO	8.09	9.18	10.27	9.34	10.84
Na$_2$O	1.48	1.83	2.18	1.70	1.98
K$_2$O	0.31	0.39	0.46	0.35	0.41
MnO	0.16	0.17	0.17	0.18	0.21
TiO$_2$	1.82	2.13	2.43	2.10	2.43
P$_2$O$_5$	0.21	0.24	0.26	0.24	0.28
φ (CaO)	–	1.13	1.27	1.15	1.34
φ (Na$_2$O)	–	1.24	1.47	1.15	1.34
φ (TiO$_2$)	–	1.17	1.34	1.15	1.34
φ (P$_2$O$_5$)	–	1.14	1.24	1.15	1.34

that the trend is related to partial melting (Green and Ringwood 1967; MacDonald 1968), while olivine control has been advocated by Irvine (1977), and compositional models based on magma mixing have been proposed by Wright and Fisher (1971).

11 Trace Element Fractionation

The trace element concentrations in magmas may afford evidence about the degrees of fractionation and partial melting, and trace elements have at least in theory a great potential in petrogenetic theory. The quantitative applications of trace element concentrations have hitherto been limited by the accuracy of analytical techniques and the absence of reliable partition coefficients. The validity of major element percentages may very roughly be controlled by their totals which should be slightly lower than 100%. There is no definite control for the analyses of trace elements, even comparison with international standards should afford good evidence about the accuracy for a specific procedure. A review of the literature will show that a given rock type of well-defined major element composition may have widely different trace element concentrations, and an estimate of a representative trace element content for a rock type is, therefore, difficult to obtain. The most applicable results are probably obtained if the same analyst investigates a rock series from a single province. This procedure will afford rather reliable estimates of the relative trace element concentrations, which are more important than the absolute values.

The second important parameter is the partition coefficients. It has generally been assumed that the partition coefficients may be considered constant. Analyses of phenocryst/matrix concentration ratios and experimental work have shown that the partition coefficient for some trace elements depends on their absolute concentrations and the compositions of the magmas. As was shown in Eq. (4) the solution to the Rayleigh integral for $k = x^l/x^s$ is given by:

$$\ln \left(\frac{W}{W^0} \right) = \frac{k}{1-k} \ln \left(\frac{x^l}{x_0^l} \right) \tag{28}$$

Fig. X-20. Trace element concentrations for different D values and degrees of fractionation $(D = x^s/x^l)$. The initial amount of liquid is 100% (W), and the initial trace element concentration is 100 ppm

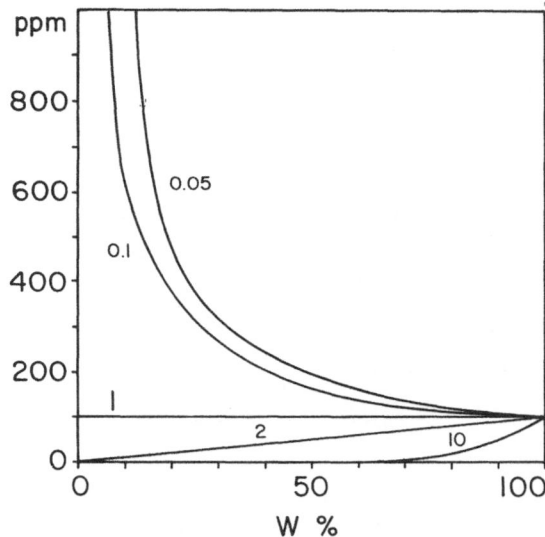

Most investigators use a distribution coefficient defined by $D = 1/k = x^s/x^l$, and we then get:

$$\ln\left(\frac{W}{W^0}\right) = \frac{1}{D-1}\ln\left(\frac{x^l}{x_0^l}\right) \tag{29}$$

The variation in trace element concentrations for some different D values is shown in Fig. X-20 and in Table X-4. The incompatible elements with $D < 1$ increase slowly in concentration during the initial stage of fractionation, but quickly during the last stage. The variation in the concentration of the refractory elements with $D > 1$ is dif-

Table X-4. Trace element concentrations for different D-values and degrees of fractionation

W%	D							
	0.05	0.1	0.2	1	2	5	10	20
100	100	100	100	100	100	100	100	100
90	111	110	109	100	90	66	39	34
80	124	122	110	100	80	41	13	1
70	140	138	133	100	70	24	4	0
60	162	158	150	100	60	13	0	0
50	193	187	174	100	50	6	0	0
40	239	228	208	100	40	3	0	0
30	314	296	262	100	30	1	0	0
20	461	426	362	100	20	0	0	0
10	891	794	631	100	10	0	0	0
5	1,722	1,482	1,099	100	5	0	0	0
1	7,944	6,310	3,981	100	1	0	0	0
0.1	70,809	50,121	25,119	100	0	0	0	0

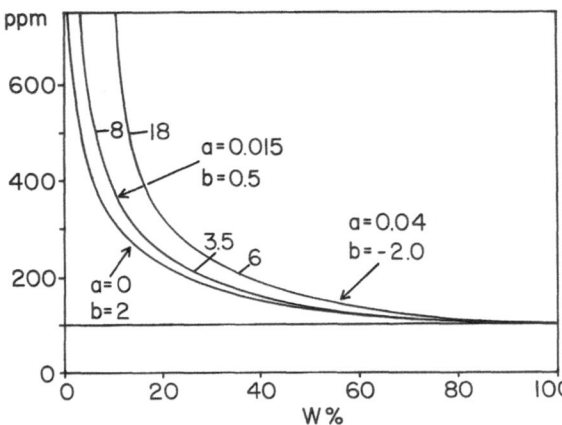

Fig. X-21. Trace element variations by varying distribution coefficients given by Eq. (25). The initial value for D is 2 for all curves. The values for *a* and *b* are shown in the diagram, and the variation in the partition coefficient is also shown for two values

ferent. The concentration of the refractory elements initially decrease quite fast, whereafter the change in concentration is small.

If the ratio x^l/x^s has been estimated as a linear function of the composition of the magma so that Eq. (25) applies then Eq. (27) may be used for an estimate of the concentrations. Some examples of concentration curves for variable distribution coefficients are shown in Fig. 21.

12 Analysis of Trends

The trend diagrams for magma suites are made by plotting the oxides or trace elements against a selected oxide or trace element. One will generally choose an oxide as abscissa that shows a large variation, for basalts, MgO is selected (Bowen diagrams), and for granites and andesites, SiO_2 is used as abscissa (Harker diagrams). If there is a systematic variation in chemical composition, it will be evidence from these plots, and the points will define trends, which may be straight lines or curves (Fig. X-1).

The trend diagrams provide a quantitative representation of the variations in chemical composition, and form the basis for an evaluation of the petrogenesis of the magma suites. The approximate composition of a given fractionate may be estimated from the trends, in principle, it is possible to estimate its composition accurately if the required distribution coefficients are known, but the scatter of analysis prevent an accurate estimate. When the approximate composition of the fractionate has been estimated, then the relative amounts of fractionate and derivative magma may be calculated using the lever rule.

The relative amounts of rock constituents were probably first related in diagrams by Reyer (1888) and Judd (1881). Their diagrams showed the variation in chemical composition as a function of rock type, and was only qualitative. The first quantitative diagrams were employed by Iddings (1892) who used the now conventional weight percentages. In these diagrams one oxide is plotted as a function of another oxide and may be called oxide–oxide diagrams. Ratio–oxide diagrams where a ratio is plotted as a function of an oxide and ratio–ratio diagrams have been used by Stanton (1967) and Pearce (1968). More complex diagrams have been devised by Thorton and Tuttle (1960) using the Thorton Tuttle index as abscissa (normative $Q+Ab+Or+Ne+Kp+Lc$) and Kuno et al. (1957) who proposed the solidification index as abscissa $(MgO \times 100)/(MgO+FeO+Fe_2O_3+Na_2O+K_2O)$. Further types of variation diagrams have been considered, however, they do not involve any major advantages, but merely represent the chemical compositions in different manners. The oxide–oxide diagrams were used extensively by Bowen (1928) and Harker (1909) and will be used in this work because they are simple, and allow the most accurate representation of chemical data.

The oxide–oxide diagrams have been rather severely criticized by Chayes (1962) and Pearce (1968). Chayes emphasized correctly that chemical analyses of rocks only result in a relative estimate of the chemical constituents, the analyses, therefore, represent a closed array, i.e., the variation of one oxide alone will induce variation in the percentage of all other oxides. He further showed that a closed array always will possess some covariance and that the value of the estimated correlation coefficients will be relatively high. These observations are perfectly correct, but do not affect the fundamental and useful properties of variation diagrams. Only the statistical implications of correlations should be evaluated with care.

When a limited number of oxides is fractionated from a magma, the percentages of all oxides will vary. The variation of the enriched oxides in oxide–oxide diagrams is, therefore, only ostensible. This relationship was considered by Pearce (1968), who, therefore, recommended a ratio–ratio diagram where the ratio oxide(A)/oxide(C) is plotted as a function of the ratio oxide(B)/oxide(C). If A and B are remaining in the magma, a horizontal line will result in this type of diagram. In oxide–oxide diagrams coordinate points (A, B) will fall along a line through the origin (0,0) with a slope equal to the ratio A/B. The same basic information will, therefore, be obtained from the two types of diagrams, it is only the graphical representations that are different. The procedure proposed by Pearce (1968) is less useful for calculatory purposes as the ratio–ratio diagrams employ three oxides, while the oxide–oxide diagrams only use two. The uncertainty of the ratio–ratio diagrams will, therefore, be larger than for the oxide–oxide diagrams.

Some simple quantitative relationships for oxide–oxide diagrams will be proven below. They have been tacitly assumed in the literature (cf. Richter and Murata 1966), but their principal aspects have not been dealt with in detail.

13 Trend and Control Lines

The material that fractionates from an initial type of magma will be called the fractionate, and the magma generated by the fractionation for the derivative, magma. The curve that displays the functional relationship between two oxides is called for the trend curve, or the trend line if the trend is fitted by a straight line (Fig. X-22). The straight line joining the coordinates for the assumed fractionate and the initial magma is called the control line. Two theorems for oxide–oxide diagrams will now be proven, a summary of used parameters is shown in Table X-5:

1. In oxide–oxide diagrams the coordinates for the fractionated magma have to be situated on the extensions of the control lines.

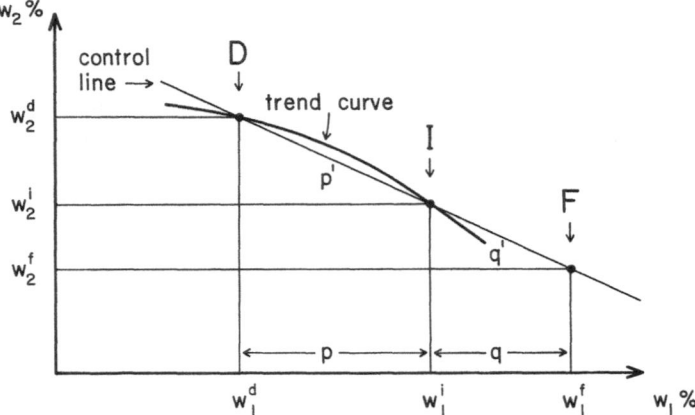

Fig. X-22. The trend curve and the control line for a magma composition and its fractionate. The initial composition of the magma is *I*, and the derived or fractionated magma has composition *F*. The geometrical relationships will show that $p/q = p'/q'$

Table X-5. Parameters used for trend analysis

D : the amount of derived magma in g
F : the amount of fractionate in g
p : is equal to $w_j^i - w_j^d$ (Fig. X-22)
q : is equal to $w_j^f - w_j^i$ (Fig. X-23)
w_j^f : the oxide percentage for the fractionate
w_j^i : the oxide percentage for the initial magma
w_j^d : the oxide percentage for the derived magma
α : see Eq. (34), the slope of the trend line
γ : see Eq. (35), a constant in the linear equation

2. If the ratio between the amounts of derivative magma and fractionate is D/F, then $D/F = q/p$, where $q = (w_j^f - w_j^i)$, and $p = (w_j^i - w_j^d)$ for an oxide with index j.

The first theorem may be proven as follows. The weight percentages of n oxides in the initial magma may be given by:

$$w_1^i, w_2^i, w_3^i, \ldots w_n^i$$

and the percentages for the derivative magma may be given by:

$$w_1^d, w_2^d, w_3^d, \ldots w_n^d$$

and similarly for the fractionate:

$$w_1^f, w_2^f, w_3^f, \ldots w_n^f$$

The absolute amount of an oxide j in the derivative magma is given by:

$$w_j^d = (w_j^i I - w_j^f F)/100 \tag{30}$$

and the concentration in wt % is given by:

$$w_j^d = (w_j^i I - w_j^f F)/D \tag{31}$$

where

$$D = I - F \tag{32}$$

If an oxide with index k is plotted as a function of the oxide with index j, then a straight line joining the two pairs of coordinates (w_j^f, w_k^f) and (w_j^i, w_k^i) will be given by the equation:

$$w_k = \alpha w_j + \gamma \tag{33}$$

the coefficient of slope will be given by:

$$\alpha = \frac{w_k^f - w_k^i}{w_j^f - w_j^i} \tag{34}$$

and the constant γ is given by:

$$\gamma = w_k^i - \frac{w_k^f - w_k^i}{w_j^f - w_j^i} w_j^i \tag{35}$$

so that Eq. (33) becomes:

$$w_k = \frac{w_k^f - w_k^i}{w_j^f - w_j^i} w_j + \frac{w_k^i w_j^f - w_k^f w_j^i}{w_j^f - w_j^i} \tag{36}$$

Using Eq. (31) the wt % of the oxides j and k for the derivative magma becomes:

$$w_j^d = (w_j^i I - w_j^f F)/D \tag{37}$$

$$w_k^d = (w_k^i I - w_k^f F)/D \tag{38}$$

If the composition of the derivative magma is situated on the control line given by Eq. (36), then w_k^d as defined by Eq. (38) should be obtained when inserting w_j^d given by Eq. (37) in Eq. (36). This is found to be the case and theorem 1 is proven.

Next the significance of the ratio p/q may be evaluated. With respect to oxide j the ratio will be given by:

$$\frac{p}{q} = \frac{w_j^i - w_j^d}{w_j^f - w_j^i} \tag{39}$$

According to Eq. (37) which defines w_j^d then:

$$\frac{p}{q} = \frac{w_j^i - \dfrac{w_j^i I - w_j^f F}{D}}{w_j^f - w_j^i} = \frac{F}{D} \tag{40}$$

Thus, the amounts of F and D are given by the ratio p/q, when the total amount of initial magma is known. In calculations it might be convenient to put I = 100, whereafter D is obtained as percentage of the initial amount of magma.

For most magma series the composition of the fractionate is unknown, but it should be situated on the line joining I and D. Some trend curves display strong curvature, in these cases, a line joining I and D will result in an average estimate for the composition of the fractionate.

The trend curves should be estimated by regression analysis. It should be noted that it is mandatory that the analyses are plotted before regression is performed, otherwise regression might be performed across abrupt changes in the trend.

For magma series the composition of the fractionate is unknown, while representative compositions of the members can be estimated from the trend curves. Thus, points like I: (w_j^i, w_k^i) and D: (w_j^d, w_k^d) will be given, and the average composition of the fractionate formed from I to D should be situated on the trend line. In case of Fig. 22, the composition of the fractionate may be situated at any given position on the trend line to the right of I. This information is fairly useless as this composition interval is large compared to the interval between I and D. However, from the minimal enrichment ratio considered next, it is possible to delimit the permissible composition interval for the fractionate quite substantially.

14 The Minimal Enrichment Ratio φ

If one of the oxide components of a magma is absent from the fractionate, then the concentration of this component will afford an exact estimate of the relative amount of the fractionate. As mentioned, the composition of the fractionate is unknown, and one cannot in this situation assume any oxide component has been absent from the fractionate. Nevertheless, the component which displays the largest degree of enrichment will indicate the minimal degree of fractionation. Consider for example P_2O_5 of the alkalic series of Hawaii. The wt% of P_2O_5 in alkalic olivine basalt is 0.44% P_2O_5, and 0.56% P_2O_5 for hawaiite. The weight percentage has increased 27%, and at least 21.26% of the alkalic olivine basalt magma must have been fractionated in order to form hawaiite.

Let the wt % of an oxide j in the initial magma be w_j^i, and let the wt % for this oxide in the derivative magma be w_j^d. Then a minimal enrichment ratio φ may be defined as follows:

$$\varphi = \frac{w_j^d}{w_j^i} \tag{41}$$

For the example just mentioned, $\varphi = 1.27$. As evident from Eq. (41), φ is the very minimum enrichment ratio as some of the oxide j may have fractionated from the initial magma. Assuming that all of oxide j is concentrated in the derivative magma, the exact degree of fractionation may be estimated. In this case from material balance:

$$100 w_j^i = D w_j^d \quad (I = 100) \tag{42}$$

and from Eq. (41):

$$D = 100 \frac{w_j^i}{w_j^d} = \frac{100}{\varphi} \tag{43}$$

Using the P_2O_5 weight percentages from the example above D may be estimated from Eq. (53) as 78.74. According to this estimate the maximal amount of hawaiite that may be formed from alkalic olivine basalt would be 78.74%.

If φ was the true enrichment ratio, the exact amount of derivative magma, D, formed from the initial magma, I, could have been estimated from Eq. (43). However, since φ either has the correct value or is too small, the estimated values of D will either be correct or too large as evident from Eq. (43). The uncertainty in the estimate of φ and D might appear a hindrance for further quantitative estimates. Fortunately, this is not the case as the minimum value of D affords valuable information with respect to the composition of the fractionate.

Since D as estimated from φ has a maximum value, it follows from Eq. (43) that the ratio:

$$\frac{p}{q} = \frac{F}{D} = \frac{100 - D}{D} = \varphi - 1 \tag{44}$$

will assume a minimum value. The exact value of p will be known from chemical analyses, and a maximal value of q can, therefore, be estimated from Eq. (44). Thus, from the above derivations the important result is obtained that the average composition of the fractionate has to be situated on the trend lines, and the compositional interval on this line will be limited by the value of q (Fig. X-22). Values for the ratio φ vary between 1.2 and 2, and q will, therefore, attain values between p and 5p. The possible composition interval for the fractionate may consequently be specified within a relatively narrow interval. As most magmas display less than four phenocryst phases (MacDonald 1949), and the composition interval for six to eight oxides can be evaluated, it will often be possible to estimate a limited number of phase assemblages for the fractionate.

The exactness of the method is restricted by the accuracy of analyses of the incompatible elements, such as TiO_2, K_2O, and P_2O_5. These elements may enter the fractionate to some degree, the relative enrichment of these elements may be evaluated by plotting them against each other (Fig. X-23). The confidence intervals s, for the minimal enrichment ratios may be estimated from:

$$s = \frac{c}{w_i^2} \sqrt{w_i^2 - w_d^2} \tag{45}$$

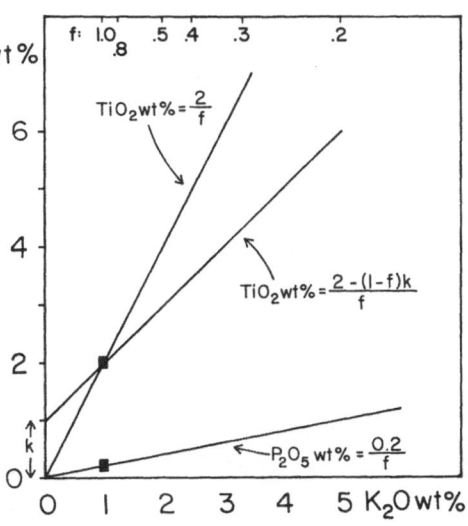

Fig. X-23. The concentrations of different incompatible elements for different degrees of fractionation (f). The initial concentrations of K_2O, TiO_2, and P_2O_5 are 1.0% K_2O, 2.0% TiO_2, and 0.2% P_2O_5, respectively. The concentration lines should intersect the origin, 0,0 if the incompatible elements totally enter the liquid. If they form part of the fractionate they will intersect the ordinate, as shown for TiO_2. The amount entering the fractionate will be represented by k that is zero for perfect incompatibility

where c is the confidence interval for an oxide in absolute wt %, w_i and w_d the weight percentages for this oxide for the initial and derivative magma, respectively (cf. Kreyszig 1970, p. 357).

In practice, different φ values will be obtained for the various oxides and trace elements, in principle, the largest value is the most significant, but the total variation in φ values and their confidence intervals should be considered before a specific value is considered the best estimate.

15 Analysis of the Fractionate

Having estimated the trend lines and their confidence intervals as well as the possible composition for a fractionate by use of the φ ratio, the next step will be an estimate of the phase assemblages that can constitute the fractionate. Three methods of different nature should be allowed for estimating the mineralogic species of the fractionate:

1. The principle of opposition, a graphical method.
2. The least square method.
3. The matrix method, a solution based on linear equations.

The principle of opposition is used when considering the trend lines for a suite (Fig. X-24). The composition of the minerals that might constitute the fractionate should be situated on opposite sides of the trend lines, or at least one of the minerals should be in opposition with respect to the trend line. This principle is simple and fairly evident, nevertheless, it might be quite powerful evaluating the possible mineral combinations for the fractionate. A plot of trend curves and mineral compositions for several oxides will in fact restrict these combinations substantially.

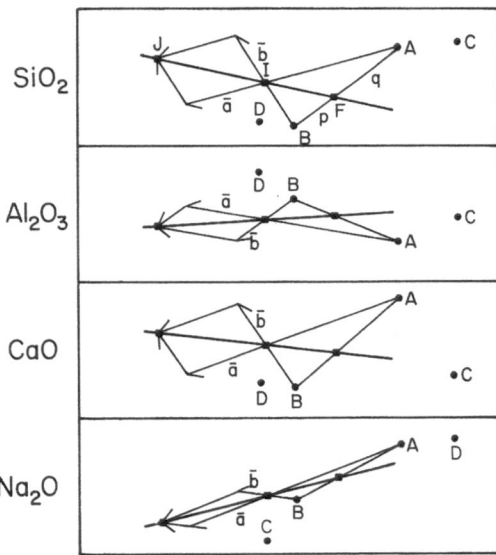

Fig. X-24. The principle of opposition illustrated for four oxides. The minerals that might constitute the fractionate are A, B, C, and D. The vectorial sum of a and b should have the vector from I to J as resultant. Thus, the ratio p/q should be the same for all oxides, if A and B constitute the fractionate

Consider as an example the four schematic trends shown in Fig. X-24. It is considered possible that the four minerals A, B, C, and D might form the fractionate. The mineralogical constituents of the fractionate formed by fractionation from 1 to 2 should now be estimated. If say A and B constitutes the fractionate, then A and B should be situated on opposite sides of the trend line, as the resultant vector of the two vectors $\bar{a}$ and $\bar{b}$ should have the direction of the trend line. Further, the resultant should extend from I to J, and the ratio p/q should be the same for all four oxides. The minerals that might constitute the fractionate should, thus, be in opposition with respect to the trend line for all oxides. The relative positions are as follows:

$$SiO_2 \quad \frac{A\ C}{B\ D}$$

$$Al_2O_3 \quad \frac{D\ B}{A\ C}$$

$$CaO \quad \frac{A}{B\ D\ C}$$

$$Na_2O \quad \frac{A\ D}{B\ C}$$

According to the first two ratios, A, B, and D might form part of the fractionate, and a similar result is also obtained from the third ratio. This possibility must be excluded from the fourth ratio, and the only possibility is then A + B.

The least square and linear equation method might be used after graphic accessment has been done. The method will result in an approximate estimate of the relative amounts of minerals forming a given fractionate. Considering that all chemical analyses are approximate estimates, the least square method is a very realistic one, because it allows some inaccuracy of analysis, even though this inaccuracy should not be too large (Reid et al. 1973).

16 The Linear Equation Method and the Least Square Method

The variation diagrams allow a determination of the possible composition range of the fractionate, and in favorable cases its composition can be determined rather accurately. The next problem is the determination of the mineralogical composition of the fractionate, both with respect to species of minerals and their compositions. This is far from a straightforward task as there are several minerals that might constitute a given fractionate. If four minerals appear possible from phenocryst assemblages, or from phase relations, then there will be one combination of four species, four combinations of three species, six combinations of two, and four combinations of one mineral, and a total of 15 combinations. Some of these might be excluded using the principle of opposition, but as the composition of the minerals are unknown parameters, there might still be a variety of possibilities. Sometimes it is considered that a magma has been fractionated by a mineral assemblage similar to the phenocryst assemblage, in that case the mineral compositions will be known, while their accurate proportions remain to be estimated. As an example, alkakic olivine basalt contains the typical phenocryst assemblage olivine + augite, as do some hawaiites. Various proportions of olivine and augite might, therefore, be substracted from the composition of alkalic olivine basalt in order to test the phenocryst assemblage as a possible fractionate. However, it is by no means granted that phenocrysts provide guiding evidence for the mineral species constituting a given fractionate. In practise, a given set of minerals with some specific compositions will be chosen, and the next step is then an estimate of their relative proportions in the fractionate.

The relative proportions may be calculated by one of two methods, the linear equation method, and the least square method. Both methods have their advantages, but the least square method is the most powerful one.

Consider the situation where a differentiated magma M_d has formed from a more primitive magma M_P. The fractionate F is considered defined by three minerals A, B, and C. Their compositions are given by the arrays a_i, b_i, and c_i, respectively. Let the weight percentages of these three minerals be x, y, and z, respectively. The following equation should then be solved for all oxides:

$$M_P = M_d + F \tag{46}$$

The linear equation method can only be used if the number of unknowns equals the number of equations, and only three oxides can, therefore, be considered in this example. The following equations should be solved:

$$a_1 x + b_1 y + c_1 z = f_1 \tag{47a}$$

$$a_2 x + b_2 y + c_2 z = f_2 \tag{47b}$$

$$a_3 x + b_3 y + c_3 z = f_3 \tag{47c}$$

where f_i is the composition of the fractionate. These three equations can easily be solved using Cramer's rule (Kreyszig 1967). However, this rule only applies for equations with two or three unknowns, and Gauss' algorithm should be used when a number of linear equations are to be solved. Gauss' method is described by Kreyszig (1967), and the method will only be demonstrated for the above example. The variables are eliminated in successive steps. First, x is eliminated from Eq. (47b), Eq. (47a) is multiplied with a_2/a_1 whereafter the multiplied equation is subtracted from Eq. (47b). The same procedure is repeated for Eq. (47c), then Eq. (47a) is multiplied with a_3/a_1, and the obtained equation is subtracted from Eq. (47c). The result is that x is eliminated from Eqs. (47b) and (47c):

$$a_1 x + b_1 y + c_1 z = f_1 \tag{48a}$$

$$k_2 y + l_2 z = m_2 \tag{48b}$$

$$k_3 y + l_3 z = m_3 \tag{48c}$$

where $k_2 = b_2 - (a_2/a_1) b_1$; $l_2 = c_2 - (a_2/a_1) c_2$; $m_2 = f_2 - (a_2/a_1) f_1$; etc. The whole procedure is then repeated for Eqs. (48b) and (48c), whereby y is eliminated from the last equation:

$$a_1 x + b_1 y + c_1 z = f_1 \tag{49a}$$

$$k_2 y + l_2 z = m_2 \tag{49b}$$

$$n_3 z = o_3 \tag{49c}$$

The variable z may now be estimated from the last equation (49 c), then knowing z, y is calculated from (49 b), and finally x is estimated from (49 a).

Since the number of equations should equal the number of unknowns, only three oxides or trace elements may be included in the present calculation. Evidently, if the compositions of the minerals are perfectly correct, then any three oxides will give the same result for x, y, and z. This will be a rare situation indeed, at least the analytical errors involved will disturb this ideal situation. The number of variables may in the present case be increased with one solving this equation:

$$A + B + C + M_d = M_p \tag{50}$$

Usually the compositions of the minerals cannot be specified, and in this case the number of oxides involved may be increased to almost any number. The minerals are solid solutions, and the end member compositions may be used rather than some specific composition. If mineral A is olivine, then the composition of A may be defined by the composition of fayalite and forsterite, and if B is clinopyroxene or garnet, then B may be estimated from three end members. The upper limit of the oxides will be given by the largest number of oxides that have been analyzed with acceptable accuracy. In this case, seven oxides will be involved:

$$A_1 + A_2 + B_1 + B_2 + B_3 + C + M_d = M_p \tag{51}$$

The solutions obtained from linear equations will be absolutely accurate for the oxides involved. The same situation does not apply for the least square method, as shown below. The particular advantage of the linear equation method is its precision for the selected oxides. On the other hand, the result obtained will be highly sensitive on analytical error, and the method is probably best suited as a control method.

The least square method is elegant and quite powerful. The critical property of the method is that it is rather diplomatic, solutions that appear acceptable, may not be as good as believed. The deviation from a perfect solution will be evident from the residuals, and if these are small for all constituents, then the obtained solution might be accepted. The least square method has been used in biological sciences for years, but was introduced into petrology by Chayes (1968) and Bryan et al. (1969), and the method was later fully developed for petrological purposes by Wright and Doherty (1970), Reid et al. (1973) and Albarede and Provost (1977).

In order to get a realistic estimate of the relative proportions of minerals in a fractionate, several oxides or trace elements should be used, and the method employed should not be too dependent on analytical error. The least square method fulfills these requirements, and the mathematical basis of the method will now be demonstrated. The present treatment is based on unpublished notes by Doherty and Wright (1975, personal communication). If three minerals are constituting a given fractionate, and five oxides are considered important variables, then the equations defining the solution are overdetermined, and generally no solution will fit all oxides. However, an approximate solution may be obtained as follows. Consider the three minerals A, B, and C that were used above, and add two oxides, then the following equations should have the same solution with respect to x, y, and z:

$$a_1 x + b_1 y + c_1 \quad = f_1$$
$$a_2 x + b_2 y + c_2 z = f_2$$
$$a_3 x + b_3 y + c_3 z = f_3 \tag{52}$$
$$a_4 x + b_4 y + c_4 z = f_4$$
$$a_5 x + b_5 y + c_5 z = f_5$$

In general, for any given values of x, y, and z, there will be a discrepancy or residual, R_i, which gives the amount by which each equation fails to be satisfied.

Rewriting the equations to include residuals, one gets:

$$a_1 x + b_1 y + c_1 z - f_1 = R_1$$
$$a_2 x + b_2 y + c_2 z - f_2 = R_2$$
$$a_3 x + b_3 y + c_3 z - f_3 = R_3 \tag{53}$$
$$a_4 x + b_4 y + c_4 z - f_4 = R_4$$
$$a_5 x + b_5 y + c_5 z - f_5 = R_5$$

Since it is usually impossible to find values of x, y, and z that will make all the R_i's zero, one must be satisfied to find a solution that will minimize some function of the R_i. The sum of squares of the R_i is usually taken for the function to be minimized, because it deals with absolute magnitudes of the residuals, and also is a differentiable function. This technique using quadratic variables will be recognized by readers familiar with regression (cf. Kreyszig 1970). Hence, the function H should be minimized with respect to x, y, and z:

$$H = R_1^2 + R_2^2 + R_3^2 + R_4^2 + R_5^2 \tag{54}$$

The objective function H will attain its minimum value where:

$$\frac{H(x,y,z)}{\delta x} = 0, \quad \text{and} \quad \frac{\delta^2 H(x,y,z)}{\delta x^2} > 0$$

$$\frac{H(x,y,z)}{\delta y} = 0, \quad \text{and} \quad \frac{\delta^2 H(x,y,z)}{\delta y^2} > 0 \tag{55}$$

$$\frac{H(x,y,z)}{\delta z} = 0, \quad \text{and} \quad \frac{\delta^2 H(x,y,z)}{\delta z^2} > 0$$

If one performs the partial differentiations indicated above and sets the first partial derivatives to zero, the result will be three linear equations which may be solved for the three unknowns x, y, and z. This solution is called the least square solution to the original overdetermined system of five equations.

H is given by:

$$\begin{aligned}
H &= R_1^2 + R_2^2 + R_3^2 + R_4^2 + R_5^2 \\
&= (a_1 x + b_1 y + c_1 z - f_1)^2 \\
&\quad + (a_2 x + b_2 y + c_2 z - f_2)^2 \\
&\quad + (a_3 x + b_3 y + c_3 z - f_3)^2 \\
&\quad + (a_4 x + b_4 y + c_4 z - f_4)^2 \\
&\quad + (a_5 x + b_5 y + c_5 z - f_5)^2
\end{aligned} \tag{56}$$

By differentiation with respect to x, one gets:

$$\frac{\delta H}{\delta x} = 2a_1 (a_1 x + b_1 y + c_1 z - f_1)$$
$$+ 2a_2 (a_2 x + b_2 y + c_2 z - f_2)$$
$$+ 2a_3 (a_3 x + b_3 y + c_3 z - f_3)$$
$$+ 2a_4 (a_4 x + b_4 y + c_4 z - f_4) \tag{57}$$
$$+ 2a_5 (a_5 x + b_5 y + c_5 z - f_5)$$

For $\delta H / \delta x = 0$, one obtains by rearrangement:

$$(a_1^2 + a_2^2 + a_3^2 + a_4^2 + a_5^2) \, x$$
$$+ (a_1 b_1 + a_2 b_2 + a_3 b_3 + a_4 b_4 + a_5 b_5) \, y$$
$$+ (a_1 c_1 + a_2 c_2 + a_3 c_3 + a_4 c_4 + a_5 c_5) \, z \tag{58}$$
$$- (a_1 f_1 + a_2 f_2 + a_3 f_3 + a_4 f_4 + a_5 f_5) = 0$$

Similar equations may be obtained by differentiation with respect to y and z. By rewriting these three equations in a more compact form:

$$(\Sigma a_i^2) \, x + (\Sigma a_i b_i) \, y + (\Sigma a_i c_i) \, z = (\Sigma a_i f_i)$$
$$(\Sigma b_i a_i) \, x + (\Sigma b_i^2) \, y + (\Sigma b_i c_i) \, z = (\Sigma b_i f_i) \tag{59}$$
$$(\Sigma c_i a_i) \, x + (\Sigma c_i b_i) \, y + (\Sigma c_i^2) \, z = (\Sigma c_i f_i)$$

The sign of $\delta^2 H / \delta x^2$ may be derived consideringEq. (57). The signs of a_i, b_i, and c_i are always positive as will be evident considering Eq. (52). Hence, differentiating Eq. (57) one more time with respect to x will result in a positive expression. Again a similar result will be obtained evaluating $\delta^2 H / \delta y^2$, and $\delta^2 H / \delta z^2$, and the values obtained for x, y, and z solving Eq. (59) will, therefore, be values that minimize $H(x,y,z)$. The set of Eq. (59) may be solved using Gauss' algorithm, whereby the wanted values for x, y, and z are obtained.

Frequently one will obtain negative values for x, y, and z, and in these cases the compositions of the minerals have to be changed. If the obtained negative values are numerically small, then they may be accepted. A better approach will be to estimate the origin of the negative values and then change the mineral compositions.

All oxides used in the above calculations have the same weight, i.e., oxides a_i, b_i, and c_i are all considered of equal importance. The accuracy by which they have been estimated will be different, and it will, therefore, be relevant to develop a method by which one may put more weight on some of the oxides. A least square program with weight factors has been developed by Reid et al. (1973), and the reader is referred to their paper for a description of the method.

17 Statistics of Chemical Data

The analysis of trends and the calculation of fractionate compositions have hitherto been considered, but the reliability of the results obtained have not been dealt with. The trend curves are based on regression of a sample of chemical analysis, which displays some scatter, and the trend curves are, therefore, only approximate representations of the true trend curves. As the fractionate compositions estimated are based on extrapolated trends, the statistical significance of the trend becomes a highly important feature, and the statistical aspects of trends should, therefore, be considered.

The scattered distribution of the chemical analyses is due to both inherent errors of analysis and an intrinsic variation in the composition of the samples. The error of analysis will depend on the analyst and the method used, and no general values for error of analysis can be given. Some evidence about the error of analysis may be obtained from the spread in chemical composition of homogenized standards analyzed by several laboratories. Histograms for such standards are shown in Fig. X-25. The analysts making these analyses knew that their analyses would be compared with the results of other laboratories, and the analyses must be considered the best obtainable. The spread shown in Fig. X-25 is, thus, to be considered minimal, and not entirely representative for a realistic estimate.

The intrinsic variation might be even larger than the error of analysis. Usually only a single sample is taken from a lava flow or a granite pluton, and the compositions of such specimens only approximate the average compositions of the rocks. Another intrinsic variation is prevalent for trends based on analyses from a whole igneous province. Lavas from an igneous province have been erupted within large periods of time of the order of a few million years, and the lavas of similar compositions most likely have undergone slightly different processes. Their small differences in composition might, thus, be significant, but difficult to evaluate.

The trend curves calculated are consequently only approximate estimates, and the fractionate compositions based on extrapolations are even more approximate. A certain deviation between a calculated fractionate composition and that estimated from a trend is, therefore, acceptable. The question is now, what degree of deviation is acceptable? This question might at first hand appear easy to answer, but a definite answer has not yet been found. Presently no method has been devised that allows an

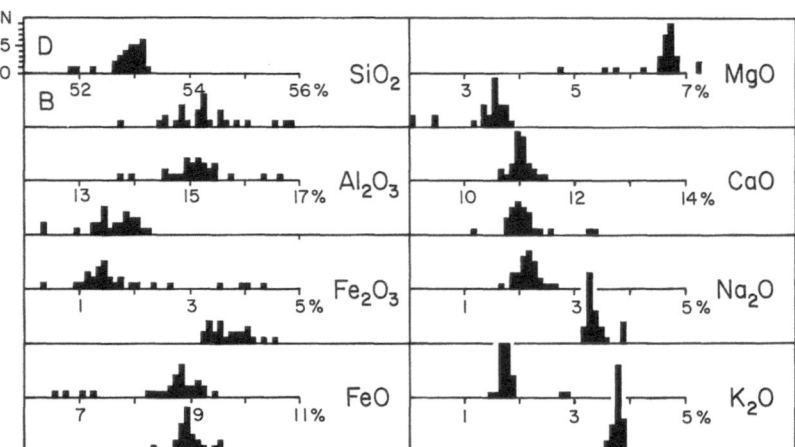

Fig. X-25. Histograms for analyses performed by several laboratories on two rock compositions, *D* (diabase, Fairbain 1951) and *B* (basalt, Flanagan 1969). The spread in analyses represents the minimum possible as the analyses were performed under optimum conditions. The spread indicates the precision of the analyses, while their accuracy will not be evident from the diagram

estimate of the true confidence interval for a trend line (Chayes 1978, pers. communication). The problems related to the statistical significance of trends were first analyzed by Chayes (1962, 1971), and his work will be dealt with here.

Before the theoretical derivations of Chayes (1962) are examined, it might be convenient to consider some statistical definitions. For an exhaustive statistical treatment the reader is referred to the excellent textbook by Kreyszig (1970).

Suppose that the variable y is a function of x:

$$y = f(x) \tag{60}$$

and that n pairs of values of (x, y) have been obtained. These pairs represent a sample of values:

$$(x_1, y_1) \; (x_2, y_2) \; (x_3, y_3) \ldots (x_n, y_n) \tag{61}$$

The mean of x is defined by:

$$\bar{x} = \frac{1}{n} (x_1 + x_2 + x_3 + \ldots x_n) \tag{62}$$

Similarly $\bar{y}$ is given by:

$$\bar{y} = \frac{1}{n} (y_1 + y_2 + y_3 + \ldots y_n) \tag{63}$$

The variance of x is s_x^2 and is defined by Eq. (64):

$$s_x^2 = \frac{1}{(n-1)} \sum_{j=1}^{n} (x_j - \bar{x})^2 \tag{64}$$

The standard deviation is the positive square root of the variance:

$$s_x = (s_x^2)^{1/2} \tag{65}$$

The variance for y is given by:

$$s_y^2 = \frac{1}{(n-1)} \sum_{j=1}^{n} (y_j - \bar{y})^2 \tag{66}$$

The covariance between x and y is given by Eq. (67):

$$s_{xy} = \frac{1}{n-1} \sum_{j=1}^{n} x_j y_j - \frac{1}{n} \left(\sum_{j=1}^{n} x_j \right) \left(\sum_{j=1}^{n} y_j \right) \tag{67}$$

The correlation coefficient r is given by:

$$r = \frac{s_{xy}}{s_x s_y} \tag{68}$$

The correlation coefficient is a measure of the linear dependence of x and y. The value of r varies between -1 and $+1$, and a perfect linear dependence is obtained when $|r| = 1$. No linear dependence is present when $r = 0$. It should be emphasized here that a trend is not necessarily linear, theoretically, it is never linear as phase boundaries never are linear, however, many trends approximate straight lines, and a good estimate of the fractionates may be obtained assuming a linear relationship. The correlation coefficient is, thus, only a measure of a linear relationship, and does not display the quality of the data when they fit a curve.

The above definitions might be used without further complications when y is a function of x, however, if x is also a function of y, then the situation becomes complicated as will be shown below.

A rock specimen consists of several components, but the concentrations of these components are not independent variables. If, say the SiO_2 content is increased, then the concentration of all other oxides will be decreased whether or not they have been subtracted. Dealing with rock analyses one can only estimate the relative changes, and one is forced to use percentages or fractions, i.e., the sum of the components should be 100 and 1, respectively. A group of variables that have a constant sum is called a closed array. The oxide percentages obtained by rock analyses are such a closed array. It is this closure constraint that makes the statistical treatment of rock analyses or trends so difficult.

Consider the N compositions consisting of M oxides in the tabulation below (N = 3; M = 4):

	M			
	10	20	30	40
N	5	50	20	25
	3	17	60	20
Average	6	29	36.7	28.3

Each of these compositions sum to 100. The average value for each oxide is shown below the tabulation. As each of the analyses sum to 100, the sum of the averages will also sum to 100. Let X_j be the percentage of oxide j in the compositions, then:

$$\sum_1^4 X_j = 100 \tag{69}$$

Let the average for each oxide be $\bar{x}_j$, then:

$$\sum_1^4 \bar{x}_j = 100 \tag{70}$$

By subtraction one obtains:

$$\sum_1^4 (X_j - \bar{x}_j) = 0 \tag{71}$$

Thus, this equation will only hold for closed arrays, and is the fundamental equation for a statistical treatment.

Let a chemical analysis of M oxides be indexed by "j", and the proportions denoted by X_j. Further, let it be assumed that there is a total of N rock analyses, and that each analysis sums to a constant K. Thus:

$$\sum_{j=1}^M X_j = K \tag{72}$$

where K is 100% if x_j is measured in wt %. The mean value of the M oxides for the analysis is given by:

$$\bar{x}_j = \frac{1}{M} \sum_{j=1}^{M} X_j \tag{73}$$

so that

$$\sum_{j=1}^{M} \bar{x}_j = K \tag{74}$$

In order to estimate the variance and covariance the deviations from the mean Δx_j should be estimated:

$$\sum_{j=1}^{M} (X_j - \bar{x}_j) = \sum_{j=1}^{M} \Delta x_j = 0 \tag{75}$$

This equation may now be multiplied with one of the Δx_j, say Δx_k:

$$\Delta x_k \sum_{j=1}^{M} \Delta x_j = 0 \tag{76}$$

isolating Δx_k^2 one gets:

$$\Delta x_k^2 + \sum_{j=1}^{M} \Delta x_k \Delta x_j = 0 \quad (j \neq k) \tag{77}$$

The variance of oxide k and the sum of covariances may now be obtained by multiplying Eq. (77) with $1/N - 1$ and by summing all terms from 1 to N:

$$\frac{1}{N-1} \left[\sum_{i=1}^{N} \Delta x_k^2 + \sum_{i=1}^{N} \sum_{j=1}^{M} \Delta x_{ik} \Delta x_{ij} \right] = 0 \quad (j \neq k) \tag{78}$$

Compared with Eqs. (75) and (73) it will be evident that the first term is the variance of oxide j, and the second term is the sum of covariances:

$$s_k^2 + \sum_{j=1}^{M} s_{kj} = 0 \tag{79}$$

This equation shows that the variance of one of the variables in a closed array is equal to the sum of covariances and opposite in sign.

Using Eq. (79) the standard deviation and the correlation coefficients may be estimated. The standard deviation s_k is given by:

$$s_k = (s_k^2)^{1/2} \tag{80}$$

By division of Eq. (79) with s_k one gets:

$$s_k + \sum_{j=1}^{M} \frac{s_{kj}}{s_k} = 0 \tag{81}$$

and by multiplication with s_j in the nominator and denominator in the last term:

$$s_k + \sum_{j=1}^{M} \frac{s_{kj} s_j}{s_k s_j} = s_k + \sum_{j=1}^{M} s_j r_{kj} = 0 \tag{82}$$

By division with s_k Eq. (82) becomes:

$$1 + \sum_{j=1}^{M} b_{kj} = 0 \tag{83}$$

where b_{kj} is the regression coefficient, i.e. the slope of the regression line. The standard deviation is always a positive number and, therefore, at least one of the correlation coefficients r_{kj} must be zero according to Eq. (82). As $r_{kj} > -1$, then it is evident from Eq. (82) that:

$$s_k \leq \sum_{j=1}^{M} s_j \tag{84}$$

so that the largest standard deviation in any closed array must be smaller than the sum of the other ones.

As mentioned above, the correlation coefficient is a measure of the magnitude of linear dependence between variables. The problem with closed arrays is that some correlation is always included, because the sum of variables is constant. This means if a group of completely arbitrary numbers are summed and each number is divided by the sum, then there will be some correlation between the fractions obtained. Consequently, there will always be some correlation between oxide percentages, and a correlation coefficient of say, 0.6 need not indicate a significant correlation. Generally, a correlation coefficient of 0.85 will be quite acceptable, but this is not necessarily so by closed arrays. One must, therefore, evaluate another test for statistical reliability.

The order of magnitude of correlation induced by the closure constraint may be estimated as follows. Consider that all the standard deviations are the same, $s_k = s_j$, then by division by s_k or s_j in Eq. (82):

$$1 + \sum_{j=1}^{M} r_{kj} = 0 \tag{85}$$

or:

$$\sum_{j=1}^{M} r_{kj} = -1 \tag{86}$$

If there are M oxides, there will be $M-1$ coefficients of covariance so that the average value for r_{kj} is given by:

$$E(r_{kj}) = \frac{1}{M-1} \tag{87}$$

where E denotes that the value obtained is the expected one, but not necessarily the real one. If $M = 10$, then $E(r_{kj}) = -0.11$. This value was obtained without any consideration of any chemical analyses, and the average correlation for an arbitrary set of ten figures with a constant sum is, thus, -0.11. It is evident from this result that correlation coefficients for trend lines of the order of 0.2–0.3 are not acceptable as the closure constraint generates a correlation of the order of 0.1. The larger M is, the smaller is $E(r_{kj})$, that is, the larger the number of oxides, the smaller is the expected value for r_{kj}.

The standard deviation for one of the oxides will frequently be larger than the other deviations. For igneous rocks the SiO_2 or MgO weight percentages may display a large variation, and the standard deviation of these oxides may, therefore, be larger than those for the other oxides. In these cases the standard deviations of the other oxides will be rather similar. Thus, assuming s_k different from s_j, and the s_j values equal, Eq. (82) becomes:

$$\sum_{j=1}^{M} s_j r_j = (M-1) s_j r_j = -s_k \tag{88}$$

so that:

$$E(r_j) = \frac{s_k}{(1-M) s_j} \tag{89}$$

The values for r_j calculated from Eq. (89) will indicate the average values of r_j that might be expected for random numbers. If the correlation coefficients for a trend have values similar to $E(r_j)$ obtained from Eq. (89), then the correlation is so poor that the chemical data should be discarded without hesitation.

The statistical parameters for a natural trend were worked out by Chayes (1961), and his results will now be considered as an example.

The trend for the Katmai province was estimated by Fenner (1926), who reported 18 analyses. The variance and correlation coefficients calculated by Chayes (1961) are shown in Table X-6. The trend is andesitic, and the SiO_2 wt%, therefore, displays a large variation with the result that the $s_{SiO_2}^2$ is high. The problem is now, are those correlation coefficients in Table X-6 indicating a significant correlation between the oxides, or are the weight percentages just random numbers? The lavas in an igneous province may have been generated by various processes and the correlation coefficients may be large or small depending on whether or not the same processes have generated a series of lavas. In order to test the possibility that the oxides constitute random numbers, one should compare the obtained correlation coefficients with the random one. There are nine oxides so from Eq. (89), $E(r_j)$ is estimated as -0.1111. The average value for $E(r_j)$ is not entirely significant, as the r_j values obtained from random numbers will display a certain variation. It will, there-

Table X-6. Variance and correlation coefficients for the Katmai andesitic series

	Variance	Correlation coefficients
SiO_2	36.3067	–
Al_2O_3	1.8497	−0.9558
Fe_2O_3	2.0058	−0.6548
FeO	2.1734	−0.6723
MgO	1.8115	−0.9476
CaO	4.2385	−0.9803
Na_2O	0.1521	0.7410
K_2O	0.4447	0.9625
TiO_2	0.0539	−0.7808

fore, be necessary to compare the r_j values of Table X-5 with the possible range of random r_j values. Assuming the r_j values are normally distributed, the confidence interval may be calculated for different confidence levels like 95% and 99%. The test to be used is the Fisher-z test. If one wants to test the significance of the obtained r_j values against the hypothesis that $r_j = 0$, then the student-t test might be used. Here, the significance of the observed r_j values should be compared with a $E(r_j)$ value different from zero, and the Fisher-z test has to be used. The theory of this test will not be considered here, but the test will be demonstrated with the above example. For a detailed account, the reader is referred to Snedecor and Cochran (1967).

Fisher introduced a variable z defined as a function of r in the following manner:

$$z = \tfrac{1}{2} (\ln (1+r) - \ln (1-r)) \tag{90}$$

and r is obtained from z by Eq. (91):

$$r = \frac{e^{2z} - 1}{e^{2z} - 1} \tag{91}$$

The value of $E(r_j)$ was estimated as −0.1111, and z is, therefore, equal to −0.1116. The range of $E(r_j)$ values may be calculated from the variance of z. The variance of z is given by $s_z = 1/(n-3)^{\frac{1}{2}}$, and as $n = 18$, $s_z = 0.2582$. The confidence level required is a somewhat subjective judgement, if 95% is chosen, then 95% of the values of $E(r_j)$ will be within the interval; and if 99% is considered relevant, then 99% of the values of $E(r_j)$ will be adopted here. The confidence interval for z may be calculated using Student's–t distribution (Snedecor 1958). The value of t will usually depend on both the sample size and the confidence interval, but as z is distributed almost normally, independent of sample size, one might use the t values for infinite samples. Some confidence levels and t values are listed below:

Confidence level	t value
50%	0.6745
90%	1.6448
95%	1.9600
99%	2.5758
99.9%	3.2905

The confidence interval for z is called c_z and is calculated from:

$$c_z = ts_z \tag{92}$$

For $t = 2.5758$ and s_z equal to 0.2582, one gets $c_z = 0.6651$. The range of z is, thus, between:

$$-0.3698 < z < 0.1466$$

and the range values are then obtained from Eq. (91):

$$-0.3538 < E(r_j) < -0.1455 \tag{93}$$

The correlation coefficients for random numbers average −0.1111 and their range is from −0.3538 to −0.1455. As evident from Table X-6 none of the correlation coefficients for the trend are within this range, and the trend might be considered significant.

The statistics of the trend were estimated in a different manner by Chayes (1961), who omitted some of the oxides calculating $E(r_j)$, but when compared with randomness all oxides are of equivalent weight, and the above method is, therefore, preferred.

The trend estimated by Fenner (1926) displays a similar scatter to most variation diagrams, and it is gratifying to find that the trend is of significance. However, what is of mandatory importance for petrogenetic purposes is an estimate of the confidence intervals for the trend curves, as the extrapolated trend curves have to be compared with the compositions of fractionates. Such confidence intervals may be calculated by ignoring the closure constraint, by assuming one of the oxides as an independent variable. This procedure is not entirely correct as the confidence intervals are a function of r_j and s_j, and as the closure constraint increases r_j above its "real" value, then the calculated confidence intervals will be too small. No standard method has yet been worked out which estimates the true confidence intervals for closed arrays, but it is possible to make an approximate estimate, and evaluate under which conditions that the closure constrain has significant consequences for the confidence interval.

For a given value of x the value of y will be defined within a certain confidence interval dependent on the goodness of fit of the regression line and the confidence level. Let the confidence interval be $k(x)$, y is then defined within the interval:

$$y - k(x) < y < y + k(x) \tag{94}$$

The confidence interval $k(x)$ is generally estimated from:

$$k(x) = c\left(\frac{h\sqrt{q}}{\sqrt{n-2}}\right) \tag{95}$$

where n is the number of analyses, and:

$$h^2 = \frac{1}{n} + \frac{(x - \bar{x})^2}{(n-1)\,s_x^2} \tag{96}$$

$$q = (n-1)\,(s_y^2 - b^2\,s_x^2) \tag{97}$$

$$b = \frac{s_{xy}}{s_x^2} \tag{98}$$

The value of c is obtained from tables of Student's t-distribution. The effect of closure on $k(x)$ can be estimated if $k(x)$ can be expressed as a function of the correlation coefficient r.

According to Eq. (95) $k(x)$ is a function of c, q, h, and n. Using the definition of r, and that of b, we may express q by:

$$q = s_y^2\,(n-1)\,(1-r^2) \tag{99}$$

By inserting q as given by Eq. (99) in Eq. (95) one obtains:

$$k(x) = cs_y\,(1-r^2)^{\frac{1}{2}}\left(\frac{n-1}{n-2}\right)^{\frac{1}{2}}\left(\frac{1}{n} + \frac{(x-\bar{x})^2}{(n-1)\,s_x^2}\right) \tag{100}$$

This expression is simplified if k (x) is estimated for the mean value of x, for $x = \bar{x}$ Eq. (100) becomes:

$$k(\bar{x}) = \frac{c}{n} \left(\frac{n-1}{n-2}\right)^{\frac{1}{2}} s_y (1-r^2)^{\frac{1}{2}} \tag{101}$$

For a given number of analyses and a given confidence level $k(\bar{x})$ will only depend on s_y and r. The evaluation of Eq. (101) would be straightforward, if $k(\bar{x})$ could be expressed as a function of r only, but that is not possible. One of the variances, in this case s_y, will form part of the expression. The next problem is, therefore, to consider the relationship between s_y and r.

The value of s_y will mainly be related to the total variation in y, as evident from Eq. (66). If the scatter of analysis becomes large compared to the total variation in y, then the value of s_y will increase, because the range of y values is increased. However, the scatter in analyses of major elements and trace elements is generally smaller than the total variation, if the trend lines have a slope near zero that will not be the case. The value of s_y may, therefore, with approximation be considered nearly independent of the correlation coefficient. The confidence interval $k(\bar{x})$ is, thus, mainly a function of r. The variation in $k(\bar{x})$ for various sample sizes and r values is shown in Fig. X-26. If the correlation coefficient has a high value, then $k(\bar{x})$ is sensitive on r. For $s_y = 5.0$, $c = 1.66$, and $n = 25$, $k(\bar{x})$ will be equal to 0.0489 for $r = 0.99$, but for $r = 0.95$, we have $k(\bar{x}) = 0.1082$. The confidence interval has, thus, increased to more than the double by a decrease in r from 0.99 to 0.95. For smaller r values the effect of closure will be less pronounced. For $M = 10$, the average expected correlation is -0.11 according to Eq. (87). This induced correlation will have a significant effect for high values of r, but will be of less importance for r values less than 0.8, when n is larger than 25.

The alkalic suite of Hawaii shown in Fig. X-1 is a fairly well-defined suite, and the $|r|$ values for the different oxides vary between 0.2 and 0.7. In order to evaluate the effect of closure for the confidence intervals, a representative value is assumed as

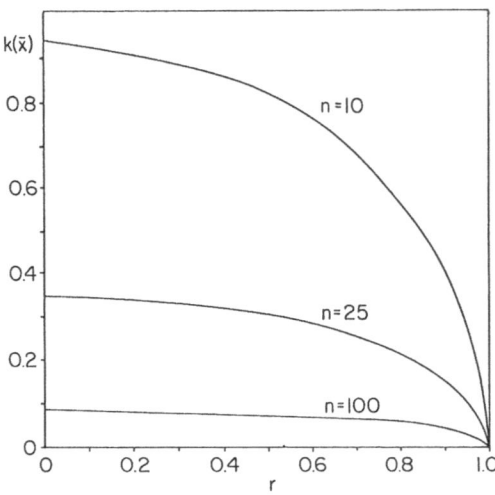

Fig. X-26. The variation in $k(\bar{x})$ as function of r for various values of sample sizes (10, 25, and 100). The confidence level used is 95%, and the s_y value is 5.0. The *curves* were calculated using Eq. (101)

0.5. The regression line for the intermediate trend from 10% to 5% MgO is based on 39 analyses. With a confidence level of 95% the value of c is 1.68 (Kreyszig 1970, p. 454), and a representative value for s_y was estimated as 2.0. Using Eq. (101) $k(\bar{x})$ is calculated as:

$$k(\bar{x}) = 0.0873(1-r^2)^{\frac{1}{2}} = 0.0756, \quad (r = 0.25) \tag{102}$$

If the average expected correlation induced is about 0.1, then a more realistic value of r is 0.4, and $k(\bar{x})$ is then calculated as 0.0800. The confidence interval has, thus, only increased with about 6% by a decrease in r of 0.1. The closure will, thus, have a small effect for the confidence intervals of the Hawaiian alkalic suite, which is a fairly typical suite. The closure will in this and most other cases cause a correction in the confidence intervals rather than a major correction. Most major element trends have correlation coefficients smaller than 0.7, and the closure effect for the confidence intervals is not significant. The correlation coefficients for trace elements are frequently higher and above 0.9, and the closure effect may have significance in that case.

If the regression coefficient between two oxides or trace elements is high, and the value of the confidence interval is considered important, then the confidence intervals may be calculated according to the following procedure.

1. Estimate r and the variance s_y^2 by conventional regression methods.
2. Choose a confidence level, typically either 95% or 99%.
3. Estimate the value of c using a table for Student's t-distribution.
4. Calculate the average expected correlation using Chayes equation, $E(r) = 1/(1-M)$, where M is the number of oxides.
5. Calculate a revised value for r: $r_c = |r| - |E(r)|$.
6. Finally, calculate $k(\bar{x})$ as given by Eq. (101).

XI Magma Kinetics

1 Diffusion

Nearly all petrological processes imply changes in the phase assemblages where new phases are formed and the old ones change their compositions. This applies for anatexis, partial melting, and fractionation as well as for metamorphic reactions. All these processes require that diffusion can take place over a short distance of the order of centimeters. The compositions of the new phases will be determined by the thermodynamic properties of the phases and the P–T conditions, but the diffusion rate will also influence the composition of the new phases as the rate of equilibration is directly dependent on the diffusion rate. Fast diffusion will, thus, ensure perfect equilibrium, while slow diffusion rates might hinder the attainment of equilibria. As an example the exsolution in solids may be mentioned. All the exposed rocks have been cooled down to temperatures near 0 °C. Nevertheless, the composition of minerals might suggest equilibration temperatures between 600° and 1200 °C. These temperatures are the temperatures at which the reaction ceased because the diffusion practically stopped. Probably the most conspicuous feature about the diffusion in magmas and rocks is the slow rate by which the diffusion occurs. As will be demonstrated later, the distances by which diffusion can occur in the span of a million years is less than a few meters. The slow diffusion consequently shows that material transport over long distances has to occur by other processes, like flow or gravitative fractionation. A comprehensive treatment of diffusion has been given by Crank (1975), while experimental aspects have been considered in detail by Frischat (1975). A detailed review of the diffusion processes in rocks and the diffusion coefficients of rocks and magmatic liquids have been given by Hofmann (1980).

2 The Diffusion Process

The diffusion rate and direction are determined by the gradient in chemical potential. The diffusion is normally related to concentration gradients, and a difference in concentration will also result in diffusion, as a difference in concentration in most cases also implies a difference in chemical potential. However, it should be emphasized that diffusion is first and foremost caused by a gradient in the chemical potential (Denbigh, 1971). The diffusion of molecules in a liquid is due to the thermal motions of the molecules. They move among each other and the collision frequency of the molecules is about 10^{12}/s. If the chemical potential of a molecule in two parts

Fig. XI-1 a, b. The relationship between chemical potential and the direction of diffusion. When the chemical potential in the two compartments A and B are the same, the molecules will diffuse in both directions with no net transfer. However, if the potentials are different, then there will be a net transfer from one compartment to the other, here from B to A

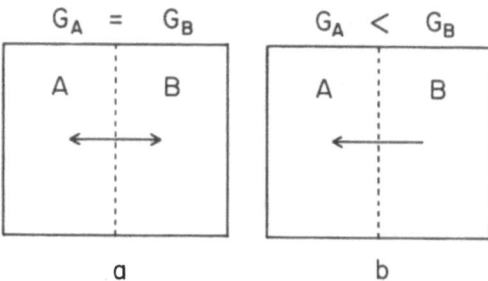

of a body is identical, then the same amount of ions will pass in opposite directions with respect to an arbitrary division plane (Fig. XI-1 a). On the other hand, if the chemical potential is smaller in one part, say A, then the molecules will move from B to A. The difference in chemical potential may be caused by a difference in both temperature and concentrations. The diffusive motion of the molecules are completely arbitrary; they move completely by chance. Let the concentration of some component in B be larger than in A. The molecules of this component will move in accidental directions in both A and B. However, the number of molecules that move to the left into A will be larger than the flux of molecules in the opposite direction, simply because there are less molecules in A.

The actual arbitrariness of the diffusion process is well illustrated from the following example (Shewmon 1963). Consider a volume of a solid with a concentration gradient along an x-axis (Fig. XI-2). The diffusion of the atoms occurs in small jumps from one position to another. The jump distance will depend on the diameter of the atoms, and is usually of the order of a few Å. Let this distance be α. Consider now two adjacent planes designated I and II a distance α apart. The number of atoms that takes part in the diffusion in plane I is n_1, and the similar number for plane II is n_2. The jumps occur with a frequency ν, so that the number of jumps per second is $n \cdot \nu$. In the time, dt, the number of jumps will be $n_1 \nu dt$ for plane I. The atoms jump in either direction and the number of atoms that jump from plane I to

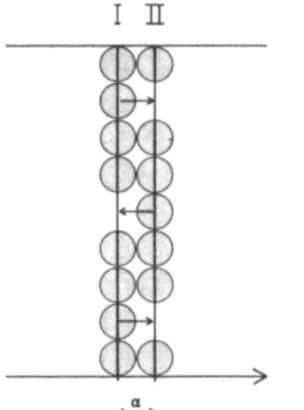

Fig. XI-2. The diffusion of molecules between two imaginary planes at a distance α from each other

II is ($\frac{1}{2}$) n_1 νdt. Similarly, the number that jumps from II to I is ($\frac{1}{2}$) n_2 νdt. The flux of atoms from I to II is given by:

$$J = \frac{1}{2}(n_2 - n_1) \tag{1}$$

The difference in the number of atoms $(n_1 - n_2)$ is related to the concentration per unit volume. If the jump distance is identical to the distance between the atoms, then the volume occupied by a plane with the unit sides is $\alpha \cdot 1 \cdot 1$. The concentrations are then given by:

$$c_1 = \frac{n_1}{\alpha} \qquad c_2 = \frac{n_2}{\alpha} \tag{2}$$

so that:

$$J = \frac{1}{2}(c_2 - c_1)\,\alpha\,\nu \tag{3}$$

The difference in concentration at planes I and II is given by:

$$\frac{dc}{dx} = \frac{1}{\alpha}(c_2 - c_1) \tag{4}$$

Thus, from Eqs. (3) and (4) we obtain:

$$J = -\frac{1}{2}\alpha^2\,\nu\,\frac{dc}{dx} \tag{5}$$

This equation is identical to Fick's first law if:

$$D = -\frac{1}{2}\alpha^2\,\nu \tag{6}$$

This equation was obtained assuming linear diffusion, i.e., that the diffusion only occurs in one dimension. If the diffusion occurs in all these dimensions we get a very similar relationship:

$$D = -\frac{1}{6}\alpha^2\,\nu \tag{7}$$

The diffusion constant is, therefore, determined by the frequency by which the atoms jump and the jumping distance. It was assumed in this derivation that the movement of the atoms is arbitrary, i.e., that the atoms jump in either of two possible directions. The diffusion of atoms in a concentration gradient does not result from any bias of the molecules to jump in any particular direction. Instead, the frequency of jumps is larger where the concentration is larger. If c_1 is larger than c_2, then the number of molecules that jump from I to II is larger than the number that jumps from II to I.

The process just described considered the diffusion in a concentration gradient. It was mentioned in the introduction that the driving force for diffusion is the chemical potential. What happens then if the diffusion occurs at constant concentration, but at a varying chemical potential due to a difference in temperature? In this case the direction of the diffusion will be determined by the variation in the frequency of jumping. Consider again planes I and II and that the temperature at plane I is higher than at II. The frequency of jumping will then be higher at I than at II so that more molecules must move from I to II than in the opposite direction.

The type of diffusion just considered is called volume diffusion. The diffusion between grain boundaries is grain boundary diffusion. This type of diffusion is frequently much faster than volume diffusion. The movement of molecules in a phase with no gradients is called self-diffusion. The diffusion in multicomponent systems, which generally involves the diffusion of several types of molecules, is termed interdiffusion.

The flux J of molecules by volume diffusion is estimated from Fick's first and second laws (Crank 1975). The first law by Fick states that:

$$J = -D\frac{dc}{dx} \tag{8}$$

where D is the diffusion coefficient, usually in the unit (cm^2/s), c the concentration in mol/cm^3, and x the coordinate in cm. The unit of J is then ($mol/cm^2 s$). This law applies for the steady state, and shows that the flux is proportional to the concentration gradient, the coefficient D being the constant of proportionality. D is typically 10^{-1} cm^2/s for gases, 10^{-7} cm^2/s for silicate liquids, and 10^{-12} cm^2/s for minerals. Readers familiar with thermodynamics may inquire here that one should use the activity α instead of the concentration c. This is correct, however, the value of D is obtained from experiments, so the deviation from ideal solution is incorporated with the value of D.

Fick's second law for diffusion deals with time dependent diffusion and is derived from the first one as follows (Fig. XI-3). Consider that some molecules move in the positive direction of x, and through the two unit planes at x and $x + \Delta x$. In the time dt, the number of moles entering the volume element from left is $J_x dt$, while the number leaving at $x + \Delta x$ is $J_{x + \Delta x}$. During an infinitesimal amount of time there will be no change in the concentration gradient, and the flux into the volume element is, therefore, given by:

$$J = -D\frac{dc}{dx} \tag{9}$$

The number of moles n, in the volume element varies with time and is determined by the difference in the flux into the element and the flux of molecules out of the element. Thus, dn is given by:

$$dn = J_x dt - J_{x+\Delta x} dt \tag{10}$$

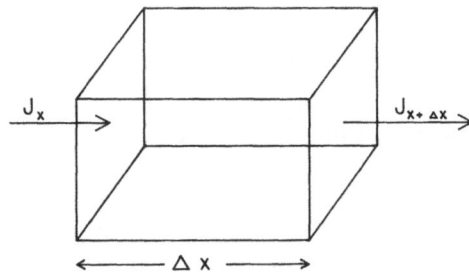

Fig. XI-3. The flux through two planes at a distance Δx from each other. The flux is J_x through the left plane and $J_{x+\Delta x}$ through the right plane

The flux at $x + \Delta x$ is given by:

$$J_{x+\Delta x} = J_x + \frac{\delta J_x}{\delta x} \Delta x \tag{11}$$

so that:

$$dn = -\frac{\delta J_x}{\delta x} \Delta x \, dt$$

The variation in the concentration in the volume element is given by $dc = dn/\Delta x$, hence, we have:

$$dc = -\frac{\delta J_x}{\delta x} dt \tag{12}$$

By division with dt we obtain:

$$\frac{dc}{dt} = -\frac{\delta J_x}{\delta x} \tag{13}$$

and by inserting J_x as given by Eq. (9) in Eq. (13) we get:

$$\frac{dc}{dt} = \frac{\delta}{\delta x} \left(D \frac{\delta c}{\delta x} \right) \tag{14}$$

This is the most general form for Fick's second law. The equation applies for the case where D is dependent on composition. If D is assumed independent on composition, as normally has to be done in petrological calculations, we get:

$$\frac{\delta c}{\delta t} = D \frac{\delta^2 c}{\delta x^2} \tag{15}$$

Both magmas and minerals consist of metallic cations and silicate anions. Diffusion of cations must occur simultaneously with the diffusion of anions. A cation which by itself has a fast diffusion rate is slowed down because the anions diffuse with a slow rate. The diffusion coefficients that are estimated for a given ion in silicate melt are, therefore, dependent on the composition of the melt. These coefficients are sometimes called EBDC (effective binary diffusion coefficients) in order to emphasize their compositional dependence (Hofmann 1980). The relationships between the interdiffusion coefficient and the separate diffusion coefficients D_j were estimated by Darken (1948)

$$D_i = \sum_{j=1}^{j=n} x_j D_j \tag{16}$$

where x_j is the mole fraction. For a binary mixture one gets:

$$D = x_1 D_1 + x_2 D_2; \quad (x_1 + x_2 = 1) \tag{17}$$

If the diffusion coefficients are dependent on composition, these equations become more complicated (Darken 1948; Onsager 1945).

3 Temperature Dependence of D

The diffusion coefficient varies a great deal with temperature, while the pressure dependence is less prounounced. Experimental investigations show that the temperature dependence is expressed by the following function which was first estimated by Arrhenius:

$$D = D_0 \exp\left(\frac{-Q}{RT}\right) \tag{18}$$

This expression was obtained from experiments, but the statistical mechanics have shown that the expression has a physical meaning. The molecules in a liquid have a wide range of kinetic energies, but the number of molecules with a specific energy is determined by the thermodynamic properties of the material. Let the total number of molecules be N and the number of molecules with free energy ΔG be n, then (Wall 1974):

$$n = N \exp\left(\frac{\Delta G}{RT}\right) \tag{19}$$

where R is the gas constant. This equation is Boltzman's distribution law in a slightly simplified form. The diffusion of molecules in a solid or liquid requires that the molecules are able to move by each other. In order that a molecule can move from one site to the next, it should possess an extra energy, the activation energy Q. The equation for n shown in Eq. (19) shows that the diffusion coefficient is proportional to the number of molecules n that possess the required free energy for the diffusion to occur. The number of molecules with high energies increases with increasing temperature, and the diffusion coefficient consequently also increases with increasing

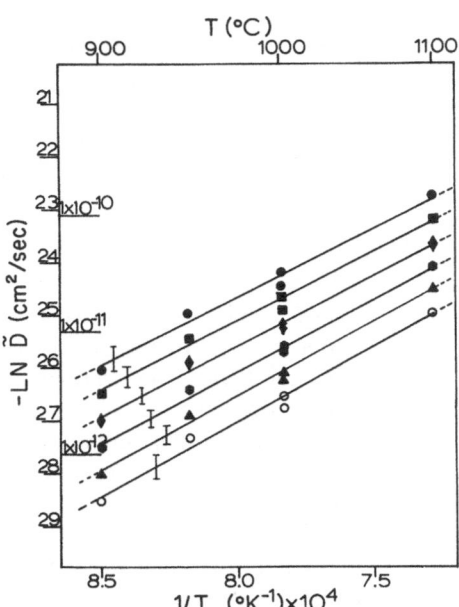

Fig. XI-4. The variation in ln D as function of 1/T for diffusion of Mg and Fe in olivine for different forsterite contents. The *upper line* is for 10 mol% Fo and the *lower line* is for 60 mol% Fo (Misener 1974)

temperature. The variation in D is estimated from experiments performed at different temperatures. The values of D^0 and Q are then estimated in the following manner. Take the natural logarithm to Eq. (18):

$$\ln D = \ln D_0 + \frac{-Q}{RT} \tag{20}$$

or by rearrangement:

$$\ln D_0 = \ln D + \frac{Q}{RT} \tag{21}$$

Thus by plotting $\ln D$ as a function of $1/T$, the values of D^0, and Q may be obtained (Fig. XI-4).

4 Pressure Dependence of D

The value of the diffusion coefficient also displays variation with pressure, but the effect of pressure is smaller than the effect of temperature. An increase in pressure will decrease the number of vacancies in the solid, and diminish the space between the molecules of a liquid, and the diffusion coefficient might, therefore, be expected to decrease with pressure. The diffusion rate will also in this case be determined by the number of molecules that possess the required free energy of activation. The variation in free energy with pressure is given by:

$$\left[\frac{\delta G}{\delta P}\right]_T = \varDelta V \tag{22}$$

The variation in D with pressure may, therefore, be expected to depend on a volume term. This volume term, called the activation volume, is not the total volume of the phase considered. The activation volume is dependent on the volume of the vacancies in the solid, and on the volume of the activated molecules in both solids and liquids (Shewmon 1963). The pressure dependence may be written in the following form (Misener 1976):

$$D = D_0 \exp\left(\frac{-P \varDelta V}{RT}\right) \tag{23}$$

The activation volume is estimated from experimental work in a similar manner as Q is estimated, an example is shown in Fig. XI-5, which shows the variation in D for olivine (Misener 1974). The average value for the activation volume is in this case, 5.5 cm³/mol. The activation volume is not only a parameter of interest in connection with diffusion, the activation volume is also an important parameter in the equations describing the creep properties of rocks (Kohlstedt et al. 1976).

The variation in D with pressure for silicate liquids is not known in detail yet. The variation in viscosity with pressure is known for both basaltic and granitic liquids, and an approximate value of D may be calculated if D is known at one pressure, and the viscosity is known at different pressures. Let η be the viscosity, r the

Fig. XI-5. The variation in ln D for the diffusion of Mg and Fe in olivine as function of pressure estimated for different forsterite contents. The value of D decreases with pressure, but the effect of pressure is relatively small. The percentages shown are mol% forsterite (Misener 1974). The values of lnD has been estimated at 900 °C (*squares*) and 1100 °C (*dots*)

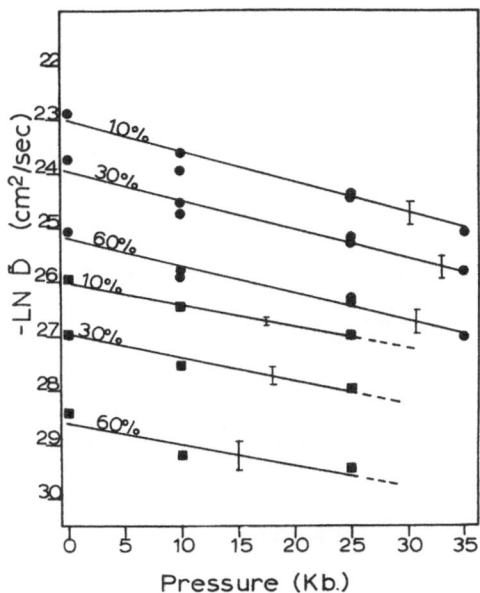

radius of the diffusing molecule, and T temperature in degrees Kelvin, then (Jost 1960, p. 462):

$$D = \frac{kT}{6\pi\eta r} \tag{24}$$

where k is Boltzman's constant. If a diffusion coefficient D_1 is known at a given temperature and pressure P_1 and the viscosities at the two pressures P_1 and P_2 are known, then the value of D_2 at pressure P_2 is estimated from:

$$\frac{D_1}{D_2} = \frac{\eta_2}{\eta_1} \tag{25}$$

5 Diffusion in Gases and Liquids

The molecules of a gas display translatory motions. They move a certain distance, collide with another gas molecule whereby their direction and velocity is changed, collide again, and then again all the time. The diffusion rate in a gas must be dependent on the velocity of the gas molecules, and it might be expected that the diffusion rate increases with the velocity of the gas molecules. The average velocity of the gas molecules can be determined assuming that the gas molecules behave as small elastic balls, and that the gas obeys Boyle-Mariotte's law:

$$PV = nRT \tag{26}$$

where P is the pressure, V the volume, n the number of moles, R the gas constant, and T the temperature. Let the average velocity be c, the kinetic energy of the gas molecules is then given by:

$$\frac{1}{2} Mc^2 = \frac{3}{2} RT \tag{27}$$

where M is the weight of 1 mol. The kinetic energy is proportional with the temperature, while the average velocity is proportional to the square root of the temperature. For water at 25 °C the velocity is estimated as 643 m/s. However, the molecules do not move in space with this velocity as the molecules collide with each other. The distance a molecule travels between the collisions is given by:

$$\lambda = \frac{RT}{d^2 N} \frac{1}{P} \tag{28}$$

where d is the diameter of the molecules, N Avogadro's number, and λ the distance. For $d = 3$ Å, $T = 298$ °K, and $P = 1$ bar, one gets $\lambda = 4.57 \cdot 10^{-5}$ cm. Using the velocity

Table XI-1. Diffusion constants

Diffusing element	Media	T°C	Typical D value	D°	Q cal/mol
H_2O	Air	20	0.24		
H_2O	Water 200 bar	800	$1.1 \cdot 10^{-3}$		
Si	Water, 1000 bar	800	$8.0 \cdot 10^{-4}$		
Si	Water, 2000 bar	800	$6.0 \cdot 10^{-4}$		
Si	Water, 1000 bar	400	$1.8 \cdot 10^{-4}$		
NaCl	Water, 1 bar	20	1.50		
Diopside	Ab_2An_1 mixture	1500	$1.7 \cdot 10^{-7}$		
Diopside	Ab_1An_1 mixture	1500	$1.6 \cdot 10^{-6}$		
Diopside	Ab_1An_2 mixture	1500	$2.3 \cdot 10^{-6}$		
Ca	$CaO\text{-}Al_2O_3\text{-}SiO_2$	1400	$6.7 \cdot 10^{-7}$		
Si	$CaO\text{-}Al_2O_3\text{-}SiO_2$	1400	$1.0 \cdot 10^{-7}$		
^{17}O	$CaO\text{-}Al_2O_3\text{-}SiO_2$	1400	$6.0 \cdot 10^{-6}$		
Ca	Olivine tholeiite	1300	$4.2 \cdot 10^{-7}$	0.54	$44 \cdot 10^3$
Sr	Olivine tholeiite	1300	$2.5 \cdot 10^{-7}$	0.28	$44 \cdot 10^3$
Ba	Olivine tholeiite	1300	$2.0 \cdot 10^{-7}$	0.06	$39 \cdot 10^3$
Eu, Gd	Basalt	1300	$1.3 \cdot 10^{-7}$	0.06	$41 \cdot 10^3$
Ba	Obsidian	900	$1.0 \cdot 10^{-10}$	0.04	$46 \cdot 10^3$
Na	Obsidian	900	$1.0 \cdot 10^{-5}$	0.04	$23 \cdot 10^3$
K	Obsidian	900	$1.0 \cdot 10^{-7}$	0.01	$26 \cdot 10^3$
Sr	Obsidian	900	$6.1 \cdot 10^{-10}$	0.06	$43 \cdot 10^3$
Na	Na	100	$4.2 \cdot 10^{-5}$	0.001	$2.4 \cdot 10^3$
^{18}O	Albite	600		$2.3 \cdot 10^{-9}$	21.3
^{18}O	Anorthite	600		$1.4 \cdot 10^{-7}$	26.2
Na	Orthoclase	800		8.9	$52.7 \cdot 10^3$
K	Orthoclase	800		16.1	$68.2 \cdot 10^3$
Rb	Orthoclase	800		38.0	$73.0 \cdot 10^3$
H_2O	Feldspar	800	$7.0 \cdot 10^{-12}$		
K	Biotite	650	$2.0 \cdot 10^{-17}$		
Ar	Phlogopite	800		0.75	$57.9 \cdot 10^3$
Mg	Crystalline olivine with 90% Fo	1400	$3.8 \cdot 10^{-11}$		

of the water molecules just calculated, the collision frequency is estimated as $1.4 \cdot 10^9$ collisions per second. The gas molecules, thus, move with a velocity of the order of 100 m/s, but their movements are incessantly interrupted by the collision with other molecules. The average distance that the molecules actually travel in space in a given time is estimated assuming that the molecules move in a completely arbitrary manner. This distance must be related to the diffusion constant, and it was shown by Einstein (1921) that D is related to this distance $\varDelta$, by the equation:

$$D = \frac{\varDelta^2}{2t} \tag{29}$$

where t is the time. The D value for the diffusion of water into air is 0.24 cm²/s, and the distance the molecules travel is then estimated as 0.69 cm/s. Despite the high velocity of the gas molecules, they do not move that far from their original position because of the high collision frequency. Some D values for gases are shown in Table 1. As evident from Eq. (28) the distance the molecules move between collisions is inversely proportional with pressure, and the D values should, therefore, decrease with pressure. In accordance with this the diffusion constant is about 1 cm²/s for water at 1 bar, while D is about $5 \cdot 10^{-4}$ cm²/s in a super critical gas (Fig. XI-6).

The above Eq. (27) and (29) are also valid for the molecules of a liquid. In the liquid the frequency of collisions will be higher than in the gas, about 10^{12} collisions per second. There is less space between the molecules in the liquid, and the frequency is, therefore, higher. The movement of the molecules in the liquid has the character of vibration rather than a translatory motion. In this respect the liquid is similar to a solid where the molecules move by vibrations. The difference between the solid

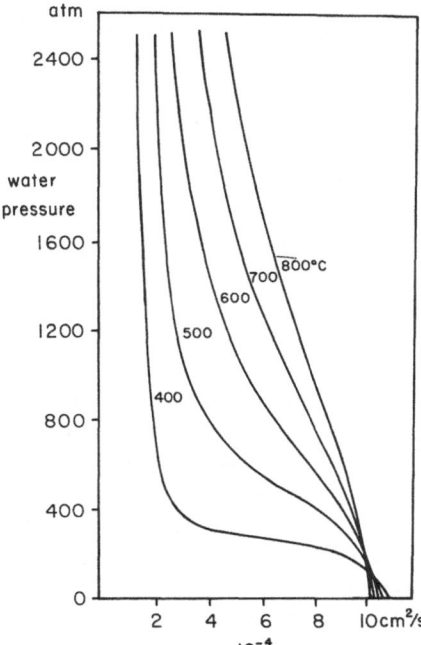

Fig. XI-6. The variation in the diffusion coefficient for the diffusion of Si^{4+} in supercritical water gas at various temperatures. The value of D decreases with increasing pressure, and increases with increasing temperature (Walton 1960)

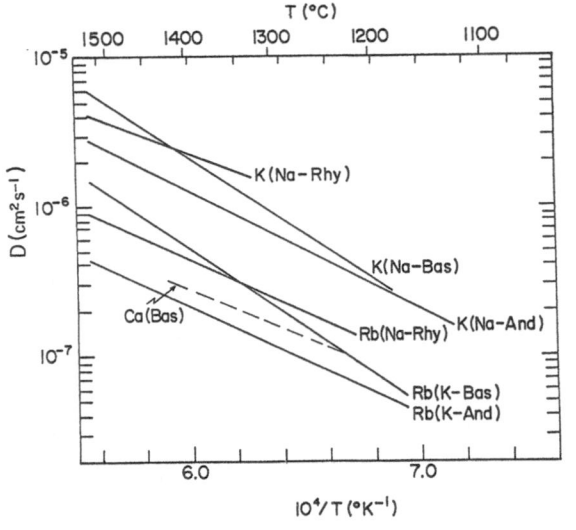

Fig. XI-7 and XI-8. D values for andesite, rhyolite and basalt at various temperatures (Hofman 1980)

Fig. XI-7.

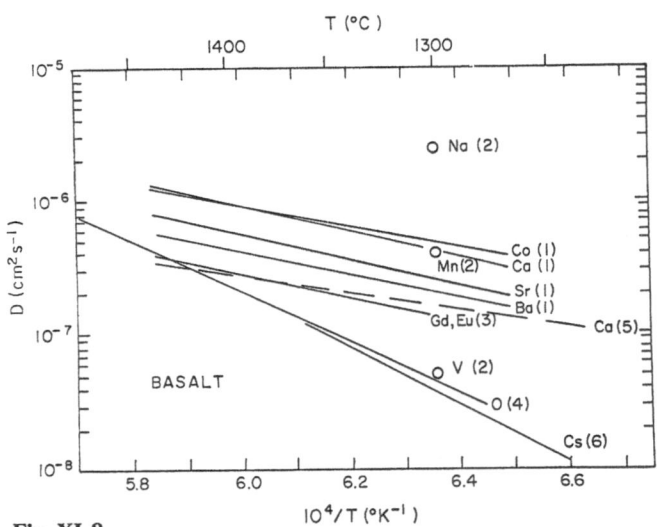

Fig. XI-8.

and the liquid is that the solid molecules move about a fixed point in the lattice, while the liquid molecules shift position all the time.

Some diffusion coefficients for silicate liquids, both synthetic and natural ones are listed in Table XI-1. A representative value for basaltic liquids is about 10^{-7} cm^2/s, while the value for granitic melts is about 10^{-8} cm^2/s (Figs. XI-7, XI-8, XI-9). However, the diffusion coefficients vary both with the metallic ions and the type of melt, as well as with the temperature. The diffusion coefficients that are estimated for a given ion in silicate melt are, therefore, dependent on the composition of the melt.

Fig. XI-9. The variation in D as function of log (Z^2R), where Z is the atomic number and R the radius of the ions. Apparently the relationship is nearly linear, suggesting that the D values for other ions can be obtained by interpolation (Hofmann 1980)

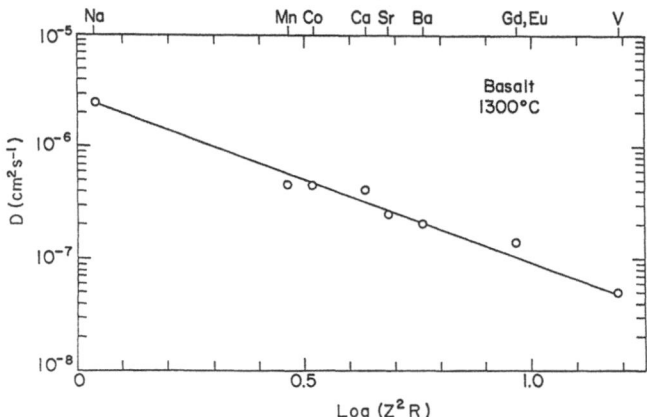

6 Diffusion in Solids

The diffusion rate in solids is much slower than in liquids, the D values for solids are about 10^5 times smaller than for liquids (Table XI-1). The diffusion rate is extremely slow in some minerals like plagioclase, and these minerals will retain their zoning during cooling. Other minerals like olivine and pyroxene have a relatively fast diffusion with the result that these minerals are rarely zoned in plutonic rocks, while their zoning may be preserved in extrusive rocks which have undergone rapid cooling.

The atoms in a solid vibrate with average frequencies between 10^{10} and 10^{13} cps, the exact value depending on both the temperature and the mass of the atoms. All the atoms do not vibrate with the same frequency. The atoms vibrate with both different frequencies and in different directions. The single atoms will sometimes possess a high vibration energy and sometimes the vibration energy will be small. The atoms may shift position in the lattice when they obtain an especially high energy, the activation energy discussed above. The diffuse movement may occur in different manners, of which the following three are the most important: (1) the vacancy mechanism, (2) the interstitial mechanism; and (3) by simultaneous rotation of three or more atoms.

In all crystals some of the lattice sites are unoccupied. These free sites are called vacancies. The number of vacancies in a crystal varies, but is typically of the order of 0.001% (Kittel 1956). Their number increases with increasing temperature, and the diffusion rate, therefore, increases with increasing temperature. The diffusion mechanism by vacancy diffusion is illustrated in Fig. XI-10. The atom A may move from its original site into site A' if it has the sufficient energy and vibrates in the right direction from A towards A'. This condition requires that the surrounding atoms have added energy to atom A, and that their movements allow the atom to pass from site A to site A'.

An atom is said to diffuse by the interstitial mechanism when it jumps from one interstitial site in the lattice to another without disturbing permanently the atoms at the lattice sites (Fig. XI-11). This movement also requires an activation energy as the atom has to pass between the other atoms.

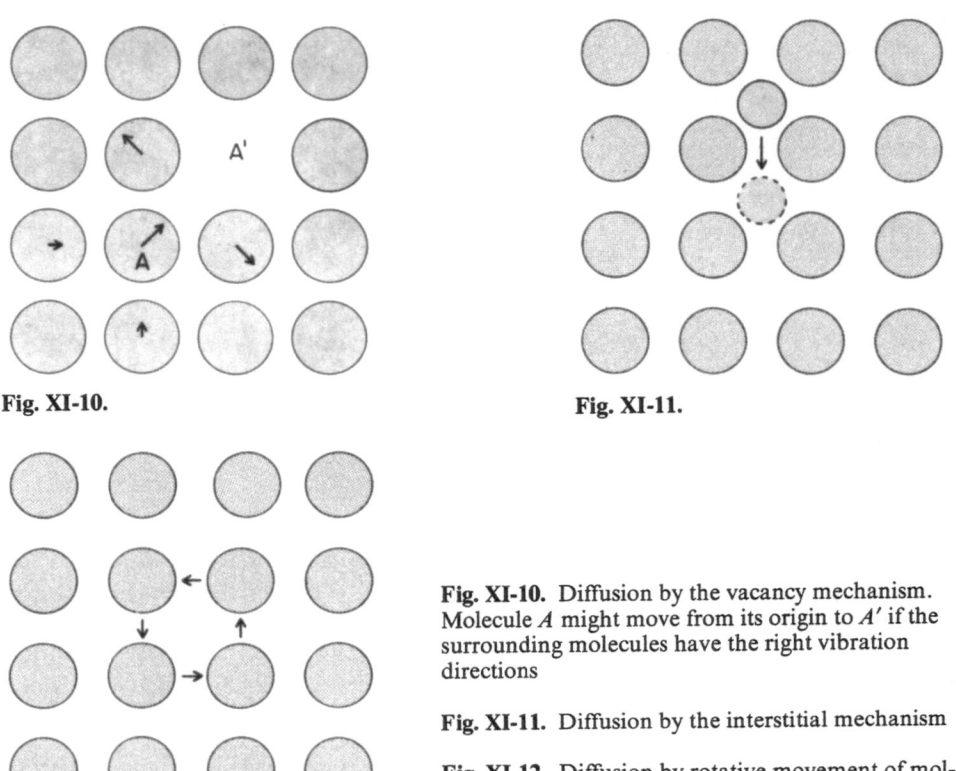

Fig. XI-10.

Fig. XI-11.

Fig. XI-12.

Fig. XI-10. Diffusion by the vacancy mechanism. Molecule A might move from its origin to A' if the surrounding molecules have the right vibration directions

Fig. XI-11. Diffusion by the interstitial mechanism

Fig. XI-12. Diffusion by rotative movement of molecules

Finally, the diffusion may occur by a rotative movement as shown in Fig. XI-12. This type of diffusion occurs only in an open lattice where the distance between the atoms is rather large. Some diffusion coefficients for solids are shown in Table XI-1.

7 Diffusion Equations

The variation in concentration with time for a given problem is estimated from the second order differential equation given by Fick's second law (Eq. 15). The estimation of solutions of such differential equations is far from straightforward, quite an amount of ingenuity is frequently required for the solution of a specific problem. The mathematical theory of differential equations may be found in Kreyszig (1967), and the solutions to diffusion problems is given by Crank (1975) and Jost (1960). The solutions of the differential equations will be dependent on the boundary conditions, the solutions for a spherical body will, thus, differ from that of a cylindrical body. The solutions can, therefore, only be stated for specific cases, and some cases with petrological applications are shown below. These solutions are of general validity for the boundary conditions given, and may be used for any particular problem. There are no approximations involved.

Fig. XI-13. The normalized error function which varies between −1.0 and +1.0. The values for the function are obtained from Table XI-2 (Kreyszig 1967)

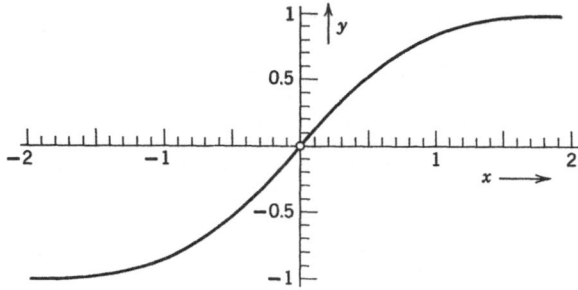

Many solutions of Fick's second law involve the normalized error function which is given by the integral (Kreyszig 1967):

$$\text{erf}(x) = \frac{2}{\sqrt{\pi}} \int_0^x e^{-t^2} dt \qquad (30)$$

This integral has no analytical solution just like the integral of $1/x$ has no analytical solution. Its value may be obtained by numerical methods in the same manner as the values of the $\ln(x)$ function are obtained. The values of the error function vary between −1 and +1 and are obtained from tables like Table XI-2 (Peirce and Foster 1963). The error function is shown in Fig. XI-13. The value of the error function for negative values of x is also estimated from Table XI-2, as the following relationship holds:

$$\text{erf}(-x) = -\text{erf}(x) \qquad (31)$$

Example 1:

Assume that a value of x has been calculated as 0.913. The value of the error function is then obtained from Table XI-2 by entering the table at 0.913 and the value of the error function is then estimated as $\text{erf}(0.913) = 0.80336$.

If x is negative, for example −2.352, then the table is entered for $x = +2.352$ and the value for $\text{erf}(2.352)$ is obtained as 0.99912, thus, according to Eq. (31): erf $(-2.352) = -0.9912$.

The solutions to Fick's second law will now be given for four sets of boundary conditions.

Linear Diffusion 1

Boundary conditions:

$$c = c_1 \quad \text{for } x < 0 \text{ at all } t$$
$$c = c_0 \quad \text{for } x > 0 \quad \text{for } t = 0$$
$$c = (x, t) \quad \text{for } x > 0 \quad \text{and} \quad t > 0$$

$$c(x, t) = c_1 + (c_0 - c_1) \, \text{erf}\left(\frac{x}{2\sqrt{Dt}}\right) \qquad (32)$$

Table XI-2. The probability integral $\left(\dfrac{2}{\sqrt{\pi}}\int\limits_{0}^{x}e^{-u^{2}}\,du\right)$

x	0	1	2	3	4	5	6	7	8	9
0.00	0.00 000	00 113	00 226	00 339	00 451	00 564	00 677	00 790	00 903	01 016
0.01	0.01 128	01 241	01 354	01 467	01 580	01 692	01 805	01 918	02 031	02 144
0.02	0.02 256	02 369	02 482	02 595	02 708	02 820	02 933	03 046	03 159	03 271
0.03	0.03 384	03 497	03 610	03 722	03 835	03 948	04 060	04 173	04 286	04 398
0.04	0.04 511	04 624	04 736	04 849	04 962	05 074	05 187	05 299	05 412	05 525
0.05	0.05 637	05 750	05 862	05 975	06 087	06 200	06 312	06 425	06 537	06 650
0.06	0.06 762	06 875	06 987	07 099	07 212	07 324	07 437	07 549	07 661	07 773
0.07	0.07 886	07 998	08 110	08 223	08 335	08 447	08 559	08 671	08 784	08 896
0.08	0.09 008	09 120	09 232	09 344	09 456	09 568	09 680	09 792	09 904	10 016
0.09	0.10 128	10 240	10 352	10 464	10 576	10 687	10 799	10 911	11 023	11 135
0.10	0.11 246	11 358	11 470	11 581	11 693	11 805	11 916	12 028	12 139	12 251
0.11	0.12 362	12 474	12 585	12 697	12 808	12 919	13 031	13 142	13 253	13 365
0.12	0.13 476	13 587	13 698	13 809	13 921	14 032	14 143	14 254	14 365	14 476
0.13	0.14 587	14 698	14 809	14 919	15 030	15 141	15 252	15 363	15 473	15 584
0.14	0.15 695	15 805	15 916	16 027	16 137	16 248	16 358	16 468	16 579	16 689
0.15	0.16 800	16 910	17 020	17 130	17 241	17 351	17 461	17 571	17 681	17 791
0.16	0.17 901	18 011	18 121	18 231	18 341	18 451	18 560	18 670	18 780	18 890
0.17	0.18 999	19 109	19 218	19 328	19 437	19 547	19 656	19 766	19 875	19 984
0.18	0.20 094	20 203	20 312	20 421	20 530	20 639	20 748	20 857	20 966	21 075
0.19	0.21 184	21 293	21 402	21 510	21 619	21 728	21 836	21 945	22 053	22 162
0.20	0.22 270	22 379	22 487	22 595	22 704	22 812	22 920	23 028	23 136	23 244
0.21	0.23 352	23 460	23 568	23 676	23 784	23 891	23 999	24 107	24 214	24 322
0.22	0.24 430	24 537	24 645	24 752	24 859	24 967	25 074	25 181	25 288	25 395
0.23	0.25 502	25 609	25 716	25 823	25 930	26 037	26 144	26 250	26 357	26 463
0.24	0.26 570	26 677	26 783	26 889	26 996	27 102	27 208	27 314	27 421	27 527
0.25	0.27 633	27 739	27 845	27 950	28 056	28 162	28 268	28 373	28 479	28 584
0.26	0.28 690	28 795	28 901	29 006	29 111	29 217	29 322	29 427	29 532	29 637
0.27	0.29 742	29 847	29 952	30 056	30 161	30 266	30 370	30 475	30 579	30 684
0.28	0.30 788	30 892	30 997	31 101	31 205	31 309	31 413	31 517	31 621	31 725
0.29	0.31 828	31 932	32 036	32 139	32 243	32 346	32 450	32 553	32 656	32 760
0.30	0.32 863	32 966	33 069	33 172	33 275	33 378	33 480	33 583	33 686	33 788
0.31	0.33 891	33 993	34 096	34 198	34 300	34 403	34 505	34 607	34 709	34 811
0.32	0.34 913	35 014	35 116	35 218	35 319	35 421	35 523	35 624	35 725	35 827
0.33	0.35 928	36 029	36 130	36 231	36 332	36 433	36 534	36 635	36 735	36 836
0.34	0.36 936	37 037	37 137	37 238	37 338	37 438	37 538	37 638	37 738	37 838
0.35	0.37 938	38 038	38 138	38 237	38 337	38 436	38 536	38 635	38 735	38 834
0.36	0.38 933	39 032	39 131	39 230	39 329	39 428	39 526	39 625	39 724	39 822
0.37	0.39 921	40 019	40 117	40 215	40 314	40 412	40 510	40 508	40 705	40 803
0.38	0.40 901	40 999	41 096	41 194	41 291	41 388	41 486	41 583	41 680	41 777
0.39	0.41 874	41 971	42 068	42 164	42 261	42 358	42 454	42 550	42 647	42 473
0.40	0.42 839	42 935	43 031	43 127	43 223	43 319	43 415	43 510	43 606	43 701
0.41	0.43 797	43 892	43 988	44 083	44 178	44 273	44 368	44 463	44 557	44 652
0.42	0.44 747	44 841	44 936	45 030	45 124	45 219	45 313	45 407	45 501	45 595
0.43	0.45 689	45 782	45 876	45 970	46 063	46 157	46 250	46 343	46 436	46 529
0.44	0.46 623	46 715	46 808	46 901	46 994	47 086	47 179	47 271	47 364	47 456
0.45	0.47 548	47 640	47 732	47 824	47 916	48 008	48 100	48 191	48 283	48 374
0.46	0.48 466	48 557	48 648	48 739	48 830	48 921	49 012	49 103	49 193	49 284
0.47	0.49 375	49 465	49 555	49 646	49 736	49 826	49 916	50 006	50 096	50 185
0.48	0.50 275	50 365	50 454	50 543	50 633	50 722	50 811	50 900	50 989	51 078
0.49	0.51 167	51 256	51 344	51 433	51 521	51 609	51 698	51 786	51 874	51 962

Table XI-2. (Continued)

x	0	1	2	3	4	5	6	7	8	9
0.50	0.52 050	52 138	52 226	52 313	52 401	52 488	52 576	52 663	52 750	52 837
0.51	0.52 924	53 011	53 098	53 185	53 272	53 358	53 445	53 531	53 617	53 704
0.52	0.53 790	53 876	53 962	54 048	54 134	54 219	54 305	54 390	54 476	54 561
0.53	0.54 646	54 732	54 817	54 902	54 987	55 071	55 156	55 241	55 325	55 410
0.54	0.55 494	55 578	55 662	55 746	55 830	55 914	55 998	56 082	56 165	56 249
0.55	0.56 332	56 416	56 499	56 582	56 665	56 748	56 831	56 914	56 996	57 079
0.56	0.57 162	57 244	57 326	57 409	57 491	57 573	57 655	57 737	57 818	57 900
0.57	0.57 982	58 063	58 144	58 226	58 307	58 388	58 469	58 550	58 631	58 712
0.58	0.58 792	58 873	58 953	59 034	59 114	59 194	59 274	59 354	59 434	59 514
0.59	0.59 594	59 673	59 753	59 832	59 912	59 991	60 070	60 149	60 228	60 307
0.60	0.60 386	60 464	60 543	60 621	60 700	60 778	60 856	60 934	61 012	61 090
0.61	0.61 168	61 246	61 323	61 401	61 478	61 556	61 633	61 710	61 787	61 864
0.62	0.61 941	62 018	62 095	62 171	62 248	62 324	62 400	62 477	62 553	62 629
0.63	0.62 705	62 780	62 856	62 932	63 007	63 083	63 158	63 233	63 309	63 384
0.64	0.63 459	63 533	63 608	63 683	63 757	63 832	63 906	63 981	64 055	64 129
0.65	0.64 203	64 277	64 351	64 424	64 498	64 572	64 645	64 718	64 791	64 865
0.66	0.64 938	65 011	65 083	65 156	65 229	65 301	65 374	65 446	65 519	65 591
0.67	0.65 663	65 735	65 807	65 878	65 950	66 022	66 093	66 165	66 236	66 307
0.68	0.66 378	66 449	66 520	66 591	66 662	66 732	66 803	66 873	66 944	67 014
0.69	0.67 084	67 154	67 224	67 294	67 364	67 433	67 503	67 572	67 642	67 711
0.70	0.67 780	67 849	67 918	67 987	68 056	68 125	68 193	68 262	68 330	68 398
0.71	0.68 467	68 535	68 603	68 671	68 738	68 806	68 874	68 941	69 009	69 076
0.72	0.69 143	69 210	69 278	69 344	69 411	69 478	69 545	69 611	69 678	69 744
0.73	0.69 810	69 877	69 943	70 009	70 075	70 140	70 206	70 272	70 337	70 403
0.74	0.70 468	70 533	70 598	70 663	70 728	70 793	70 858	70 922	70 987	71 051
0.75	0.71 116	71 180	71 244	71 308	71 372	71 436	71 500	71 563	71 627	71 690
0.76	0.71 754	71 817	71 880	71 943	72 006	72 069	72 132	72 195	72 257	72 320
0.77	0.72 382	72 444	72 507	72 569	72 631	72 693	72 755	72 816	72 878	72 940
0.78	0.73 001	73 062	73 124	73 185	73 246	73 307	73 368	73 429	73 489	73 550
0.79	0.73 610	73 671	73 731	73 791	73 851	73 911	73 971	74 031	74 091	74 151
0.80	0.74 210	74 270	74 329	74 388	74 447	74 506	74 565	74 624	74 683	74 742
0.81	0.74 800	74 859	74 917	74 976	75 034	75 092	75 150	75 208	75 266	75 323
0.82	0.75 381	75 439	75 496	75 553	75 611	75 668	75 725	75 782	75 839	75 896
0.83	0.75 952	76 009	76 066	76 122	76 178	76 234	76 291	76 347	76 403	76 459
0.84	0.76 514	76 570	76 626	76 681	76 736	76 792	76 847	76 902	76 957	77 012
0.85	0.77 067	77 122	77 176	77 231	77 285	77 340	77 394	77 448	77 502	77 556
0.86	0.77 610	77 664	77 718	77 771	77 825	77 878	77 932	77 985	78 038	78 091
0.87	0.78 144	78 197	78 250	78 302	78 355	78 408	78 460	78 512	78 565	78 617
0.88	0.78 669	78 721	78 773	78 824	78 876	78 928	78 979	79 031	79 082	79 133
0.89	0.79 184	79 235	79 286	79 337	79 388	79 439	79 489	79 540	79 590	79 641
0.90	0.79 691	79 741	79 791	79 841	79 891	79 941	79 990	80 040	80 090	80 139
0.91	0.80 188	80 238	80 287	80 336	80 385	80 434	80 482	80 531	80 580	80 628
0.92	0.80 677	80 725	80 773	80 822	80 870	80 918	80 966	81 013	81 061	81 109
0.93	0.81 156	81 204	81 251	81 299	81 346	81 393	81 440	81 487	81 534	81 580
0.94	0.81 627	81 674	81 720	81 767	81 813	81 859	81 905	81 951	81 997	82 043
0.95	0.82 089	82 135	82 180	82 226	82 271	82 317	82 362	82 407	82 452	82 497
0.96	0.82 542	82 587	82 632	82 677	82 721	82 766	82 810	82 855	82 899	82 943
0.97	0.82 987	83 031	83 075	83 119	83 162	83 206	83 250	83 293	83 337	83 380
0.98	0.83 423	83 466	83 509	83 552	83 595	83 638	83 681	83 723	83 766	83 808
0.99	0.83 851	83 893	83 935	83 977	84 020	84 061	84 103	84 145	84 187	84 229

Table XI-2. (Continued)

x	0	1	2	3	4	5	6	7	8	9
1.00	0.84 270	84 312	84 353	84 394	84 435	84 477	84 518	84 559	84 600	84 640
1.01	0.84 681	84 722	84 762	84 803	84 843	84 883	84 924	84 964	85 004	85 044
1.02	0.85 084	85 124	85 163	85 203	85 243	85 282	85 322	85 361	85 400	85 439
1.03	0.85 478	85 517	85 556	85 595	85 634	85 673	85 711	85 750	85 788	85 827
1.04	0.85 865	85 903	85 941	85 979	86 017	86 055	86 093	86 131	86 169	86 206
1.05	0.86 244	86 281	86 318	86 356	86 393	86 430	86 467	86 504	86 541	86 578
1.06	0.86 614	86 651	86 688	86 724	86 760	86 797	86 833	86 869	86 905	86 941
1.07	0.86 977	87 013	87 049	87 085	87 120	87 156	87 191	87 227	87 262	87 297
1.08	0.87 333	87 368	87 403	87 438	87 473	87 507	87 542	87 577	87 611	87 646
1.09	0.87 680	87 715	87 749	87 783	87 817	87 851	87 885	87 919	87 953	87 987
1.10	0.88 021	88 054	88 088	88 121	88 155	88 188	88 221	88 254	88 287	88 320
1.11	0.88 353	88 386	88 419	88 452	88 484	88 517	88 549	88 582	88 614	88 647
1.12	0.88 679	88 711	88 743	88 775	88 807	88 839	88 871	88 902	88 934	88 966
1.13	0.88 997	89 029	89 060	89 091	89 122	89 154	89 185	89 216	89 247	89 277
1.14	0.89 308	89 339	89 370	89 400	89 431	89 461	89 492	89 522	89 552	89 582
1.15	0.89 612	89 642	89 672	89 702	89 732	89 762	89 792	89 821	89 851	89 880
1.16	0.89 910	89 939	89 968	89 997	90 027	90 056	90 085	90 114	90 142	90 171
1.17	0.90 200	90 229	90 257	90 286	90 314	90 343	90 371	90 399	90 428	90 456
1.18	0.90 484	90 512	90 540	90 568	90 595	90 623	90 651	90 678	90 706	90 733
1.19	0.90 761	90 788	90 815	90 843	90 870	90 897	90 924	90 951	90 978	91 005
1.20	0.91 031	91 058	91 085	91 111	91 138	91 164	91 191	91 217	91 243	91 269
1.21	0.91 296	91 322	91 348	91 374	91 399	91 425	91 451	91 477	91 502	91 528
1.22	0.91 553	91 579	91 604	91 630	91 655	91 680	91 705	91 730	91 755	91 780
1.23	0.91 805	91 830	91 855	91 879	91 904	91 929	91 953	91 978	92 002	92 026
1.24	0.92 051	92 075	92 099	92 123	92 147	92 171	92 195	92 219	92 243	92 266
1.25	0.92 290	92 314	92 337	92 361	92 384	92 408	92 431	92 454	92 477	92 500
1.26	0.92 524	92 547	92 570	92 593	92 615	92 638	92 661	92 684	92 706	92 729
1.27	0.92 751	92 774	92 796	92 819	92 841	92 863	92 885	92 907	92 929	92 951
1.28	0.92 973	92 995	93 017	93 039	93 061	93 082	93 104	93 126	93 147	93 168
1.29	0.93 190	93 211	93 232	93 254	93 275	93 296	93 317	93 338	93 359	93 380
1.30	0.93 401	93 422	93 442	93 463	93 484	93 504	93 525	93 545	93 566	93 586
1.31	0.93 606	93 627	93 647	93 667	93 687	93 707	93 727	93 747	93 767	93 787
1.32	0.93 807	93 826	93 846	93 866	93 885	93 905	93 924	93 944	93 963	93 982
1.33	0.94 002	94 021	94 040	94 059	94 078	94 097	94 116	94 135	94 154	94 173
1.34	0.94 191	94 210	94 229	94 247	94 266	94 284	94 303	94 321	94 340	94 358
1.35	0.94 376	94 394	94 413	94 431	94 449	94 467	94 485	94 503	94 521	94 538
1.36	0.94 556	94 574	94 592	94 609	94 627	94 644	94 662	94 679	94 697	94 714
1.37	0.94 731	94 748	94 766	94 783	94 800	94 817	94 834	94 851	94 868	94 885
1.38	0.94 902	94 918	94 935	94 952	94 968	94 985	95 002	95 018	95 035	95 051
1.39	0.95 067	95 084	95 100	95 116	95 132	95 148	95 165	95 181	95 197	95 213
1.40	0.95 229	95 244	95 260	95 276	95 292	95 307	95 323	95 339	95 354	95 370
1.41	0.95 385	95 401	95 416	95 431	95 447	95 462	95 477	95 492	95 507	95 523
1.42	0.95 538	95 553	95 568	95 582	95 597	95 612	95 627	95 642	95 656	95 671
1.43	0.95 686	95 700	95 715	95 729	95 744	95 758	95 773	95 787	95 801	95 815
1.44	0.95 830	95 844	95 858	95 872	95 886	95 900	95 914	95 928	95 942	95 956
1.45	0.95 970	95 983	95 997	96 011	96 024	96 038	96 051	96 065	96 078	96 092
1.46	0.96 105	96 119	96 132	96 145	96 159	96 172	96 185	96 198	96 211	96 224
1.47	0.96 237	96 250	96 263	96 276	96 289	96 302	96 315	96 327	96 340	96 353
1.48	0.96 365	96 378	96 391	96 403	96 416	96 428	96 440	96 453	96 465	96 478
1.49	0.96 490	96 502	96 514	96 526	96 539	96 551	96 563	96 575	96 587	96 599

Table XI-2. (Continued)

x	0	2	4	6	8	x	0	2	4	6	8
1.50	0.96 611	96 634	96 658	96 681	96 705	2.00	0.99 532	99 536	99 540	99 544	99 548
1.51	0.96 728	96 751	96 774	96 796	96 819	2.01	0.99 552	99 556	99 560	99 564	99 568
1.52	0.96 841	96 864	96 886	96 908	96 930	2.02	0.99 572	99 576	99 580	99 583	99 587
1.53	0.96 952	96 973	96 995	97 016	97 037	2.03	0.99 591	99 594	99 598	99 601	99 605
1.54	0.97 059	97 080	97 100	97 121	97 142	2.04	0.99 609	99 612	99 616	99 619	99 622
1.55	0.97 162	97 183	97 203	97 223	97 243	2.05	0.99 626	99 629	99 633	99 636	99 639
1.56	0.97 263	97 283	97 302	97 322	97 341	2.06	0.99 642	99 646	99 649	99 652	99 655
1.57	0.97 360	97 379	97 398	97 417	97 436	2.07	0.99 658	99 661	99 664	99 667	99 670
1.58	0.97 455	97 473	97 492	97 510	97 528	2.08	0.99 673	99 676	99 679	99 682	99 685
1.59	0.97 546	97 564	87 582	97 600	97 617	2.09	0.99 688	99 691	99 694	99 697	99 699
1.60	0.97 635	97 652	97 670	97 687	97 704	2.10	0.99 702	99 705	99 707	99 710	99 713
1.61	0.97 721	97 738	97 754	97 771	97 787	2.11	0.99 715	99 718	99 721	99 723	99 726
1.62	0.97 804	97 820	97 836	97 852	97 868	2.12	0.99 728	99 731	99 733	99 736	99 738
1.63	0.97 884	97 900	97 916	97 931	97 947	2.13	0.99 741	99 743	99 745	99 748	99 750
1.64	0.97 962	97 977	97 993	98 008	98 023	2.14	0.99 753	99 755	99 757	99 759	99 762
1.65	0.98 038	98 052	98 067	98 082	98 096	2.15	0.99 764	99 766	99 768	99 770	99 773
1.66	0.98 110	98 125	98 139	98 153	98 167	2.16	0.99 775	99 777	99 779	99 781	99 783
1.67	0.98 181	98 195	98 209	98 222	98 236	2.17	0.99 785	99 787	99 789	99 791	99 793
1.68	0.98 249	98 263	98 276	98 289	98 302	2.18	0.99 795	99 797	99 799	99 801	99 803
1.69	0.98 315	98 328	98 341	98 354	98 366	2.19	0.99 805	99 806	99 808	99 810	99 812
1.70	0.98 379	98 392	98 404	98 416	98 429	2.20	0.99 814	99 815	99 817	99 819	99 821
1.71	0.98 441	98 453	98 465	98 477	98 489	2.21	0.99 822	99 824	99 826	99 827	99 829
1.72	0.98 500	98 512	98 524	98 535	98 546	2.22	0.99 831	99 832	99 834	99 836	99 837
1.73	0.98 558	98 569	98 580	98 591	98 602	2.23	0.99 839	99 840	99 842	99 843	99 845
1.74	0.98 613	98 624	98 635	98 646	98 657	2.24	0.99 846	99 848	99 849	99 851	99 852
1.75	0.98 667	98 678	98 688	98 699	98 709	2.25	0.99 854	99 855	99 857	99 858	99 859
1.76	0.98 719	98 729	98 739	98 749	98 759	2.26	0.99 861	99 862	99 863	99 865	99 866
1.77	0.98 769	98 779	98 789	98 798	98 808	2.27	0.99 867	99 869	99 870	99 871	99 873
1.78	0.98 817	98 827	98 836	98 846	98 855	2.28	0.99 874	99 875	99 876	99 877	99 879
1.79	0.98 864	98 873	98 882	98 891	98 900	2.29	0.99 880	99 881	99 882	99 883	99 885
1.80	0.98 909	98 918	98 927	98 935	98 944	2.30	0.99 886	99 887	99 888	99 889	99 890
1.81	0.98 952	98 961	98 969	98 978	98 986	2.31	0.99 891	99 892	99 893	99 894	99 896
1.82	0.98 994	99 003	99 011	99 019	99 027	2.32	0.99 897	99 898	99 899	99 900	99 901
1.83	0.99 035	99 043	99 050	99 058	99 066	2.33	0.99 902	99 903	99 904	99 905	99 906
1.84	0.99 074	99 081	99 089	99 096	99 104	2.34	0.99 906	99 907	99 908	99 909	99 910
1.85	0.99 111	99 118	99 126	99 133	99 140	2.35	0.99 911	99 912	99 913	99 914	99 915
1.86	0.99 147	99 154	99 161	99 168	99 175	2.36	0.99 915	99 916	99 917	99 918	99 919
1.87	0.99 182	99 189	99 196	99 202	99 209	2.37	0.99 920	99 920	99 921	99 922	99 923
1.88	0.99 216	99 222	99 229	99 235	99 242	2.38	0.99 924	99 924	99 925	99 926	99 927
1.89	0.99 248	99 254	99 261	99 267	99 273	2.39	0.99 928	99 928	99 929	99 930	99 930
1.90	0.99 279	99 285	99 291	99 297	99 303	2.40	0.99 931	99 932	99 933	99 933	99 934
1.91	0.99 309	99 315	99 321	99 326	99 332	2.41	0.99 935	99 935	99 936	99 937	99 937
1.92	0.99 338	99 343	99 349	99 355	99 360	2.42	0.99 938	99 939	99 939	99 940	99 940
1.93	0.99 366	99 371	99 376	99 382	99 387	2.43	0.99 944	99 942	99 942	99 943	99 943
1.94	0.99 392	99 397	99 403	99 408	99 413	2.44	0.99 944	99 945	99 945	99 946	99 946
1.95	0.99 418	99 423	99 428	99 433	99 438	2.45	0.99 947	99 947	99 948	99 949	99 949
1.96	0.99 443	99 447	99 452	99 457	99 462	2.46	0.99 950	99 950	99 951	99 951	99 952
1.97	0.99 466	99 471	99 476	99 480	99 485	2.47	0.99 952	99 953	99 953	99 954	99 954
1.98	0.99 489	99 494	99 498	99 502	99 507	2.48	0.99 955	99 955	99 956	99 956	99 957
1.99	0.99 511	99 515	99 520	99 524	99 528	2.49	0.99 957	99 958	99 958	99 958	99 959
2.00	0.99 532	99 536	99 540	99 544	99 548	2.50	0.99 959	99 960	99 960	99 961	99 961

Table XI-2. (Continued)

x	0	1	2	3	4	5	6	7	8	9
2.5	0.99 959	99 961	99 963	99 965	99 967	99 969	99 971	99 972	99 974	99 975
2.6	0.99 976	99 978	99 979	99 980	99 981	99 982	99 983	99 984	99 985	99 986
2.7	0.99 987	99 987	99 988	99 989	99 989	99 990	99 991	99 991	99 992	99 992
2.8	0.99 992	99 993	99 993	99 994	99 994	99 994	99 995	99 995	99 995	99 996
2.9	0.99 996	99 996	99 996	99 997	99 997	99 997	99 997	99 997	99 997	99 998
3.0	0.99 998	99 998	99 998	99 998	99 998	99 998	99 998	99 998	99 999	99 999

These boundary conditions apply for the case when a body has a constant composition (Fig. XI-14). The body situated to the left retains a constant composition and material from this body diffuses to the right. Consider that the body to the left is a granite magma and that the body to the right is a host rock. The diffusion in the magma will be much faster than in the rock, so that the concentration near the contact may be considered as approximately constant. For $D = 10^{-10}$ cm²/s one obtains the concentration curves shown in Fig. XI-15. Now let:

$D = 10^{-10}$

$x = 100$ cm

$c_1 = 100\%$

$t = 10^6$ y $= 3.1536 \cdot 10^{13}$ s

For $x = 100$ cm one gets:

$$\frac{x}{2 \sqrt{Dt}} = 0.8904$$

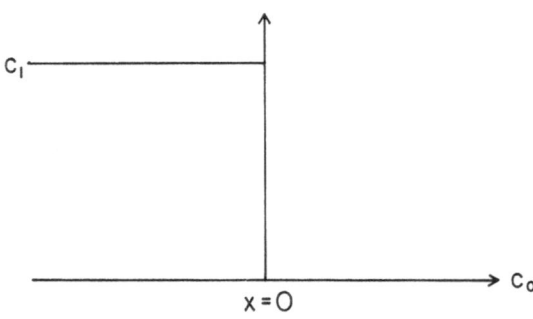

Fig. XI-14. Boundary conditions for example 1

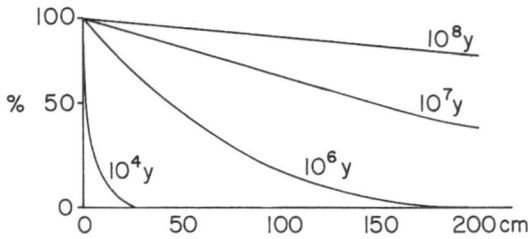

Fig. XI-15. Calculated concentration curves for example 1

Using Table XI-2 the value of the error function is estimated as 0.79184, thus:

c (100 cm, 10^6 yr) = 20.82%

The concentration has increased from 0% to 20.82% after 1 m. y. This example suggests that diffusion processes occur at a very slow rate in rocks. Comparing the values of D for silicate liquids and solids given in Table XI-1, it is apparent that material transport by diffusion can only occur over very short distances of the order of a few meters even on a geological time scale. The calculation here is simplified in one respect, the diffusion is assumed to be volume diffusion. In the natural rocks the diffusion will also occur along grain boundaries, and this type of diffusion is somewhat faster than volume diffusion, but the diffusion rate is still very small.

Linear Diffusion 2

Boundary conditions:

$c = c_1$ for $x < 0$ and $t = 0$
$c = c_0$ for $x > 0$ and $t = 0$
$c = f(x, t)$ for all x and $t > 0$

Equation:

$$c(x, t) = \frac{c_1 - c_0}{2}\left(1 - erf\left(\frac{x}{2\sqrt{Dt}}\right)\right) \tag{33}$$

Let us apply the equation using the same value for D as in case 1:

$c_1 = 100\%$
$c_0 = 0\%$
$x = 100$ cm
$D = 10^{-10}$ cm^2/s
$t = 10^6$ yrs

Thus, erf(0.8904)=0.79184, and c(100 cm, 10^6 yr)= 10.408%. The concentration obtained is smaller here because the concentration at x=0 cm decreases with time,

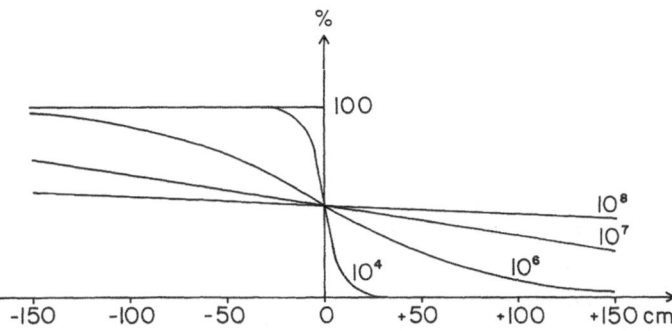

Fig. XI-16. Calculated concentration curves for example 2

while it was constant in the former case (Figs. XI-15, XI-16). It might be noted that the concentration at $x=0$ is equal to $\frac{1}{2}(c_1 - c_0)$ for an extended period of time. This model assumes that the diffusion coefficient D is the same in both bodies. This approximation is fairly valid if they both are rocks at the same temperature. The equation can be used if one wants to calculate the diffusive exchange of material between two rocks that undergo metamorphism, or two types of magmas that are in contact.

Linear Diffusion 3

Boundary conditions:

$c = c_1$ for $|x| < h$,   and $t = 0$
$c = c_0$ for $|x| > h$,   and $t = 0$
$c = f(x, t)$ for all x   .and $t > 0$

Equation:

$$c(x, t) = \frac{c_1}{2}\left(\text{erf}\left(\frac{h + x}{2\sqrt{Dt}}\right) + \text{erf}\left(\frac{h - x}{2\sqrt{Dt}}\right)\right)$$

(34)

This equation is used for a slab formed body like a dyke or a sill (Fig. XI-17). The variation in concentration for $D = 10^{-10}$ is shown in Fig. XI-18.

Radial Diffusion

Boundary conditions:

Radius of sphere: $r = a$
$c = c_1$ for $r < a$ and $t = 0$
$c = c_0$ for $r > a$ and all t
$c = f(x, t)$ for $r < a$ and $t > 0$ (Fig. XI-19)

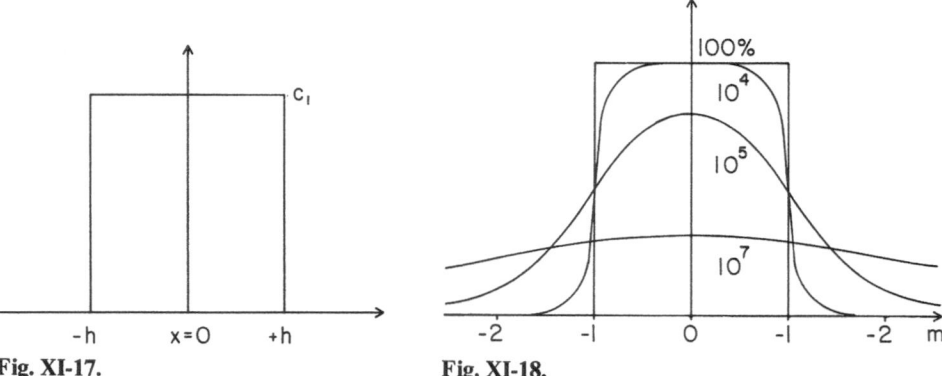

Fig. XI-17.   Fig. XI-18.

Fig. XI-17. Boundary conditions for example 3
Fig. XI-18. Calculated concentration curves for linear diffusion, example 3. The concentration curves have been calculated for periods of time of 10^4 yr, 10^5 yr, and 10^7 yr

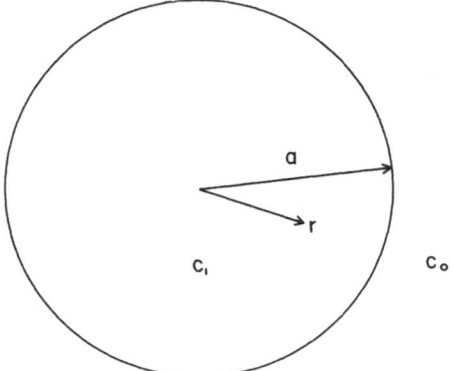

Fig. XI-19. Boundary conditions for spherical diffusion. The concentration is initially c_1 within the sphere and c_0 outside the sphere

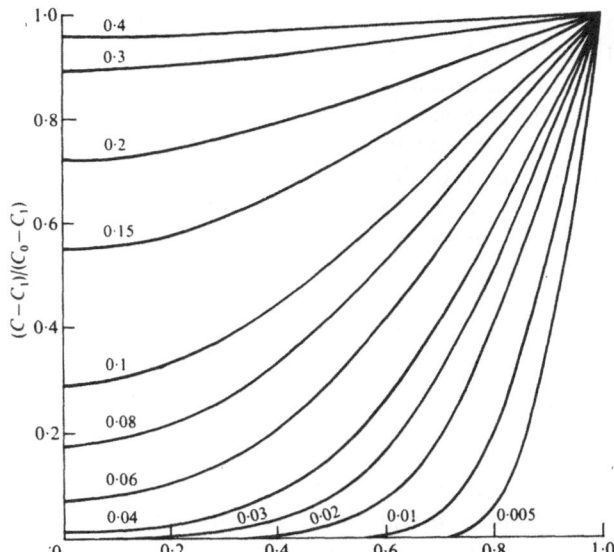

Fig. XI-20. The curves used for solutions to Eq. (35). The *abscissa* is the ratio r/a, while the *curves* are for different values of Dt/a^2

Equation (Crank 1975):

$$\frac{c - c_1}{c_0 - c_1} = 1 + \frac{2a}{\pi r} \sum_{n=1}^{\infty} \frac{(-1)^n}{n} \sin\left(\frac{n \pi r}{a}\right) \exp\left(\frac{- D n^2 \pi^2 t}{a^2}\right) \qquad (35)$$

This equation may be solved with the required degree of accuracy by calculating the sufficient number of terms in the summation. A quite accurate solution is obtained using the curves in Fig. XI-20. The right side of Eq. (35) is estimated from the ratio r/a and the ratio Dt/a^2. An example is shown in Fig. XI-21 which shows the variations in concentration for a spherical particle with a radius of 1 mm and a D value of 4.10^{-11} cm²/s. The boundary conditions imply that the concentration around the sphere is constant. The model may be used for a calculation of the change in chemical composition of crystals or gas bubbles. The minerals in a rock undergoing metamorphism may either become unstable or adjust their composition when the P–T

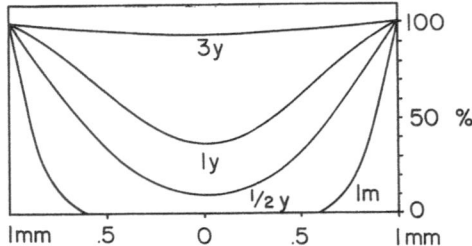

Fig. XI-21. Concentration curves for diffusion in olivine at 1400 °C after 1 mo, ½ yr, 1 yr, and 3 yr. The percentages are for the relative difference. The forsterite content of the spherical olivine crystal is 90 mol% Fo, and the D value used is $4 \cdot 10^{-11}$

conditions change. In the latter case the minerals will exchange materials with their surroundings, and Eq. (35) will give a determination about the time required for equilibration. As an example we might consider the processes anatexis and partial melting which typically occur at temperatures near 900 °C and 1400 °C, respectively. The D value for orthoclase, one of the mineral constituents of gneiss is about $5 \cdot 10^{-11}$ at 900 °C, and the value for olivine with 90% Fo at 1400 °C is $4 \cdot 10^{-11}$ (cf. Table XI-1). Some petrogenetic theories assume that these two processes do not occur at equilibrium conditions. Considering that both take place during a considerable time interval, we may ask what is actually the time required for equilibration. Consider a crystal with a radius of 1 mm, and a D value of $4 \cdot 10^{-11}$. The variation in concentration at different times is shown in Fig. XI-21, which was estimated using the curves of Fig. XI-20. The initial difference is assumed as 100 relative percent. This 100% may be equivalent to a real weight percentage of say 5 or 30 wt%. If the relative percentage is estimated at 25% and the total real difference is 10 wt%, then the percentage is estimated at 2.5 wt%. When is equilibrium then obtained? Strictly speaking, the perfect equilibrium will first be obtained after an infinite amount of time has passed. The accuracy of chemical analyses is about 2%, so we may say that equilibrium has been obtained when the composition of the minerals is within 2% of 100%. Using this approximate, but realistic criterion for the equilibrium composition, it is evident from Fig. XI-21 that equilibrium has been obtained after about 5 yr. Thus, a grain of alkali feldspar will adjust its composition within a few years, and this mineral will, therefore, equilibrate during anatexis which occurs during a period of at least some m. y. The same applies for an olivine crystal in the mantle. The mantle material undergoing partial melting ascends with a velocity of the order of 10 cm/y, and equilibration is, therefore, assured as it only takes a few years for the equilibration. Calculations like this one do not give an accurate result, but afford a valuable evaluation of one of the parameters involved, in this case the time required for equilibration.

8 Interfacial Energy

The minerals of a rock are separated from each other by their surfaces. The surface layer which separates the crystals is only a few atomic diameters thick, and constitutes a very small part of the crystals. The surface layer between a phenocryst and its enclosing magmatic liquid is also extremely thin, and of the same order of magnitude. The layer of atoms which separates a crystal from other crystals or a liquid is called the interfacial layer. In both cases the structure of the interfacial layer of the crystals differs from that of the interior part of the crystal and its thermodynamic properties are, therefore, different. The interfacial energy of the minerals does not influence the phase equilibria in any significant manner, and may consequently be neglected insofar the majority of calculations deal with thermodynamic equilibria. The interfacial energy plays a major role in the nucleation and growth kinetics of minerals and for textural features like grain habits as well as for the distribution of the melts formed by anatexis and partial melting.

The atoms of a crystalline surface are more loosely bonded than the atoms in the interior of the crystal. They are in a state which is intermediate between that of the crystal and the liquid. The atoms in the interfacial layer, therefore, possess a higher energy than the atoms in the bulk part of the crystal, and the formation of a surface consequently requires energy. This relationship is clearly demonstrated by a soap bubble. The surface tension of a soap bubble will tend to contract the spherical surface so that the surface of the bubble becomes as small as possible. This contraction will increase the pressure inside the bubble, and the air pressure is, therefore, larger inside the bubble than on the outside. If the size of the bubble is increased, the pressure inside the bubble must be increased by an amount determined by the surface tension of the bubble. The energy relationships between a surface and its energy can be evaluated by considering a liquid film (Fig. XI-22).

Experiments show that the energy required to increase the area of the film, x1 is proportional with the area. The energy is given by $F\,dx$, where F is the force. The increase in area is given by $1 \cdot dx$. As the film has two surfaces, the area should be multiplied by a factor 2. We, thus, have:

$$\frac{F\,dx}{21\,dx} = \gamma \tag{36}$$

where γ is the surface tension per unit area given in dyn/cm^2. The interfacial energy (erg/cm^2) is generally identical to the surface energy, the two parameters differ only at very low temperatures. The interfacial energy is a material constant which depends on pressure and temperature as well as the phase involved. The interfacial energy between two crystals will not only depend on their composition, but also on their mutual orientations. The interfacial energies between the minerals of a rock will, therefore, not be constant, but will display some variation depending on the orientation of the grains. The interfacial energies for materials of petrological interest have only been measured for very few cases. Some values for metals and minerals are listed in Table XI-3.

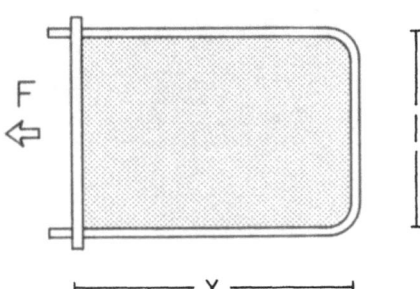

Fig. XI-22. The liquid film has the area lx and the force F required to increase the area depends on the surface tension

Table XI-3. Interfacial energies (Chalmers 1964; Cooper and Kohlstedt 1982)

Material	Solid-vapor erg/cm²	Solid-liquid erg/cm²	Solid/solid erg/cm²
Cu	1700	177	
Ag	1200	126	
Au	1400	132	
Hg		24.4	
Pb		33	
Ni		255	
Co		234	
Fe		204	
Pt		240	
Olivine/basalt Melt		500	
Olivine (010)			1600
Olivine (010)			

9 Latent Heat and Interfacial Energy

The atoms of the surface layer of a crystal are in a state which are intermediate between the crystalline and the liquid states. The bonds between the atoms are not as strong as in the solid, but are, on the other hand, somewhat stronger than those of the liquid. The energy of the atoms of the surface must, thus, be less than the latent heat per atom, which is the total amount of energy required to move one atom from the interior of the crystal to the liquid. If the latent heat/mol is L, then the latent heat per atom, λ is given by L/A, where A is Avogadros' number. The energy of the atoms in the interfacial layer is, thus, higher than the energy of the atoms of the solid, but less than λ. The transition from the interior of the crystal to the liquid may be considered to occur in two steps. The first step consists of moving the atom from the interior part of the crystal to the surface. The second step is the transition of the atom from the surface layer to the liquid. The energy relationships of the total transition may be demonstrated considering a cubic close-packed crystal. The atoms of

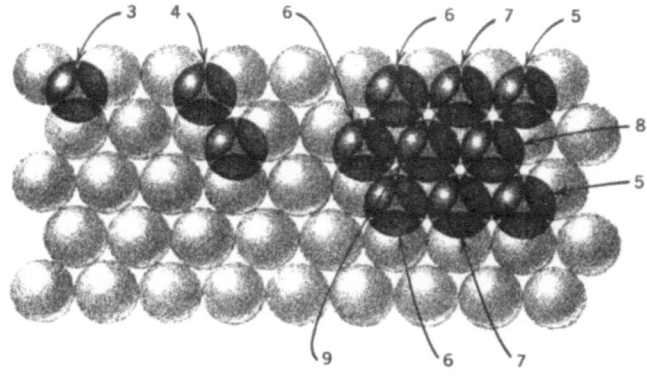

Fig. XI-23. Atoms on a cubic close-packed surface. The atoms on the surface may be surrounded by three to nine other atoms. The atoms within the crystal are surrounded by 12 atoms (Chalmers 1964)

the surface may be neighbored by three to nine other atoms, while the atoms of the interior part of the crystal are surrounded by 12 other atoms (Fig. XI-23). The energies of the atoms are related to the number of bonds they form with the surrounding atoms. The removal of the atoms from the crystal, therefore, requires the breaking of 12 bonds and the energy required per bond is on the average $1/12\ \lambda$. The energy of the atoms in the surface which is surrounded by nine atoms is, therefore, $(3/12)\ \lambda$ higher than those of the interior part of the crystal. These surface atoms may enter the liquid if the restoring nine bonds are broken which requires an additional ener-

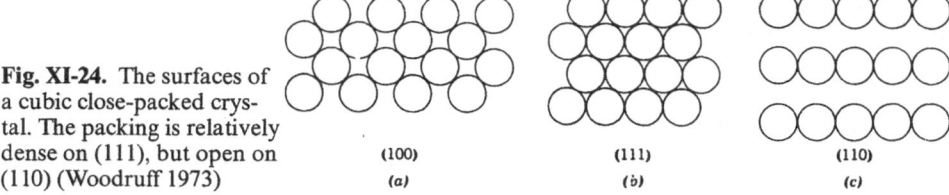

Fig. XI-24. The surfaces of a cubic close-packed crystal. The packing is relatively dense on (111), but open on (110) (Woodruff 1973)

(100)   (111)   (110)
(a)   (b)   (c)

gy of $(9/12)\ \lambda$ per atom. The atoms of the surface gain this energy from the surrounding atoms of the solid and liquid which hit the surface atoms by their thermal motion. The formation of the surface, thus, requires an energy of $\frac{1}{4}\ \lambda$ per atom and the transition from the surface to the liquid requires an energy of $\frac{3}{4}\ \lambda$ per atom. The surface energy is, thus $0.25\ \lambda$ according to this simple model. The surface energies for the solid–liquid interfaces have mostly been derived from nucleation experiments, and the results show that the surface energy rather is about $0.4\ \lambda$. The difference in theoretical and experimental value is caused by the presence of atoms at edges and corners on the crystalline nuclei which have higher energies. At the edge of a cube, an atom is surrounded by seven atoms, and at the corner it is surrounded by five atoms. The energy of an edge atom is $(5/12)\ \lambda$ or $0.42\ \lambda$ and the energy of a corner atom is $(7/12)\ \lambda$ or $0.58\ \lambda$. The surface energies obtained by the nucleation method must, therefore, be considered as being slightly too high.

The interfacial energy of a liquid drop is the same in all directions because the atoms of the liquid are distributed in an arbitrary manner. For crystals the situation is different, even for cubic crystals, as the atoms display a regular arrangement, and the density of atoms will be different on the different faces as shown in Fig. XI-24, which shows the distribution of atoms of a face-centered cubic structure. The interfacial energy of the (111) face will be smaller than that of the (100) or (110) face because the atoms of the (111) face are more closely packed. The interfacial energy will, therefore, vary from face to face. The interfacial energies obtained from the nucleation experiments are, therefore, average values for the nuclei as a whole.

10 Supercooling

The nucleation of crystals in a melt can only occur if the melt is supercooled and the growth of the crystals has to occur also at supercooled conditions. The supercooled condition denotes a deviation from true thermodynamic equilibrium conditions, and the problem, therefore, arises as to whether or not the experimentally determined phase relations can be applied directly for a determination of the phase relations of rocks. Fortunately, this appears to be the case. The amount of supercooling involved by the crystallization of plutonic rocks appears to be small, less than 20–50 °C and the nucleation and growth of crystals is controlled by the phase relations by small degrees of supercooling, because the free energy difference between magma and crystals which governs the kinetics of the crystallization is related to the amount of supercooling. The phase relations will, therefore, also exsert control on the crystallization by supercooled conditions.

The quantitative amount of supercooling may be expressed as the difference between the actual temperature T_a, and the equilibrium temperature T_e. The supercooling, ΔT is then given by:

$$\Delta T = T_e - T_a \tag{37}$$

This method is the most convenient and direct one, and is used in this book. Another method consists of the determination of the relative supercooling ψ. Let the total crystallization interval of a liquid be from T_e to T_s, where T_s is the solidus temperature. The relative supercooling is then:

$$\psi = \frac{\Delta T}{T_e - T_s} \tag{38}$$

The prevalence of supercooled crystallization conditions may be demonstrated considering the crystallization conditions of a mineral like plagioclase. Consider first that the crystallization occurs at the equilibrium temperature T_e (Fig. XI-25 a). A plagioclase liquid of composition l_e has been cooled and reaches the liquidus curve at T_e. According to the equilibrium conditions, the first crystals should form at T_e.

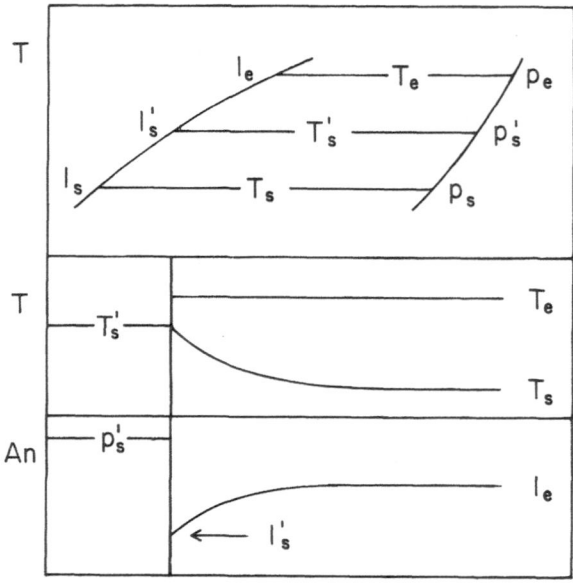

Fig. XI-25. Temperature and compositional relationships by supercooled or supersaturated crystallization demonstrated by the plagioclase system

However, this is impossible if the temperature of the liquid remains at T_e. The crystallization can only occur if the crystals can dissipate their heat into the liquid. As the temperature of the liquid is T_e and the crystals are unstable at any temperatures above T_e, there can be no crystallization, and no crystals will form at T_e.

Now let the temperature of the liquid drop a little, and let the first crystals nucleate at T_s. The nuclei will now grow by dissipation of heat into the surrounding liquid. In addition, the composition of the liquid near the crystals must be changed, because the crystals contain more anorthite than the liquid (Fig. XI-25 b). The composition of the liquid at some distance from the crystal is still l_e, but l'_s at the surface of the crystal. The actual temperature of the liquid increases toward the surface of the plagioclase crystals, and is T'_s at the surface. The composition of the liquid will be somewhere between l_s and l_e, and the composition of the plagioclase between p_s and p_e. The compositions of the liquid and the plagioclase will be determined by the cooling rate and the rate of heat dissipation and the diffusion rate in the liquid. Consequently, the compositions of phenocrysts are to some degree determined by the cooling rate of the magma. The compositions of phenocryst in lavas and dykes may, thus, deviate somewhat, but probably not much from the compositions estimated from phase equilibria.

11 Nucleation

Crystals are formed in two consecutive processes, nucleation and growth. The formation of crystals from a melt begin with the formation of small clusters of molecules, called nuclei. They have a crystalline structure, and their diameter is of the order of 100 Å. These nuclei may then grow and form crystals of macroscopical size. The basic theory for nucleation is well-known, but the theory involves such simplifications that it should be considered an ideal model. The theory cannot yet be applied quantitatively to petrogenetic problems because the physical constants controlling the process are not known. However, considering that nearly all rocks consist of crystals, the basic theory for crystallization should be known by the petrologist. The textural features of rocks are mainly related to the kinetics of crystallization, and it is likely that a firm understanding of the crystallization theory will contribute to the development of petrogenetic theory. It was not until the late 1970's that work has begun in this field. This is a natural consequence of the development of petrology. The fundamental phase relations of rocks constitutes the most significant research for petrogenetic theory. After the phase relations have been estimated, a natural next step is an estimate of the kinetics involved in phase changes. Only a few investigations have been done, most notably by Gibb (1974) and Donaldson (1979).

12 Thermodynamics of Nucleation

The kinetic relationships of nucleation are estimated from the variation in the free energies involved. We will now consider the process where a small amount of liquid forms a crystal nuclei. Let the free energy per volume unit of the liquid be γ_1 and that of the solid γ_s. Assuming that the crystal has a spherical shape, its volume is given by $(\frac{4}{3}) \pi r^3$, where r is the radius of the sphere. If a nuclei of radius r_1 is formed, then the total free energy of the liquid will decrease with an amount ΔG_1 given by:

$$\Delta G_1 = \frac{4}{3} \pi r_1^3 \gamma_1 \qquad (39)$$

The free energy related to the solid will stem from two terms, one related to the free energy of the bulk of the solid, and another related to the interfacial energy of the solid, σ. The nuclei with radius r_s will bring about the following change in free energy:

$$\Delta G_s = \frac{4}{3} \pi r_s^3 \gamma_s + 4 \pi \sigma r_s^2 \qquad (40)$$

where σ_s is the surface free energy for the nuclei. The total change in free energy caused by the formation of the nuclei is then given by the subtraction of Eq. (39) from (40):

$$\Delta G = \frac{4}{3} \pi (r_s^3 \gamma_s - r_1^3 \gamma_1) + 4 \pi r_s^2 \sigma \qquad (41)$$

When dealing with this equation, it is generally assumed that $r_s = r_1$. It is, thus, assumed that the liquid and the solid have the same density. The difference in density between a silicate and its melt is typically about 10%, which is equivalent to a difference in radius of 3%. Thus, assuming $r_s = r_1$ one gets:

$$\Delta G = \frac{4}{3} \pi r^3 \Delta G_v + 4 \pi r^2 \sigma \qquad (42)$$

The first term shows the change in free energy related to volume, and is negative below the equilibrium temperature, and positive above. The second term, the surface term, is always positive. The equation as a whole shows the variation in free energy as function of the radius of the nuclei (Fig. XI-26). For $r = 0$, ΔG is also zero. As r increases the value of ΔG increases until a maximum is reached, whereafter ΔG decreases in value and becomes negative. The free energy at the maximum is called the critical free energy, ΔG^*. Generally, a reaction will only occur from state I to state II if the free energy of state II is less than that of I. As evident from Fig. XI-26, the formation of nuclei causes an increase in the free energy for r values between zero and r^*. This is caused by the dominance of the surface term for small values of r, at larger values the volume term is the dominant one. According to this result it is somewhat surprising that nuclei might form at all. However, the total change in free energy is negative for r values larger than r_0, and the free energy is decreasing for r values larger than r^*. The initial increase in free energy is, thus, a reaction barrier which delays the process. The particles of a thermodynamic system are not in the same state of energy, the energies of the particles in a liquid will be distributed over a wide range of energy levels. Some of the liquid molecules will have such low energies that they cluster together and the size of these clusters increases with decreasing temperature. At temperatures below the liquidus temperature some of these clusters

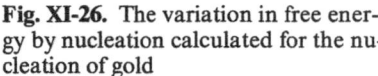

Fig. XI-26. The variation in free energy by nucleation calculated for the nucleation of gold

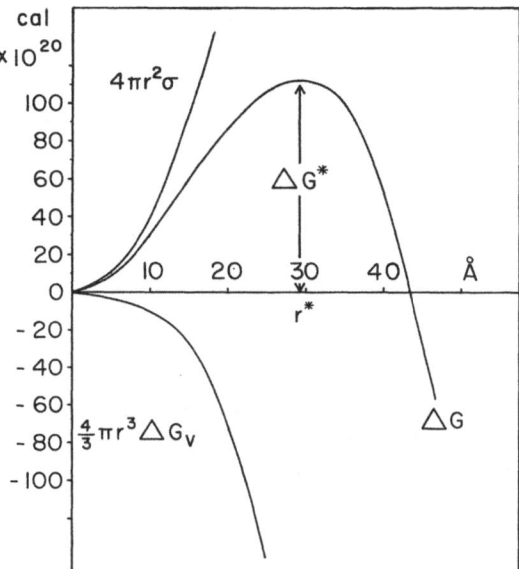

will attain the size of the critical nucleus and result in the crystallization of the liquid. The clusters with a smaller size than r* are called embryos, while those with sizes between r*and r_0 are called nuclei. It is, thus, clear that the rate of nuclei formation will be dependent on the concentration of the nuclei with the critical size. For $\Delta G_v = 0$ the right side of Eq. (42) is always positive and there can be no nucleation. The embryos that form at the liquidus temperature will be unstable and unable to attain a large size. The condition for nucleation to proceed is that the liquid is supercooled or that ΔG_v has become negative.

The nucleation process in a cooling melt proceeds as follows: Above the melting temperature the embryos in the liquid are small and few in number. As the melting temperature is approached they increase both in size and number. At the liquidus temperature there will be large embryos in the liquid, but none of them will reach the critical size. When the temperature decreases below the liquidus, the first embryos will attain the critical size and the first nuclei will be formed. A few tens of degrees below the liquidus the nucleation rate will be extremely small. For silicate melts a supercooling of 5° to 30°C is required before the nucleation proceeds at a measurable rate. If the temperature remains just below the liquidus, only a few nuclei per volume unit will be formed and the crystallizate becomes coarse grained. At large degrees of supercooling the nucleation rate increases and the crystallizate becomes more fine-grained. This variation in crystal size is observed in dykes, where the chilled margins either are glassy or fine-grained, while the center of the dykes is medium-grained.

Returning now to Eq. (42) we will derive some parameters related to the nucleation process. At the critical radius there is a maximum in the ΔG curve so that:

$$\frac{\delta \Delta G}{\delta r} = 0 \tag{43}$$

Hence, from Eq. (42):

$$\frac{\delta \Delta G}{\delta r} = 8 \pi r \sigma + 4 \pi r^2 \Delta G_v \tag{44}$$

so that:

$$r^* = \frac{-2\sigma}{\Delta G_v} \tag{45}$$

and:

$$\Delta G^* = \frac{16 \pi \sigma^3}{3 (\Delta G_v)^2} \tag{46}$$

The variation in ΔG_v may be calculated accurately from the variation in Δc_p (Lewis and Randall 1961), but an approximate value near T_f is estimated in the following manner. At equilibrium the condition is:

$$\Delta G = \Delta H - T \Delta S = 0 \tag{47}$$

At the melting temperature T_f the difference in enthalpy is equal to the latent heat of melting ΔH_f, and the difference in entropy is then:

$$\Delta S = \frac{\Delta H_f}{T_f} \tag{48}$$

Hence,

$$\Delta G_v = \frac{\Delta H_f \Delta T}{T_f} \tag{49}$$

The value of r for which ΔG becomes zero is estimated from:

$$\Delta G = \frac{4}{3} \pi r_0^3 \Delta G_v + 4 \pi r_0^2 \sigma = 0 \tag{50}$$

thus:

$$r_0 = \frac{-3\sigma}{\Delta G_v} = \frac{3}{2} r^* \tag{51}$$

13 The Nucleation of Gold

The above theory for nucleation will now be demonstrated considering the nucleation of gold. Gold is one of the few minerals for which all the required constants are known, and the theory will, therefore, be applied for this mineral rather than a silicate. We will calculate the variation in r^* and ΔG as function of supersaturation, ΔT. The constants required for this calculation are shown in Table XI-4. We will first have to calculate ΔG_v using Eq. (49) and the values given in Table XI-4, the result being:

$$G_v = 2.21 \cdot \Delta T \text{ (cal/°K mol)} \tag{52}$$

Table XI-4. The physical constants for gold

$Q = 53,000$ cal/mol
$\Delta H_f = 2,955$ cal/mol
$\sigma_{sl} = 132$ erg/cm^2
$T_f = 1,336\,°K$
$\bar{v}_s = 10.215$ cm^3 (molar volume of solid)
$\bar{v}_l = 10.736$ cm^3 (molar volume of liquid)
1 cm $= 10^8$ Å
1 erg $= 2.39006 \cdot 10^{-8}$ cal
$k = 3.2995697 \cdot 10^{-24}$ cal/degree (Boltzmans' constant)
$h = 6.63 \cdot 10^{-34}$ joule $\cdot$ s (Plancks' constant)
1 cal $= 4.184$ joule
$r_{Au} = 1.439$ Å (atomic radius of gold atom)
$v = 16.9614$ Å^3/atom the volume per atom in solid gold
$N = 0.0561$ atom/Å^3, the number of atoms per Å^3 in liquid

Before this result can be applied we will have to estimate the value of ΔG_v for a volume unit. Let the unit of volume be 1 Å^3 then:

$$\Delta G_v = 2.165 \cdot 10^{-25}\,\Delta T \text{ (cal/°K Å}^3) \tag{53}$$

Having obtained the dependence of ΔG_v on ΔT we can calculate r* from Eq. (45), and get for $\sigma = 3.1549 \cdot 10^{-22}$ cal/Å^2:

$$r^* = \frac{2914}{\Delta T} \text{ (Å °K)} \tag{54}$$

so that for $\Delta T = 100\,°K$ we have r $= 29.14$ Å (Fig. XI-26). According to Eq. (51), the value for r_0 is given by:

$$r_0 = \frac{4371}{\Delta T} \text{ (Å °K)} \tag{55}$$

Both r* and r_0 are inversely proportional with the degree of supercooling, and they decrease with decreasing temperature. The variation in $\Delta G*$ is estimated from Eq. (46), and we get:

$$G^* = \frac{1.122 \cdot 10^{-14}}{(\Delta T)^2} \text{ cal} \tag{56}$$

For $\Delta T = 100\,°K$ we get $\Delta G* = 112.2 \cdot 10^{-20}$ cal (Fig. XI-26). Since the critical free energy is inversely proportional with the square of the degree of supersaturation its value decreases rapidly with increasing supercooling.

The variation in ΔG for an embryo or nuclei with varying radius is calculated from Eq. (42) and the value for ΔG_v obtained from Eq. (49), and the result is for $\Delta T = 100\,°K$:

$$\Delta G = 3.96456 \cdot 10^{-21}\,r^2 - 9.06998 \cdot 10^{-26}\,r^3 \tag{57}$$

The curve for ΔG is shown in Fig. XI-26 which also shows the variation in the two terms of Eq. (57).

The present calculations afford equations which determine some of the parameters specifying the nucleation process. However, the kinetics of nucleation cannot be estimated from these equations, and we will, therefore, deal with this subject now.

14 Kinetics of Nucleation

The nucleation rate for phase transformations was first estimated for the vapor/liquid transformation by Volmer and Weber (1926), and their model was later improved by Becker and Døring (1935). The nucleation rate for condensed systems was derived by Turnbull and Fisher (1949), and their theory is briefly sketched below using the simplified representation given by Walton (1969).

The thermodynamic derivations considered above have defined the conditions for nucleation to occur, and allow a calculation of the critical parameters if the appropriate constants are known. In order to estimate the rate by which the nucleation occurs, we will have to estimate two additional parameters, one being the number of nuclei with critical size and the other the frequency of transformation of critical nuclei to stable and growing nuclei.

The embryos increase in size by a series of successive additions of molecules to the embryos. First, two molecules are joined together, whereby the first embryo is formed:

$$a + a = A_1$$

This process is reversible as A_1 may split into two molecules again. By the addition of a third molecule to A_1 a second embryo is formed, which may increase in size by the addition of a fourth molecule, and so on:

$$a + A_1 = A_2$$
$$a + A_2 = A_3$$
$$a + A_3 = A_4$$
$$a + A_4 = A_5$$
$$a + A_n = A_{n+1}$$

Below the liquidus these reactions will proceed to the right with a faster rate than to the left, large embryos being produced, some of which attain the critical size. The nucleation rate is then determined by rate of generation of postcritical nuclei:

$$a + A^* = A_{i+1}$$

Let R be the rate by which a molecule crosses an interface, and let S^* be the surface area of a critical nuclei. The frequency of transformation of critical nuclei is then given by (Walton 1969):

$$\nu = R S^* \tag{58}$$

The number of critical nuclei per unit volume is given by the number N^*. The rate of transformation of critical nuclei must be given by the product:

$$I = \nu N^* \tag{59}$$

or from Eq. (58):

$$I = R S^* N^* \tag{60}$$

The parameters of this equation will now be estimated. The frequency of molecular collisions is given by:

$$R = \frac{kT}{h} \exp\left(\frac{-Q}{kT}\right) \tag{61}$$

where:

    k: Boltzmans' constant
    h: Plancks' constant
    Q: the activation energy of diffusion

The surface area of the critical nuclei is estimated from the critical radius r:

$$S^* = 4\pi (r^*)^2 = \frac{16\pi\sigma^2}{\Delta G_v^2} \tag{62}$$

The number of critical nuclei is estimated from Boltzmans' distribution law (Wall 1974). This law is based on statistical mechanics and defines the number of thermo-dynamic units which attain a certain energy under the condition that the entropy should be at a maximum. Let N be the number of liquid molecules per unit volume. If the difference in free energy between the embryo consisting of i molecules and the molecules of the liquid is ΔG^*, then the number of critical nuclei is given by:

$$N^* = N \exp\left(\frac{-\Delta G^*}{kT}\right) \tag{63}$$

The nucleation rate will, thus, be determined by a function which contains the product of two exponential functions:

$$I = \frac{kT}{h} N \frac{16\pi\sigma^2}{\Delta G_v^2} \exp\left(\frac{-Q}{kT}\right) \exp\left(\frac{-\Delta G^*}{kT}\right) \tag{64}$$

The nucleation rate will be extremely sensitive on the two exponents, and the nucleation rate can only be determined satisfactorily accurate if the parameter of the exponents are known with a high degree of accuracy.

The value obtained so far for I has been based on equilibrium thermodynamics as we have used Boltzmans' distribution law for an estimate of N^*. This procedure is not strictly correct, as we are dealing with a situation where a reaction occurs, that is, when I is different from zero, then there is no equilibrium. The value for I should, therefore, be corrected, and the correction can be done by multiplying the right side of Eq. (64) with the so-called Zeldovich factor, Z (Walton 1969) which is given by:

$$Z = \frac{v\,\Delta G_v^2}{8\pi\sigma\,(\sigma kT)^{\frac{1}{2}}} \tag{65}$$

where:

    v: the volume per atom in the solid
    σ: the interfacial energy

By multiplying I with Z we finally get:

$$I = \frac{2v\,(kT\,\sigma)^{\frac{1}{2}}}{h} N \exp\left(\frac{-Q}{kT}\right) \exp\left(\frac{-\Delta G^*}{kT}\right) \tag{66}$$

The nucleation rate of gold may now be calculated using Eq. (64) and the data given in Table XI-4, the result being:

$$I = 3.8776 \cdot 10^{35} (T^{\frac{1}{2}}) \exp(-2.6671 \cdot 10^4/T) \exp(-3.4004 \cdot 10^9/\Delta T^2 T) \tag{67}$$

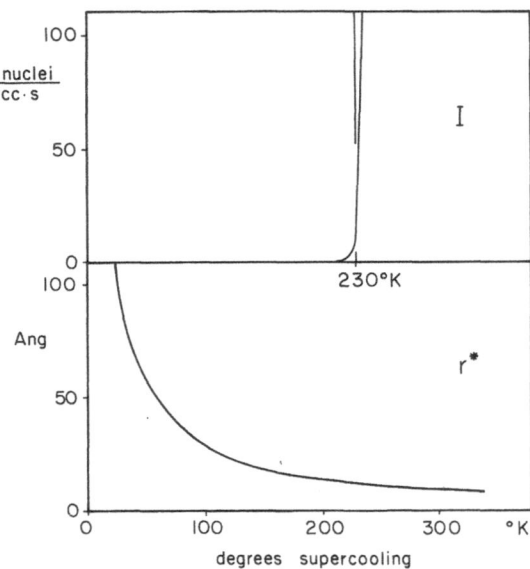

Fig. XI-27. The nucleation rate for gold (*I*) and the variation in the critical radius of the critial nuclei (*r**). The nucleation temperature estimated from experiments is 230 °K

where the unit of I is nuclei/cm³ s. The exponential form of this equation implies that I is very sensitive on the degree of supercooling and the temperature. There will be virtually no nucleation at small degrees of supercooling and nucleation does not occur at a measurable rate before the supercooling has reached 230 °K, which is the experimentally estimated degree of supercooling (Fig. XI-27). At this degree of supercooling the nucleation rate increases abruptly and becomes extremely large when the supercooling reaches 250 °K. The radius of the critical nuclei is about 10 Å for a supercooling of 230 °K, and the nuclei contains about 250 gold atoms. The agreement between the calculated nucleation rate and the experimental result for the supercooling is not completely accidental since the surface energy is estimated from nucleation experiments.

15 Nucleation in Magmas

The theory for nucleation considered so far describes the nucleation process under ideal circumstances, that is, under the condition that the nucleation only depends on the free energy and surface tension of the nuclei. The theory can be considered an ideal model for the nucleation process in the same manner as the Boyle-Mariotte's law applies for an ideal gas. The nucleation of magmatic liquids does not generally occur at such large degrees of supercooling as the ideal nucleation law predicts, because the nucleation is catalyzed by the presence of impurities and the wide variety of cations and polymeric units in the magmatic liquids. The effect of the catalyzing agents is to decrease the surface tension of the nuclei, and the degree of supercooling required for nucleation will, therefore, decrease. The degree of supercooling required for nucleation to occur will be different for the different minerals, but will probably be of the order of 1 °–50 °C in most cases. The variation in nucleation rate with increasing supercooling to be expected in magmatic liquids is shown in Fig. XI-28.

The nucleation rate is infinitely small just below the liquidus. It will increase with increasing degree of supercooling and reach a maximum a few hundred degrees below the liquidus, whereafter it will decrease and become zero several hundred degrees below the liquidus. The maximum in the nucleation curve is caused by the decrease in the diffusion rate with decreasing temperature. Consider the two exponential functions in Eq. (64). Just below the liquidus both ΔG and Q will have relatively small values. Initially the value of ΔG increases faster than the activation energy for diffusion, Q, but at still lower temperatures Q increases and the function $\exp(-Q/kT)$ decreases, and will be several orders of magnitude smaller than $\exp(\Delta G/kT)$.

The crystal size of the consolidated magma will depend on the temperature history of the magma, i.e., on much time it spends at the various values of ΔT. The very rapidly cooled rocks, like chilled margins, will be glassy because the extremely fast cooling moves the magma below the nucleation region before any nucleation can occur. The margin of a dyke may, therefore, be glassy. Normally, only the outermost millimeters or the first centimeter will be glassy. The crystal size of the dyke will increase towards the center because the cooling rate gradually becomes smaller whereby the nucleation rate decreases and the time available for growth increases. The variation in the number of crystals per cubic centimeter for a 106 m broad dyke consisting of olivine tholeiite is shown in Fig. XI-29. The crystal size decreases very rapidly near the margin, and becomes more or less constant about 10 m from the margin.

The number of crystals in a rock may be described using two different parameters, the crystallinity, and the crystal index. The crystallinity depends on the mode of the mineral, while the crystal index only depends on the average crystal size. These two parameters are estimated as follows.

Let the total number of crystals in a thin section be N and the area of the thin section A, the number of crystals per cm² is then N/A. Assuming that the crystals are equidimensional and evenly distributed in the rock, the number per cubic centimeter of the rock is given by:

$$C = (N/A)^{3/2} \tag{68}$$

Here C is the crystallinity of the rock. This parameter was originally defined by Wager (1961), but he did not mention his calculation procedure, but his numbers

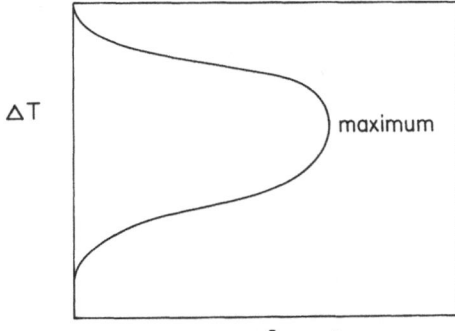

Fig. XI-28. The variation in nucleation rate with supercooling for a silicate melt. The *curve* shows the general variation that one might expect for a melt where the nucleation is catalyzed by impurities so that the critical free energy is small

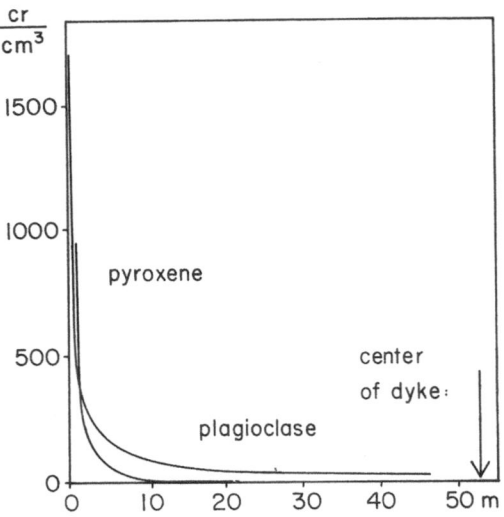

Fig. XI-29. The variation in the crystal index for a 106 m broad dyke with tholeiitic composition and phenocrysts of plagioclase and pyroxene grown in situ. The index decreases very rapidly at the margin and is nearly constant in the central part of the dyke

suggest that he calculated C in the above manner. Evidently C is dependent on the mode of the rock, so it will, therefore, be appropriate to define a parameter which is independent of the mode. If the mode in percent is M then the crystal index is calculated as follows:

$$n = \frac{100\,(N/A)^{3/2}}{M} \tag{69}$$

The total range of the crystal index for the Skaergaard intrusion is from 20 to 63,000, while the average is 2000 (Maaløe 1976). Both the crystallinity and the nucleation index depend on the habit of the crystals.

The nucleation rate of magma is difficult to estimate accurately by experimental methods because the cooling rate of magma chambers is very slow. Some data has been obtained from lava lakes, e.g., Kirkpatrick (1976) investigated the nucleation

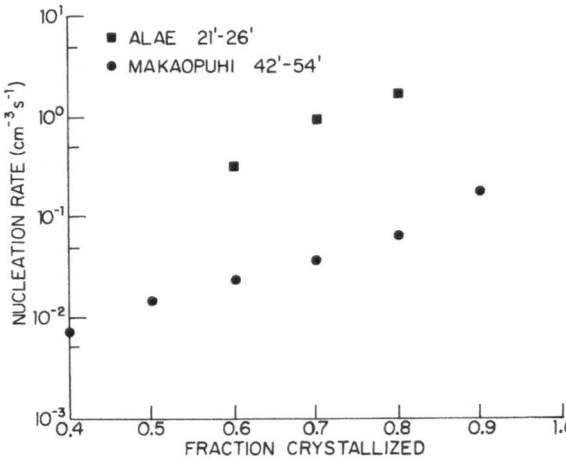

Fig. XI-30. The nucleation rate for lava lakes on Hawaii. The nucleation rate apparantly increases as the crystallization proceeds (Kirkpatrick 1976)

within a lava lake on Hawaii and obtained a nucleation rate of about 10^{-1} crystals/ cm³ s (Fig. XI-30). The average crystallinity of the Skaergaard intrusion is 2000, a number which suggests that the nucleation rate for this intrusion has been about $1.7 \cdot 10^{-3}$ crystals/cm³ s (Maaløe 1976).

16 Experimental Results

The degree of supercooling required for the nucleation of minerals will depend on their intrinsic properties like crystal structure and composition, and the cooling rate. The cooling rate will probably have the greatest significance when it is large, while the individual properties of the minerals become important when the cooling rate is small. The effect of the cooling rate for the nucleation of olivine was investigated by Donaldson (1979) who performed a series of nucleation experiments on the nucleation of olivine in basaltic melts. The diagram in Fig. XI-31 shows the relationship between the supercooling, ΔT, and the nucleation temperature where the first nuclei become visible. If the melt is cooled 60 °C below the liquidus, the first nuclei will appear after 2 h. A much longer period is required if the supercooling is only 20 °C, a period of at least some days is required before the first nuclei will appear. These results indicate that the nucleation will occur at a very slow rate in the large intrusions where the cooling rate is small. Another result from this diagram to be considered is the equilibrium time required for true thermodynamic equilibrium in experimental investigations. By far the majority of experimental runs are performed in less than 24 h, and the diagram shows that one can easily obtain an apparent equilibria which is from 20 ° to 50 °C below the true equilibrium temperature if there are no seeds in the charge. If the charge consists of the minerals which are in equilib-

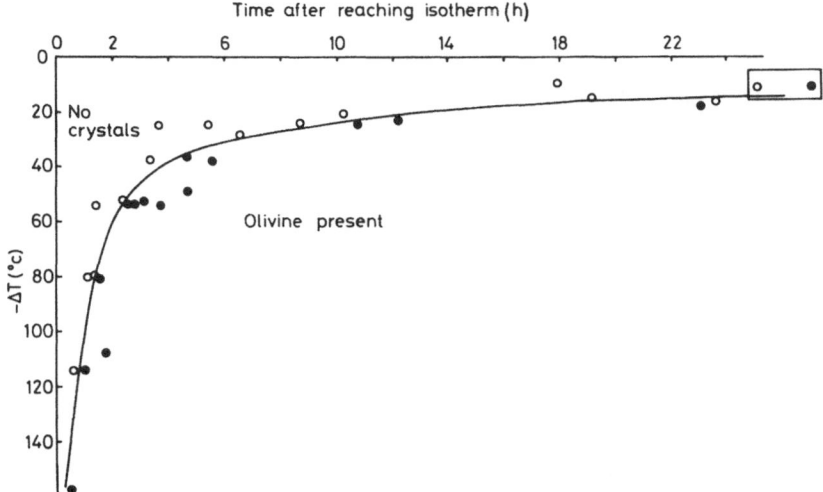

Fig. XI-31. The nucleation temperatures for olivine in a basaltic melt at different times. The supercooling is strongly dependent on time for periods of less than 2 h, and the supercooling becomes nearly constant for long periods of time (Donaldson 1979)

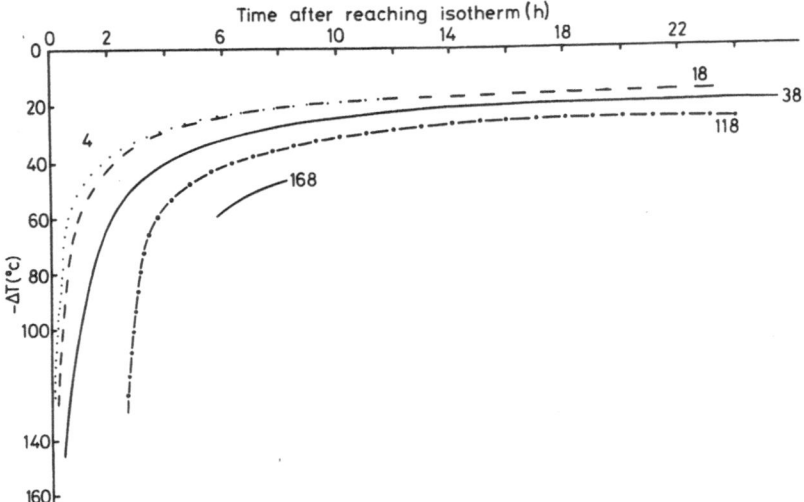

Fig. XI-32. The variation in degree of supercooling as function of superheating and time. As the initial superheating is increased the degree of supercooling required for nucleation increases. The initial superheating is shown by *curves* (Donaldson 1979)

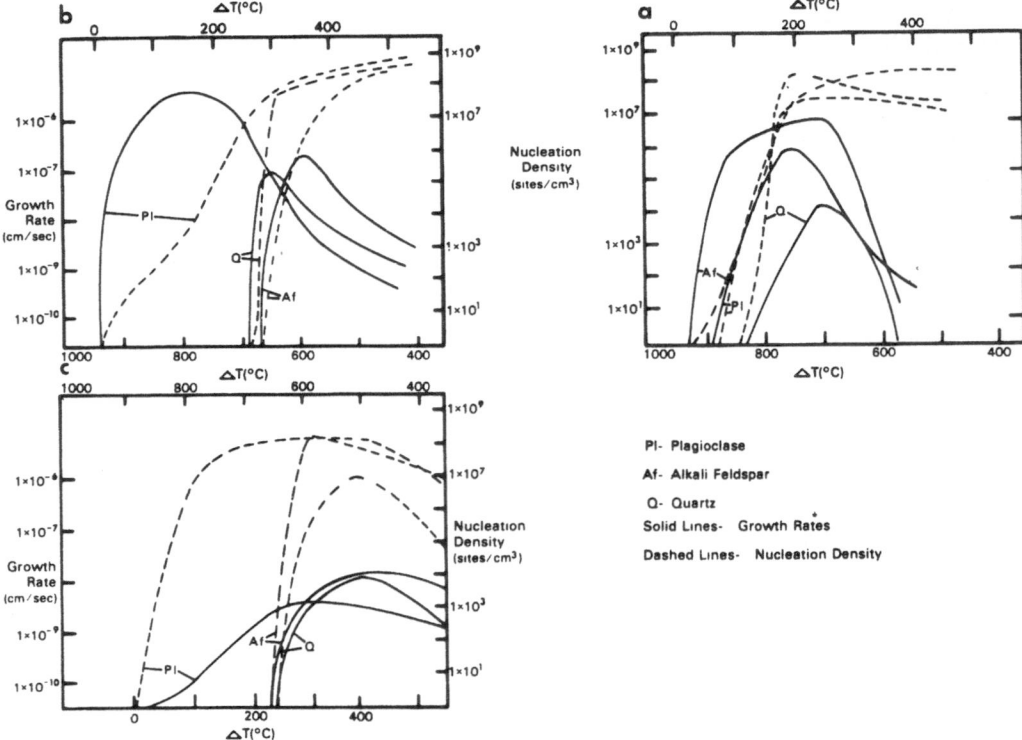

Fig. XI-33a–c. The number of crystals per cubic centimeter and the growth rates for minerals in **a** synthetic granite plus 3.5% H_2O; **b** synthetic granodiorite plus 6.5% H_2O; and **c** synthetic granodiorite plus 12% H_2O. Note the difference in rates by different water contents (Swanson 1977)

rium at the given P–T conditions, then there will be no reason for concern, but if a new mineral is stable then one might expect supercooling for this mineral. The problem can be circumvented if the charge is held below the melting temperature of the mineral for some time, and thereafter raised in temperature. This will ensure that the nuclei for the mineral have been formed so that the mineral actually is present in the charge near its melting temperature.

The nucleation rate will not only depend on the supercooling of the charge, but also on the history of the melt. The silicate melts consist of polymeric units and the degree of polymerization will depend on the time the melt has spent at a certain temperature. The polymerization of a melt which has been say 1 h at a given temperature will not be the same if the melt has been at the same temperature for 5 h. This is evident from Fig. XI-32 which shows the variation in nucleation temperature for different superheatings. A high degree of superheating implies a high degree of supercooling because the embryos in the melt are disintegrated by the superheating. The embryos will reappear after the liquid has been supercooled for some time, and the effect of the superheating is, therefore, largest by short nucleation periods of time.

The nucleation in granitic melts have been investigated by Swanson (1977) and some of the results are shown in Fig. XI-33. The number of crystals per cm³ increases fast just below the liquidus temperature and remains constant about 200 °C below the melting temperature of the minerals. The growth rates are maximal 100 °–150 °C below the melting temperature. These experiments were of less than 1 week duration so the results suggest that the nucleation rates in granitic melts must be very small.

The nucleation and growth rates in granitic melts must be much smaller than in the basaltic melts, nevertheless it is interesting to note that the crystal size of gabbros and granites are quite similar, the sizes being of the order of 1–5 mm. Considering the different kinetic properties of the melts this result is somewhat surprising, but the similar crystal sizes must be due to the longer cooling times of most granitic plutons. The nucleation rate of a granitic melt is smaller than that of a basaltic one, but if the cooling period of time of the grantic melt is much longer, the same number of nuclei might form in the granitic melt with the result that the number of crystals per cubic centimeter becomes the same.

17 Crystal Growth

The processes controlling the growth of silicate crystals are not yet known in much detail. The research dealing with crystal growth has mainly been within metallurgy because of the industrial aspects involved. Some research has been initiated, mainly by Kirkpatrick (1974, 1975), Lofgren (1974), Uhlman (1972), Swanson (1977), and Wagstaff (1967), but the values of the growth parameters for minerals still remain to be estimated. However, the general principles of crystal growth are well-known and will be dealt with here.

Both theory and experiments show that there must be three different main types of growth processes. The process that controls the growth will be dependent on the thermodynamic properties of the phases involved, the growth conditions, and the degree of supercooling. The three different types of growth are summarized in Fig. XI-34 which also shows the qualitative relationship between supercooling and growth type.

The interface reaction controlled growth occurs at small degrees of supercooling and is probably the most important type of growth for phenocrysts in magmas. The high free energy of the interfacial layer represents a kinetic barrier for growth because only a small fraction of the molecules or ions in the liquid has the required small energy which allows them to be attached to the surface. The crystals that are formed by interface reaction controlled growth will be euhedral and faceted. As shown in Fig. XI-34 there are various types of interface reactions. These are dealt with subsequently.

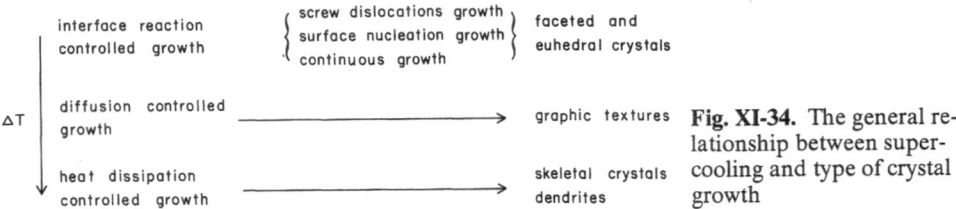

Fig. XI-34. The general relationship between supercooling and type of crystal growth

The second process, the diffusion controlled growth, occurs at slightly larger supercoolings. Here, it is the diffusion rate that delays the growth. The difference in free energy between the new phase and unstable phase is now so large that the free energy of the interfacial layer plays a subordinate role for the growth kinetics. The growth of crystals with compositions different from that of the liquid requires that material transport takes place, and the diffusion rate, therefore, begins to control the growth rate when the supercooling becomes large enough. Crystals with skeletal habit like some megacrysts and phenocrysts in magmas and crystals with graphic texture have grown under diffusion controlled conditions.

At very high cooling rates the supercooling becomes so large that the rate of heat dissipation mainly controls the growth. This type of growth results in dendritic textures like the spirifer textures of komatiites and the dendrites of pillow lavas.

The thermodynamic properties of the crystals determine to a large degree the nature of the crystalline surface, which again have a major influence on the growth process. The nature of the crystalline surfaces will, therefore, be considered first, whereafter the growth processes are described.

18 The Crystal Surface

The interfacial layer of the surface of a crystal which is neither strictly crystalline nor part of the liquid is not more than a few Å thick, and may attain two different forms.

By the atomically smooth interface essentially one atomic layer forms the boundary between the crystal and the melt. The smooth surface is only interrupted by a limited number of single steps of the type shown in Fig. XI-35 a. The other type of interfacial layer is the diffuse layer, which is characterized by an irregular distribution of the surface atoms. This layer consists of several atomic layers and is ragged on an atomic scale (Fig. XI-35 b). These two types of layers are the extreme types of the interfacial layers. The real crystals may attain surfaces which vary between these two types. The growth mechanism is to a large degree related to the type of interfacial layer. The smooth layer is prevalent by the interface reactions, where surface nucleation and screw dislocations occur. The interfacial layer is diffuse by diffusion and heat dissipation controlled growth.

The nature of the interfacial layer was investigated by Jackson (1958), who considered the free energy of the interfacial layer on the basis of the free energies of

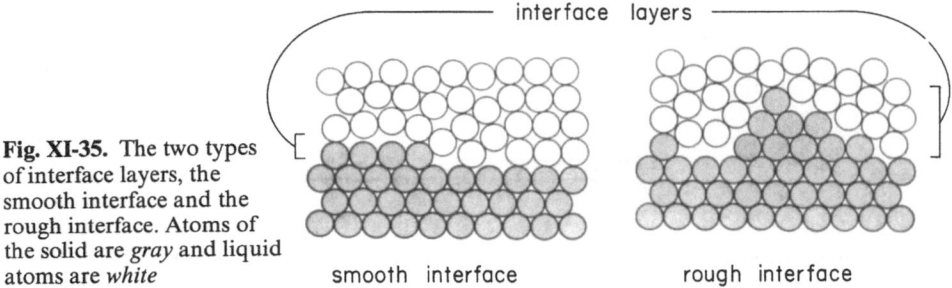

Fig. XI-35. The two types of interface layers, the smooth interface and the rough interface. Atoms of the solid are *gray* and liquid atoms are *white*

interface layers

smooth interface   rough interface

solids and liquids and the entropy of the layer. His model implies some simplifications, but displays the main features and has been quite successful. He showed that a single parameter α may define the nature of the interfacial layer. The value of α is given by (Chalmers 1964):

$$\alpha = \left(\frac{L}{RT} - 1\right)\frac{n}{v} \tag{70}$$

where:

L: latent heat of melting
R: the gas constant
n: coordination number for the solid
v: coordination number for the liquid

As the heat of melting L, as well as n and v all are intrinsic properties, then α will be a constant for a given material. It was shown that if α is less than 2, then the interfacial layer is diffuse, and for values larger than the interfacial layer, is atomically smooth. The fraction L/RT is equal to S/R where S is the entropy of melting. Thus, for materials with small entropies of melting, the interfacial layer is diffuse. Metals have entropies of melting near 2 cal/°K mol. For the cubic close packing we have n = 12, and v = 6 so that $\alpha = (2/R - 1) \approx 0$. The interfacial layer of the cubic metals is, therefore, diffuse. The latent heats of melting of minerals varies between 10,000 and 25,000 kcal/mol, and are generally higher than those for metals. The value of α will either be about 2 or somewhat higher than 2. Minerals should, therefore, mostly have atomically smooth interfaces.

The derivations of Jackson (1958) involve the simplifying assumption that the atomic forces between the solid atoms, ε_{ss} and the liquid ones, ε_{ll} are identical. A

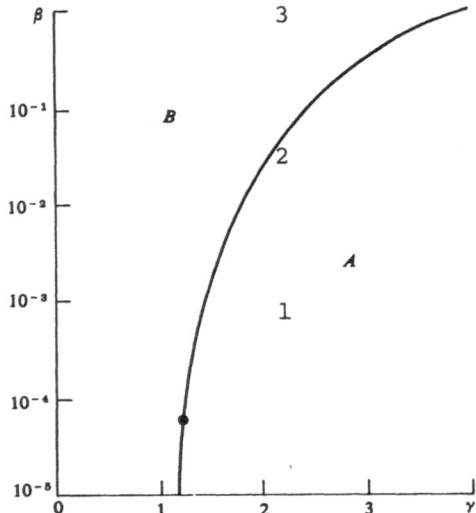

Fig. XI-36. The regions of smooth and rough interfaces. The interface is smooth in region A and rough in region B. The value of β increases with increasing degree of supercooling so a given melt will move from 1 to 3 by increasing supercooling (Woodruff 1973)

more realistic model was worked out by Temkin (1966) which involves both ε_{ss}, ε_{ll}, and ε_{sl}, the latter being the forces between the solid and liquid atoms. His model has two different parameters which define the interfacial layer (Woodruff 1973):

$$\beta = \left(\frac{L}{kT}\right)\left(\frac{\varDelta T}{T_f}\right) \tag{71}$$

and:

$$\gamma = \frac{4}{kT}\left(\varepsilon_{sl} - (\varepsilon_{ss} - \varepsilon_{ll})/2\right) \tag{72}$$

The parameter β is dependent both on the latent heat of melting, L and the supercooling given by $\varDelta T$. The other parameter γ is mainly dependent on the material as T does not vary much for small degrees of supercooling. The values of β and γ which define the two types of interfaces are evident from Fig. XI-36. The interfacial layer is atomically smooth at 1, obtains an intermediate state at 2, and becomes diffuse at 3. By comparing the variation in values of β with Eq. (71), it is evident that the degree of supercooling increases from 1 to 3 and that the nature of the layer varies with the degree of supercooling. This relationship was not evident from Jackson's model, which implies that the interfacial layer is only an intrinsic property.

19 Surface Nucleation

The interfacial layer is atomically smooth by the growth controlled by surface nucleation. A single atomic layer forms the boundary between the crystal and the liquid. The growth of the crystal occurs by the successive addition of new layers to the different faces of the crystal. A new layer is initiated by the addition of a single atom, which may attach anywhere on the surface. The atoms that become attached to the surface are sometimes called adatoms. The first atom will lose an amount of energy which is dependent on the packing of the surface. If the surface is closely packed then the adatom will come in contact with three other atoms and it will lose an amount of energy equal to $\frac{3}{12}\lambda$ (Fig. XI-23). The next atom may either attach isolated on the surface just as the first one, or attach in contact with the first adatom. The last situation is the most energetically favorable, as it will lose energy equal to $\frac{4}{12}\lambda$. Subsequent adatoms will join the first ones. As soon as seven adatoms have joined on the surface a small atomic island has been formed, which is surrounded by six edges. The energy loss for an atom being attached to an edge is $\frac{5}{12}\lambda$. It is, therefore, easier for the atoms of the liquid to be attached to an edge than to form the first adatom. The first adatom may form the nuclei for the whole surface, however, the surface of a crystal is much larger than the dimension of an atom, and several single atoms may stick to the surface. Each of these atoms form the nuclei for an atomic island. The atomic islands grow by the addition of atoms and meet at some stage whereafter an entire new layer of atoms has been formed. As the addition of new atoms to the edge of the islands requires less energy than the attachment of the first single atom, some supercooling is required before a new layer may be formed.

The conditions for the formation of a new layer on the surface of a crystal may be further illustrated by considering the kinetics of the molecules. The motion of liquid molecules and ions occurs by thermal or Brownian motion. It was shown by Einstein (1921) that the relationship between the average distance that a molecule moves, Δ and time, t, is given by:

$$\Delta = (2 D t)^{\frac{1}{2}} \tag{73}$$

where D is the diffusion constant. In order to move from the liquid to the surface of the crystal a molecule should move a distance of the order of 3 Å. The D value for silicate melts is typically 10^{-7} cm^2/s. The time required for this distance is, thus, about $5 \cdot 10^{-9}$ s. The atoms of the crystalline surface is, therefore, under constant bombardment of the liquid molecules, and an atom on the surface will be hit about $2 \cdot 10^8$ times per second. The atoms of the crystal also display thermal motions, they vibrate with a incredible frequency of 10^{13} cps (Kittel 1956). A molecule that approaches the solid surface will therefore, be hit by the solid atoms. If the translatory energy of the molecule is small, it may become attached to the crystal. If its energy is high, it will be pushed away. If the energy of the liquid molecules on the average is high then they will transfer some of their energy to the atoms of the crystal so that they will detach them from the crystal, i.e., the crystal will melt. If the kinetic energy of the liquid molecules are small, then they will be attached to the surface of the crystal which then grows.

The growth rate for surface nucleation may be estimated considering the free energy of an atomic island. The transfer of atoms from the liquid to the solid surface results in a decrease in the free energy. Let n be the number of atoms of such an island, and ΔG_a be the difference in free energy between an atom in the liquid and on the surface. The different faces of a crystal will have different free energies, and this difference may be represented by a shape factor a_1. The difference in free energy for n atoms is, thus, given by $a_1 n \Delta G_a$. This product represents the free energy change caused by the formation of a new layer consisting of n atoms. However, the free energy change of the atoms on the edges of the new island has been neglected.

Note here that the new island creates the same amount of surface as it covers. The interfacial energy is not changed except for the energy related to the edges. The edge energy is proportional to the circumference of the island which is proportional to $\sqrt{n}$. Let σ_a be the interfacial free energy per atom and α_2 a factor related to the material. The increse in free energy caused by the island is then given by $\alpha_2\,\sigma_a\,\sqrt{n}$ and the total difference is estimated from:

$$\varDelta G = -\alpha_1\,n\,\varDelta G_a + \alpha_2\,\sigma_a\sqrt{n} \tag{74}$$

The first term is proportional to n, while the second term is only proportional to $\sqrt{n}$. There will, therefore, be a decrease in $\varDelta G$ as the size of the island increases. By differentiation it is easily shown that the island has reached a critical size for n given by:

$$n^* = \frac{\alpha_2}{\alpha_1}\,\frac{\sigma_a^2}{2\,\varDelta G_a} \tag{75}$$

Assuming as an approximation that $\varDelta G_v = \dfrac{L\,\varDelta T}{T_f}$, then:

$$n^* = \frac{\alpha_2}{\alpha_1}\,\frac{T_f\,\sigma_a^2}{2L\,\varDelta T} \tag{76}$$

and

$$\varDelta G^* = \frac{\alpha_2^2}{4\,\alpha_1}\,\frac{\sigma_a^2\,T_f}{L\,\varDelta T} \tag{77}$$

It may be noted that this derivation is very similar to the derivation for the critical nuclei. According to Boltzman's statistics the number of monolayer islands is proportional to $\exp\left(-\dfrac{\varDelta G^*}{kT}\right)$. If a is the thickness of a monolayer, and I the

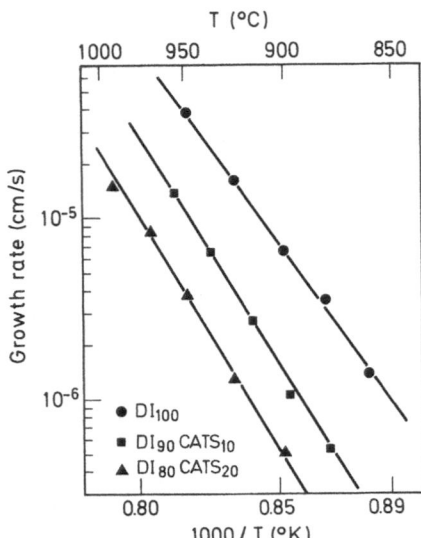

Fig. XI-37. Surface nucleation controlled growth of diopside in melts consisting of diopside and diopside and the Ca-Tschermack molecule. As ln R is a linear function of 1/T the growth is surface-controlled (Kirkpatrick 1975)

nucleation rate of islands, then the growth rate R is given by:

$$R = a I \tag{78}$$

As I is proportional to $\exp\left(\dfrac{\varDelta G^*}{kT}\right)$ then:

$$R \cong \exp - \frac{\alpha_2^2}{4\,\alpha_1}\,\frac{\sigma_a^2\,T_f}{L\,kT\,\varDelta T} \tag{79}$$

This equation shows the qualitative relationship between R and the various parameters. The main significance of Eq. (79) is the functional relationship between growth rate and temperature. According to Eq. (84) $\ln(R)$ is proportional to $1/\varDelta T$, a relationship which is shown in Fig. XI-37. Thus, if it is found by experiments that $\ln R$ is proportional to $1/T$, then it is likely that the growth is controlled by surface nucleation. It is, thus, possible from experiments to estimate the growth process although the parameters controlling the growth are unknown.

20 Screw Dislocations

The early experiments on crystal growth showed that growth occurs at smaller degrees of supercooling than would be expected from the theory based on surface nucleation. This problem was solved by Frank (1949) who suggested that the surface atoms of a crystal do not form a perfect plane layer. Instead the surface contains steps due to the presence of dislocations in the lattice of the crystal (Fig. XI-38). These dislocations are due to imperfect stacking of the atoms caused by vacancies and impurities. The growth by screw dislocations is essentially lateral growth, but while surface nucleation requires the nucleation of a new layer by the attachment of the first adatom, growth by screw dislocations starts at the edge of the dislocations. The free energy required for growth in the presence of screw dislocations is, therefore, essentially smaller than by surface nucleation controlled growth. The phenocrysts in a magma grow from a melt which contains widely different species of atoms, and dislocations in minerals should, therefore, be the rule rather than the exception. Consider an olivine phenocryst. The nominal olivine crystals have the for-

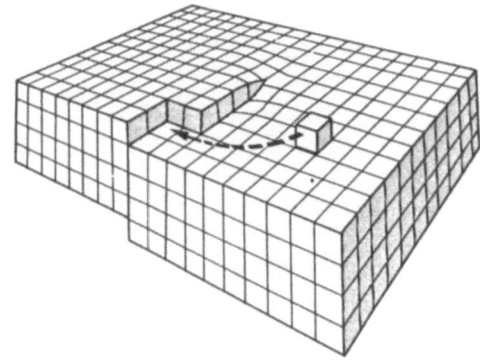

Fig. XI-38. The formation of steps on the crystal surface. Once a step has been formed by an impurity it will persist, so the number of steps will increase as the crystal increase in size. Here it is shown how atoms are added to a step (Woodruff 1973)

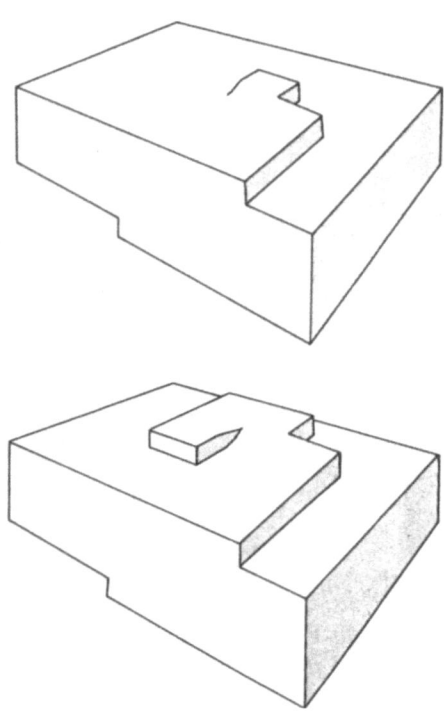

Fig. XI-39. The growth of spirals. The *upper diagram* shows the first stage of formation of a spiral, and the *lower figure* a more advanced stage (Woodruff 1973)

mula $(Fe, Mg)_2SiO_4$, but the olivine crystals also contain ions of Ni^{++}, Mn^{++}, and Ca^{++}. Thus, the lattice cannot be perfectly regular, but must be deformed in the vicinity of these ions.

The steps formed on the surface by dislocations are self-perpetual, no matter how many layers are deposited the step will persist. By continuous growth the atoms added to the step will form a screw or spiral around the step. This is shown in Fig. XI-39, which shows the central part of a spiral after several layers have been deposited. The surface of a natural crystal contains several dislocations, and several spirals will, therefore, form on the surface (Fig. XI-40). The equation for the growth

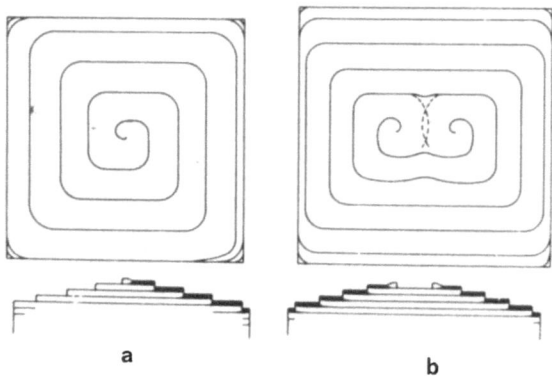

a   b

Fig. XI-40a, b. The growth of screw dislocations. A single dislocation is shown in **a,** and the inteference of two dislocations is shown in **b**

rate by screw dislocation controlled growth is not readily obtained, but the theory shows (Burton et al. 1950):

$$R \simeq K(\Delta T)^2 \tag{80}$$

The growth rate is, thus, proportional to the quadratic of the degree of supercooling. The relationship between $\ln R$ and ΔT for screw dislocation controlled growth differs from that of nucleation controlled growth [see Eq. (79)]. Experimental investigations of the growth rates will, therefore, reveal the type of growth.

Both large igneous bodies and metamorphic rocks undergo slow cooling, and it is likely that screw dislocation controlled growth is the dominant growth form for the minerals of plutonic and metamorphic rocks.

21 Continuous Growth

The surface nucleation controlled growth and the screw dislocation controlled growth occur by the lateral spread of new layers on the surface of the crystal. The energy of the liquid molecules are in both cases too high for all the molecules to stick to the surface. By the continuous type of growth the energy of the molecules has become so small that a relatively large fraction of the molecules can stick to the surface, which can then grow anywhere. The surface becomes diffuse as the molecules are now added anywhere to the surface. The rate of continuous growth also depends on the supercooling, but in a different manner than the two former types of growth. It can be shown that R displays the following dependence on ΔT (Woodruff 1973):

$$\cdot R \cong \left(1 - \exp\left(\frac{L \Delta T}{kT_f T}\right)\right) \tag{81}$$

where:

   L: latent heat of melting
   T: temperature
   k: a constant
   T_f: melting temperature

We have now considered the various types of interface reaction controlled growth forms and will now turn to the shape of the crystal controlled by this type of growth.

22 Crystal Habits

The crystals that are formed by small degrees of supercooling generally tend to be euhedral with a faceted habit. By large degrees of supercooling the growth rate is no longer dominated by the surface free energy, and the crystal may develop habits which are controlled by diffusion and heat dissipation. The habits are skeletal or dendritic, the skeletal habit being transitional between the faceted one and the dendritic one. As long as the surface free energy controls the growth rate it will also influence the habit of the crystals. The crystals will be bounded by surfaces whose free energies are the minimal possible ones. Consequently, the number of different forms will tend to be small. Cubic crystals have closed forms and the cubic crystals will frequently have faces which belong to single forms.

The faceted crystals formed in magma are surrounded by a limited number of plane faces. A planar interface between a crystal and its melt is apparently a stable configuration. Let us consider this relationship in detail. The temperature relationships for a phenocryst are shown in Fig. XI-41 using plagioclase as an example. The temperature of the interior part of the crystal is constant and above that of the melt

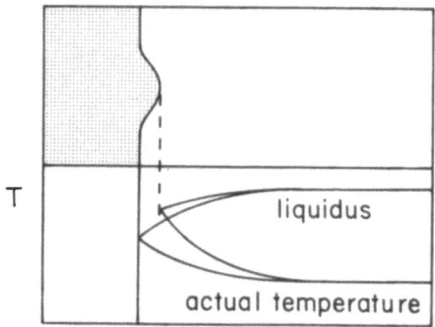

Fig. XI-41. The stability of a plane interface by crystal growth. By equilibrium crystallization at small degrees of supercooling, the temperature of the liquid will be smaller than that of the solid and protuberance is, therefore, unstable

because the crystal dissipates latent heat (cf. Fig. XI-41). The temperature of the liquid decreases away from the crystal and is smaller than the liquidus temperature. The liquid is, thus, supercooled in front of the crystal, a condition required for the dissipation of latent heat from the crystal. Consider the situation if a small pertubation extends from the solid into the liquid. The concentration of albite in the liquid will increase in the front of this protuberance if it starts to grow. However, because of its convex form the concentration of albite will be smaller than ahead of the plane interface. The liquidus temperature will, therefore, be higher in front of the pertubation than at the planar interface. The protuberance will not be able to grow as the rest of the plagioclase surface grows at the maximal possible temperature. So, the protuberance will not start to grow before it is at the same level as the planar interface. A planar interface is, therefore, the stable shape for a crystal face at small supercoolings (Chalmers 1964).

The majority of faceted crystals have only developed a few forms. However, the tendency for simplicity not only relates to the forms, their indices also tend to be simple. Forms like [010], [100], and [111] occur much more frequently than forms like [743]. Again, this feature is explained by the surface energy. As was shown in a former section the surface free energy will be smaller, the smaller the distances are between the surface atoms. The cubic close-packed surface [100] has a smaller energy than (110), which has a more open structure. Let us now consider the relationship between the orientation of a crystal surface and its free energy. The surface that has no steps is completely flat, the atoms form a single continuous layer. If a surface is inclined to the planar one, it will contain a number of steps, the number increasing with the inclination θ (Fig. XI-42). If it contains kinks in addition to steps it will have indices of the type (hkl). If there only are steps, the indices are of type (0hl).

Fig. XI-42a, b. The development of faces with high indices requires the formation of ledges and kinks

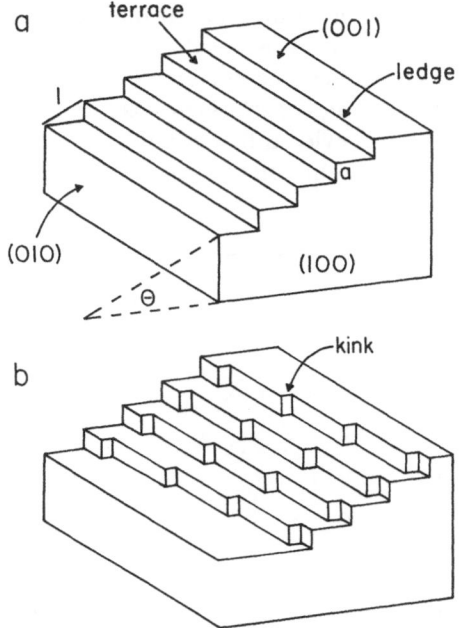

The free energy of the surface increases with the number of steps and kinks, and is estimated as follows. Let the steps be of average height a, and the average distance between the steps be 1, as shown in Fig. XI-42. The inclination angle is then given by (Woodruff 1973):

$$\sin \theta = \frac{a}{1} \tag{82}$$

The surface free energy of the surface (001) is σ_0. The energy per unit length of the steps is β. The free energy of each of the terraces is 1 $\sigma_0 \cos \theta$. The height of the steps is in average a, so that the number of atomic steps n, is given by sin (θ), hence, we have from Eq. (82):

$$n = \frac{\sin (\theta)}{a} = \frac{1}{1} \tag{83}$$

The energy of each step is β so that the total free energy for an unit area with length 1 is:

$$\sigma_0 \cos \theta + \frac{\beta}{1} = \sigma_\theta \tag{84}$$

or:

$$\sigma_\theta = \sigma_0 \cos \theta + \frac{\beta}{\alpha} \sin \theta \tag{85}$$

The variation in σ_θ is shown in Fig. XI-43 for $\sigma_0 = 1$ and $\beta/a = 1$. There is a minimum or a cusp on the curve for $\theta = 0$, so if the angle of inclination is zero, then the free energy is minimal. If the plane (001) belongs to a form with the minimal energy, then it is obvious that other surfaces with different inclinations and indices will have higher free energies and, therefore, be unstable with respect to (001). If the calculation is performed including the energy of kinks, the same result is obtained, and it is evident that a kink free surface has a lower free energy than a surface with the same amount of steps and additional kinks.

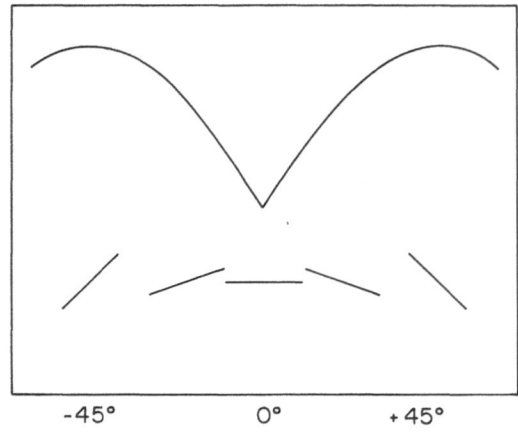

Fig. XI-43. The variation in σ_θ for various values of θ. The value of γ is minimal for $\theta = 0$. The inclination of the crystal surface is shown beneath the γ curve

The forms that will constitute the surface of a given crystal may be estimated using the construction proposed by Wulff (1901). His proof for the construction has been shown to be incorrect (Christian 1965), but his construction method is correct and will be demonstrated here from a single example shown in Fig. XI-44a. The normals are drawn to the all possible crystal faces from the center of the crystal, making the length of the normal proportional to the surface energy of the corresponding face. The normals to ($\bar{1}0$), ($\bar{1}2$), and ($\bar{1}1$) are shown in Fig. XI-44a. When this surface is inclined, the surface free energy will increase as was shown in Fig. XI-43.

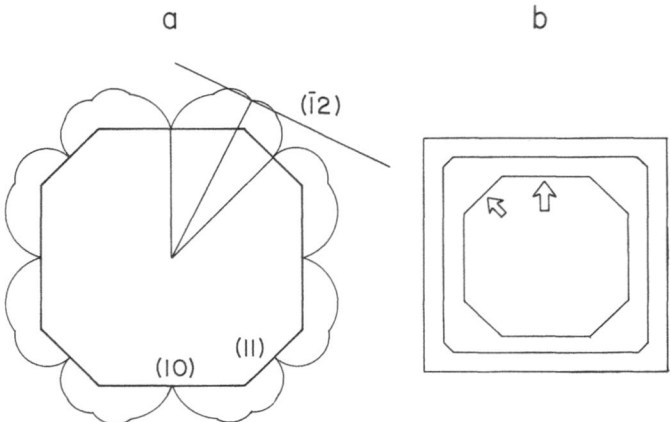

Fig. XI-44a, b. a Wulff's plot for a crystal with two stable forms; **b** Here it is shown how the faces of a form may disappear if they grow with a smaller rate than the other faces

Fig. XI-45. Growth interference of plagioclase from the layered series of the Skaergaard intrusion. The crystals that grow
towards the (010) and (0$\bar{1}$0) faces of the large
crystal dissolve the plagioclase because these
faces are less stable

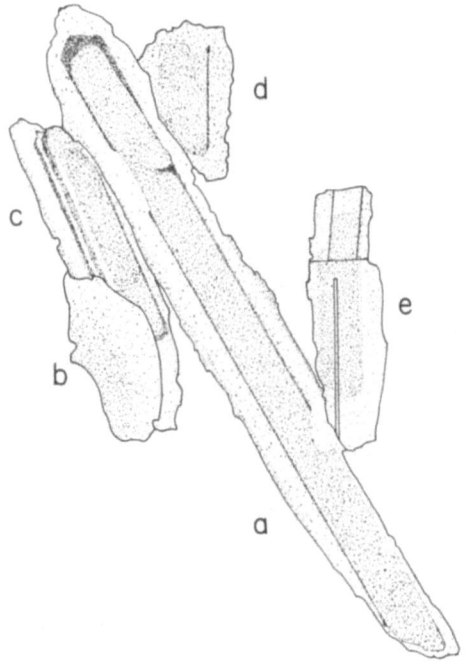

These curves for σ_θ are also shown. The obtained plot is a polar diagram for the variation in free energy. Wulff's theorem then states that the equilibrium shape is obtained by tracking the inner envelope of the normals. The figure obtained in this manner defines the equilibrium shape for the crystal. For the crystal shown in Fig. XI-44 forms [10] and [11] have a lower free energy than form [12] and will constitute the surface of the crystal. Most minerals crystallize with a characteristic habit. Plagioclase crystals are typically tabular, while micas are sheet-like. The prevalence of these habits is due to the difference in surface energies between the various forms.

The free energy of the surfaces not only defines the stable surfaces, but also their growth rate. The form with the lowest free energy will display the fastest growth, and some forms can, therefore, disappear during growth as sketched in Fig. XI-44 b. The difference in stability of the different faces of crystals is sometimes evident in plutonic rocks, an example is shown in Fig. XI-45, which shows some tabular plagioclase crystals from the cumulate of the layered series of the Skaergaard intrusion. These crystals have grown simultaneously in the cumulate from the intercumulus liquid. The tabular crystals have large (010) and (0$\bar{1}$0) faces. This suggests that one of the forms [100], [001], or [0kl] have grown with a larger rate than the form [010]. These faces must, therefore, be more stable than [010] and possess a lower free energy than [010]. This relationship is evident from the resorption that has occurred along the (010) or (0$\bar{1}$0) faces where other crystals are situated with a semi-perpendicular orientation.

The habit of the phenocrysts in magmas is mainly determined by the properties of the crystals themselves. The metamorphic minerals recrystallize in a solid matrix,

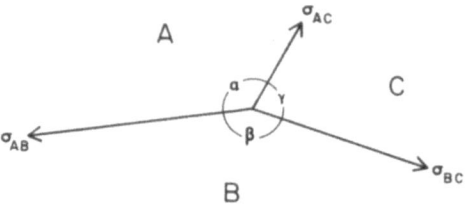

Fig. XI-46. The relationship between the dihedral angle and the interfacial energy, σ

and the shape of one mineral is, therefore, dependent on the surrounding minerals. Here it is the difference in surface-free energy between the different minerals that determines their shape. We have been dealing here mainly with the free energy of the surface. This energy is due to the surface tension, which we will consider in order to estimate the configuration of crystal boundaries. The triple junction between three solids is shown in Fig. XI-46, where the surface tensions are represented by vectors, their directions being given by the three angles α, β, and γ. These angles will tend to adjust themselves into the lowest energy possible. There is a boundary between the A and B faces with a surface tension σ_{AB}. This tension is acting in such a way to reduce the length of the boundary between the two phases A and B. This force is represented by the vector acting along the A–B boundary. Similar vectors exist for the two other boundaries. These forces are in a stable condition when:

$$\frac{\sigma_{AB}}{\sin\alpha} = \frac{\sigma_{BC}}{\sin\beta} = \frac{\sigma_{AC}}{\sin\gamma} \tag{86}$$

Assuming that all grain boundaries have the same surface tension the angles must be equal and 120°. Similarly four grains meet at an angle of 109°28'. No plane faced polyhedron can conform exactly with this geometrical requirement, but the truncated octahedron nearly fulfills this condition (Fig. XI-47). This polyhedron does not possess the minimum surface and the angles at the corners are 120° and 90°. However, a distorted version of the α-polyhedron, the β-tetrakaidecahedron minimizes the interfacial energy and fills the space by stacking. This polyhedron is the stable shape when all the solids have the same interfacial energy. The interfacial energies of the neighboring crystals will depend on the mutual orientation of the crystals, and the various faces will, therefore, not have exactly the same interfacial energy. The habit of minerals must differ to some degree from the β-tetrakaidecahedron, but, the grain boundary angles of metamorphic rocks show a tendency to

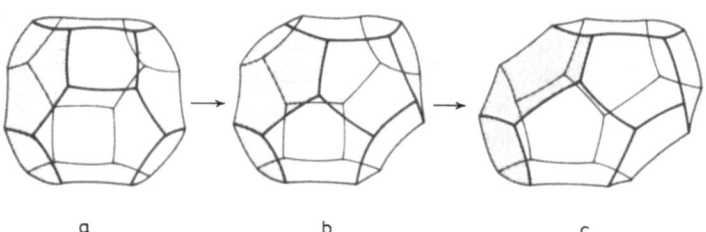

a   b   c

Fig. XI-47a–c. The α-tetrakaidekahedron (**a**) and the β-tetrakaidecahedron (**c**), and a transitional form (**b**) (Williams 1968)

Fig. XI-48a, b. Tabular (a) and mosaic (b) texture of lherzolite (Mercier and Nicolas 1975). Olivine: *blank;* orthopyroxene: *dashes;* clinopyroxene: *heavy contours* and *heavy dots;* and spinel: *black.* The dihedral angle between the grains average about 120°

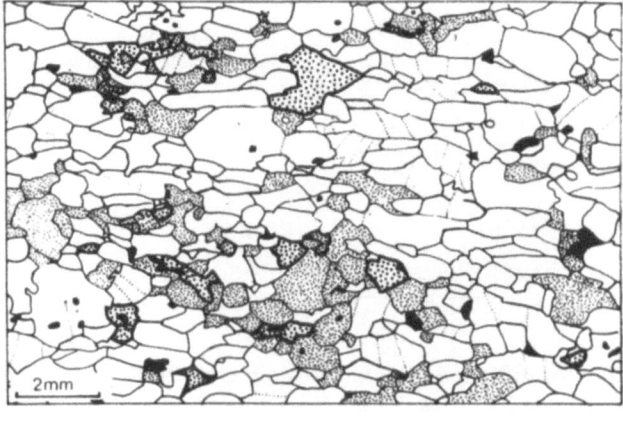

a

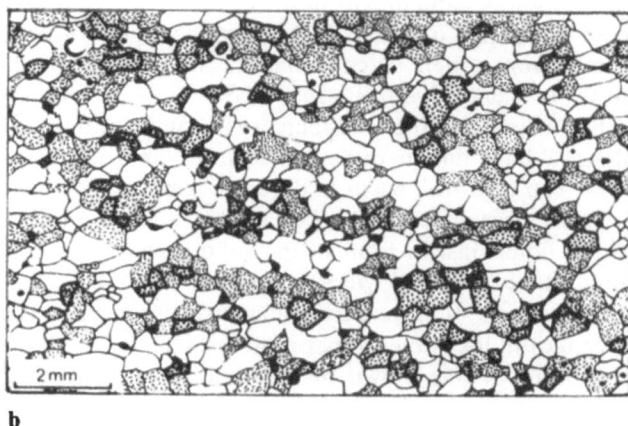

b

cluster around 120°, so the difference in interfacial energies must be small. The polyhedral texture is especially prevalent for lherzolites (Fig. XI-48), which have recrystallized at high temperatures and which, therefore, have approached an equilibrium condition.

We have now considered nucleation and growth controlled by the surface properties of the minerals and will now look into diffusion controlled growth.

23 Diffusion Controlled Growth

When the growth rate becomes so large that all the molecules of the liquid that hit the surface of the crystal become attached, then it is the diffusion rate of molecules in the liquid that controls the growth rate. This will happen at degrees of supercooling larger than those required for surface nucleation controlled growth. The difference in free energy between the liquid and the crystal is now so large that the free energy of the interfacial layer cannot retard the growth. It is now the activation

A

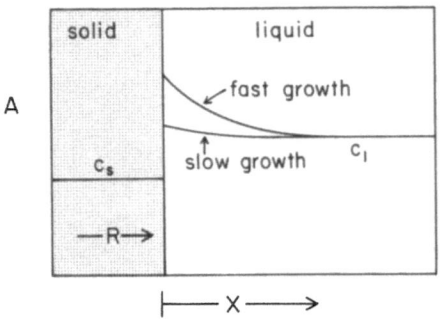

Fig. XI-49. The variation in concentration ahead of a growing face at different crystallization rates. By slow growth the concentration will approximate the equilibrium concentration, but by fast growth the concentration will deviate strongly from the equilibrium concentration

energy for diffusion that controls the growth rate. The growth of plagioclase requires that ions of Ca^{++}, Na^+, and K^+ are available, while ions like Fe, Mg should be removed. Thus, the surface of the plagioclase crystal cannot advance faster than the diffusive transport of these ions. This relationship may be examined considering the concentration of components in the liquid near the solid surface. The solid shown in Fig. XI-49 consists of A and B, A being the high temperature component. The solid is formed from a melt which contains both A and B. By the fast advance of the solid front the concentration of B must increase in front of the surface (Fig. XI-49). If the growth rate is small, then the increase in concentration of B will be small as the molecules of B have time to diffuse away from the front (Fig. XI-49). The crystallization in these two cases will occur at different temperatures. The temperature will be lower by slow crystallization than by fast crystallization, as the solid is in equilibrium with the liquid just in front of the solid surface. Also, it should be noted that the composition of the solid will vary with the crystallization rate. By fast crystallization it will contain more B than by a slow growth rate. This relationship may explain the discrepancy in the values of partition coefficients obtained by different investigators. For example, it was shown by Leeman and Lindstrom (1978) that an estimate of partition coefficients for olivine require equilibration periods of about 24 h. The same type of experiments have been made by Mysen (1976) who used shorter equilibration times, and obtained awkward results due to lack of true equilibration.

The rate of spherical crystal growth controlled by diffusion is obtained from the differential equation for diffusion, the result being (Nielsen 1964; Christian 1965):

$$R = v D (c_\infty - c_r)/r \tag{87}$$

where v = molar volume, r = the radius of the sphere, D = the diffusion coefficient, c = the concentration of a component in the bulk liquid, and c_r = the concentration at the surface of the sphere. The radius is estimated by integration of Eq. (87), the result being:

$$r = (2 v D (c_\infty - c_r) t)^{1/2} \tag{88}$$

where t is the time. The concentration at the interface may be estimated if the distribution coefficient is known. Let $k = c_s/c_r$, then $c_r = c_s/k$. As was shown above, k will depend on the growth rate and the supercooling. If a steady state condition is obtained, then the concentration in the solid is equal to c, and the concentration in the liquid is given by (Chalmers 1964):

$$c_l = \frac{c_\infty}{k} \exp\left(-\frac{R}{D} x\right) + c_\infty \tag{89}$$

where R is the advance rate of the solid front and x is the distance from the solid front (Fig. XI-49).

24 Constitutional Supercooling

The growth of a new crystal requires a certain amount of supercooling, as both the attachment of new ions and molecules to the crystalline surface and the dissipation of latent heat demand that the temperature of the old phase is lower than the equilibrium temperature. The heat generated by the growth of a phenocryst or a metamorphic mineral must be dissipated away from the crystal, and this can only happen by the conduction of heat via the material surrounding the crystal. In some cases the dissipation of latent heat may occur via the crystallizing solid, and this situation will result in what is called constitutional supercooling. This type of crystallization may occur near the margin of igneous bodies where the heat is conducted away by the margin itself. Consider a dyke where the heat is conducted via the already solidified marginal rock. The constitutional supercooling causes the development of graphic textures and will, therefore, be described shortly below.

The situation by constitutional supercooling is shown in Fig. XI-50. The liquid is cooled by the solid which has a lower temperature than the liquid. The condition for constitutional supercooling to occur is that the temperature gradient in the solidified material is larger than in the liquid as shown in Fig. XI-50. The solid front moves to the right whereby the composition of the liquid is changed in front of the solid. As the solid contains more of the high temperature component than the liquid, the liquid must be enriched in the low temperature component, and the liquidus temperature of the liquid consequently decreases towards the solid front. The surface of the solid is in equilibrium with the liquid situated just at the front, and has the same temperature as the liquid. Because heat is conducted from the liquid to the solid, the

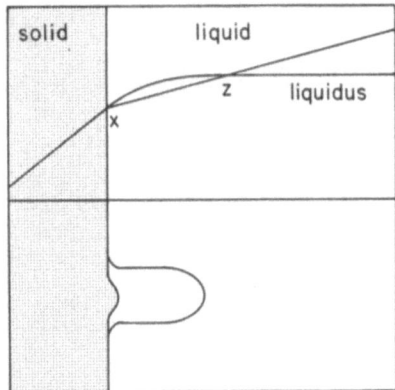

Fig. XI-50. Constitutional supercooling. The solid is more stable than the liquid between x and z and a protuberance will, therefore, form where the solid surface extends into the liquid

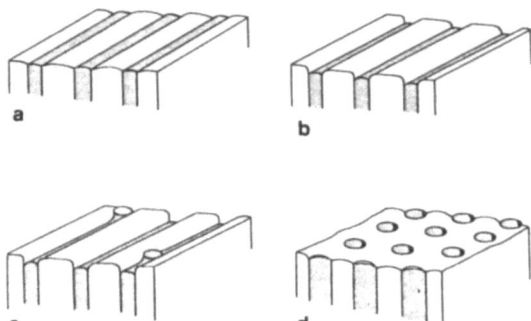

Fig. XI-51 a–d. The formation of cellular structures which are similar to the graphical texture. Initially lamellae will be formed and these may change in shape and become rods (Woodruff 1973)

temperature of the liquid decreases towards the solid, and is, therefore, lower than the liquidus temperature of the liquid in the region ahead of the solid front. Thus, the liquid is metastable between x and z. If the surface contains a small protuberance on the surface, it will grow into the liquid with a larger rate than the plane front. This type of growth results in a cellular substructure, which frequently has been observed in freezing metals (Fig. XI-51). Similar textures have been observed in pegmatites which contain graphic granite (Maaløe 1973).

25 Heat Dissipation Controlled Growth

At large degrees of supercooling of more than about 50° to 100 °C, the crystal growth becomes controlled by the rate of heat dissipation. The crystal will grow in the direction in which the latent heat can be conducted away with the largest rate. Consider a cubic crystal as the one shown in Fig. XI-52 b. The corner of the cube is surrounded by more liquid than the edge, which again is in contact with more liquid than the face. The heat formed by crystal growth can easily be dissipated from the corners, while the dissipation must occur at a slower rate in front of the faces. The growth will, therefore, occur at a higher rate at the corners and the edges than at the faces. The crystals formed by heat dissipation controlled growth obtain a skeletal texture of the type shown in Fig. XI-52 a, b and Fig. XI-53 and XI-54. It is fre-

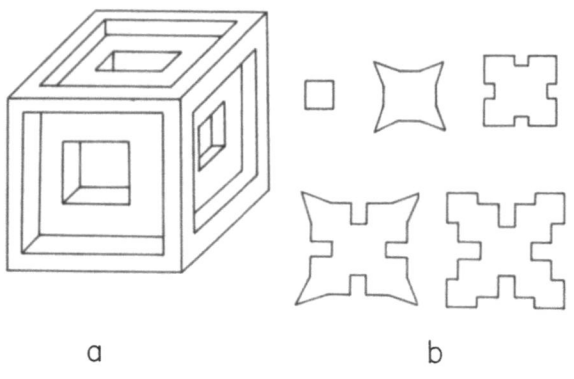

a   b

Fig. XI-52 a–b. a A cubic dendritic crystal during a stage of its growth; **b** five successive stages during the dendritic growth. The crystal starts as a small cube whereafter the edges grow faster than the faces. The last figure of **b** shows a cross-section of the crystal shown in **a**

Fig. XI-53 A–I. Skeletal textures of olivine (Donaldson 1976). The olivine crystals were formed by the crystallization of a basaltic melt

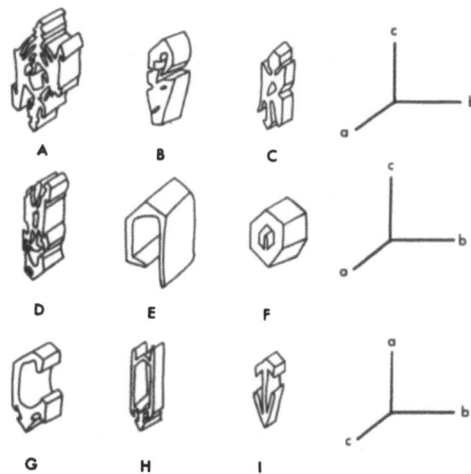

quently observed that phenocrysts and megacrysts in lavas have this texture. The most likely explanation for their occurrence is the fast cooling occurring in dykes and subvolcanic magma chambers. The magma ascending in a dyke undergoes fast cooling, which favors the development of skeletal growth.

At the most extreme supercoolings another texture is developed, the dendritic texture, which is observed in pillow lavas and in komatiites where it is called the

Fig. XI-54. Skeletal diopside phenocryst from an ankaramite from Jan Mayen. The crystal is about 1 cm large and contains three small olivine crystals (Maaløe et al. 1985)

1 mm

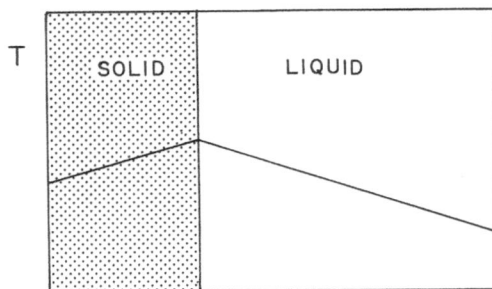

Fig. XI-55. The temperature relations by dendrite formation. The temperature is largest near the solid-liquid interface and decreases away from the interface. By surface controlled crystallization the temperature would be nearly constant within the solid, and larger than that of the liquid

spirifer texture. Both these rock types have suffered extreme cooling during their consolidation. The temperature relations by dendrite formation are shown in Fig. XI- 55. The temperature is highest at the solid–liquid interface where the crystallization occurs, and the temperature decreases away from the solid. A small protuberance of the solid into the liquid will, therefore, advance faster than the plane solid surface. A plane surface is, therefore, unstable under these conditions, it breaks up into a series of projections extending into the liquid. These projections are called dendrites, their idealized shape is shown in Fig. XI-56. The dendrites develop branches because the liquid around them is also unstable and the whole crystal gets a habit like a Christmas tree. Most metals crystallize with a dendritic texture, but this texture is rarely observed in rocks. Dendritic tectures are frequently observed in the charges from experimental runs that have been cooled quite fast, the cooling rate being of the order of 100 °C/s.

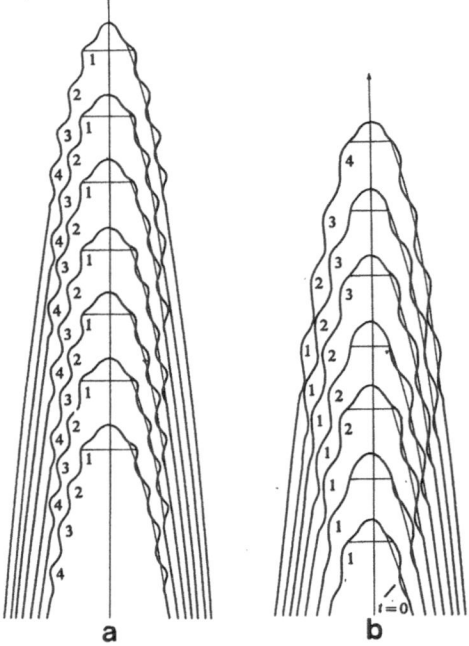

Fig. XI-56 a, b. The formation of dendrites, two types are shown. The crystals grow upwards and form small protuberances which may form new dendrites so that the crystals obtain the shape of a christmas tree (Woodruff 1973)

XII Magma Dynamics

1 Introduction

The compositions of magmas are governed by a series of processes which occur at different pressures and temperatures. The very first stage in the development of magmas is the formation of the initial magma by the partial melting of some source rock. The initial magma may then either ascend to the surface of the Earth without any fractionation or it may undergo one or several stages of fractionation. At any stage during its development the composition of the magma is controlled by the phase relations and the P–T conditions. The P–T conditions are governed by the dynamic processes, and it is, therefore, the dynamic processes which exert the ultimate control on the composition of magmas. The period of time that the magma stays at different P–T conditions before its final eruption is related to the geodynamic processes, and a determination of the petrogenesis of magmas, therefore, implies estimates of both phase relations and dynamic processes. Until the 1970's the greatest importance was attached to the phase relations of magmas, and the petrogenetic theories were mainly dealing with phase relations, while the dynamic aspects were dealt with in less detail. After the fundamental phase relations were determined, the dynamic aspects gradually began to attract attention, and the relationships between the dynamic processes and the compositions of magmas became more into focus. The present chapter gives an introduction to some of the more basic aspects of magma dynamics and will deal with the pressure relationships of magmas, the flow of magma in dykes, the ascent of plumes and convection currents, and the flow of magma in a permeable mantle. The density and viscosity of magmas control the dynamic behavior of melt accumulation and the flow of magma in dykes and we will, therefore, briefly consider these features first.

2 Densities of Rocks and Glasses

The densities of rocks vary from 2.5 g/cc to 3.4 g/cc at 1 bar, the densities of various igneous rock types are shown in Table XII-1 (Clark 1966). The densities of some rocks and their glasses are shown in Table XII-2, and Table XII-3 shows the densities of different glasses.

The density of magmatic liquids decreases with temperature, however, the decrease is small as evident from Fig. XII-1. This diagram was estimated by Murase and McBirney (1973), who made a systematic and comprehensive investigation of the physical properties of magmatic liquids.

Table XII-1. Average densities of holocrystalline igneous rocks (Clark 1966)

Rock	Number of samples	Mean density
Granite	155	2.667
Granodiorite	11	2.716
Syenite	24	2.757
Quartz diorite	21	2.806
Diorite	13	2.839
Norite	11	2.984
Gabbro	27	2.976
Diabase	40	2.965
Peridotite	3	3.234
Dunite	15	3.277
Pyroxenite	8	3.231
Anorthosite	12	2.734

Table XII-2. Density of crystalline rock and corresponding glass (Clark 1966)

Rock type	Rock density	Glass density
Granite	2.656	2.446
Granite	2.630	2.376
Syenite	2.724	2.560
Tonalite	2.765	2.575
Diorite	2.833	2.680
Diorite	2.880	2.710
Gabbro	2.940	2.791
Olivine dolerite	2.889	2.775
Dolerite	2.800	2.640
Dolerite	2.925	2.800
Diabase	2.975	2.761
Diabase	2.960	2.760
Eclogite	3.415	2.746

The density of magmatic liquids increases with increasing pressure, but the variation is fairly small, at least up to 25 kbar (Fig. XII-2). These density data suggest that the compressibility of silicate melts is small. The compressibility is defined by:

$$\varkappa = -\frac{1}{V} \left(\frac{dV}{dP}\right)_T \tag{1}$$

where V is volume and P is pressure. By integration we obtain:

$$\varkappa (P_2 - P_1) = \ln \left(\frac{V_1}{V_2}\right) \tag{2}$$

Assuming a linear variation in the density of the olivine tholeiite shown in Fig. XII-2, we obtain $\varkappa = 2.08 \cdot 10^{-6}$ bar^{-1} between 0 and 20 kbar.

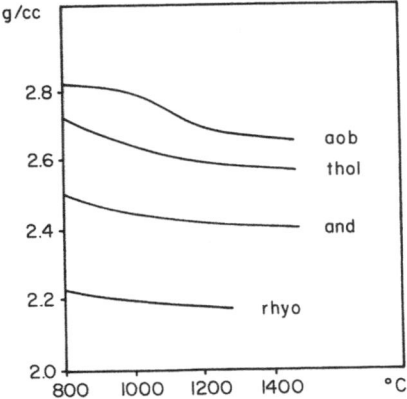

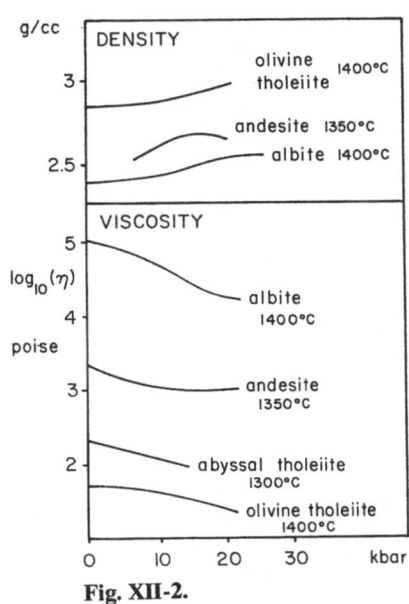

Fig. XII-1. The variation in density with temperature for four different compositions; alkaline olivine basalt (*aob*), tholeiite (*thol*), andesite (*and*), and rhyolite (*rhyo*) (Murase and McBirney 1973)

Fig. XII-2. The variation in density and viscosity with pressure for some magmatic compositions and albite liquid (Scarfe et al. 1979)

Fig. XII-2.

Table XII-3. Average densities of natural glasses (Clark 1966)

Glass	Number of determinations	Mean density
Rhyolite	15	2.370
Trachyte	3	2.450
Andesite glass	3	2.474
Basalt glass	11	2.772

3 Viscosity of Magmas

The viscosity of magmatic liquids, which are not generally simple Newtonian liquids, may vary with the stress applied. We will first consider some general relationships of viscosity and thereafter deal with the experimental estimates of the viscosity.

The viscosity is an expression for the resistance to flow, a high viscosity implies a high resistance deformation of the liquid. For a Newtonian liquid, like water, the

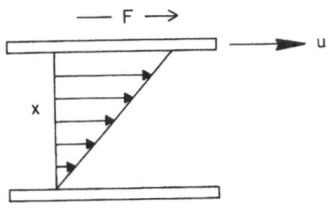

Fig. XII-3. The definition of the viscosity. The *upper plate* moves to the right with the velocity u, and the force F acts on this plate. The area of the plate is A, and the distance between the two plates is x

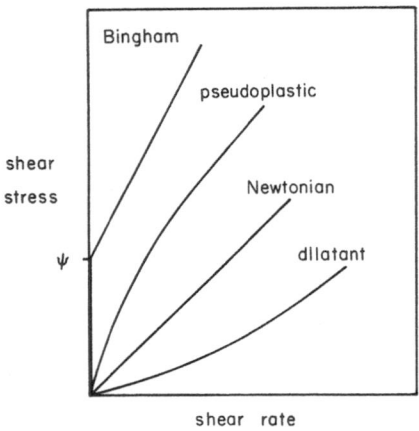

Fig. XII-4. The different types of rheological behavior displayed by liquids and solids. The yield stress for Bingham bodies is shown by ψ

shear stress, τ, is proportional to the velocity gradient (Fig. XII-3):

$$\tau = \frac{F}{A} = \eta \left(\frac{du}{dx} \right) \tag{3}$$

The constant of proportionality, η (g/cm · s), is the viscosity. The kinematic viscosity $\nu = \eta/\varrho$ is the viscosity divided by the density ϱ. The viscosity is independent of the stress by the Newtonian liquids, but varies with the stress for the non-Newtonian liquids.

The behavior of some non-Newtonian liquids is shown in Fig. XII-4 (Wilkinson 1960). The Bingham body will not be deformed at low stress, rather deformation will not begin before a certain stress, the yield stress ψ, is attained. If a magmatic liquid behaves like a Bingham body, then the phenocrysts should attain a certain size before they can settle in the liquid. The pseudoplastic body displays an intermediate behavior between that of a Newtonian liquid and a Bingham liquid. The rheology of the mantle is somewhat similar to that of a pseudoplastic material. The dilatant behavior was originally observed in suspensions of solids. The presence of suspended particles has minor effects for the flow at low flow rates, but at high flow rates the viscosity increases. The dilatant viscosity is observed in lavas with high phenocryst contents (Shaw 1969).

Magmatic liquids may also behave like thixotropic liquids. The viscosity of these liquids decreases after some flow has occurred, but the viscosity increases again after the flow has stopped. These liquids, thus, have a viscosity that varies with the flow rate and the time, and experimental work on magmatic liquids shows that these liquids display thixotropy at temperatures near or below their liquidus temperature.

4 Viscosity–Temperature Dependence

The viscosity of magmatic liquids decreases by increasing temperature. The variation in viscosity with temperature is shown in Fig. XII-5 for a tholeiite, andesite, alkalic olivine basalt, and a rhyolite.

Fig. XII-5. The variation in viscosity with temperature for rhyolite (*rhyo*), andesite (*and*), tholeiite (*thol*), and alkaline olivine basalt (*aob*) (Murase and McBirney 1973)

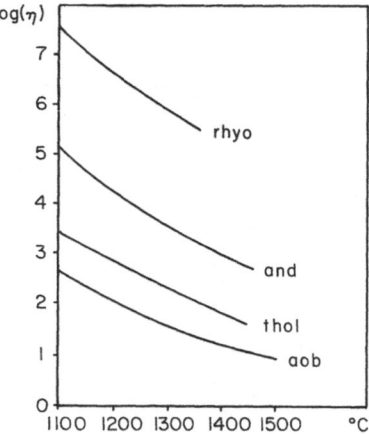

5 Viscosity–Pressure Dependence

The physical effect of temperature and pressure is generally opposite, and one would expect that the viscosity increases with increasing pressure. It was first shown by Kushiro (1976) that the viscosity of dry melts decreases with increasing pressure, and this result has later been substantiated by Scarfe et al. (1979), whose results are shown in Fig. XII-2. The decrease in viscosity is explained by a change in the polymerization of the silicate liquids. Granitic melts that contain water also display a decreasing viscosity with increasing pressure, however, the variation is here caused by the increased solubility of water with increasing pressure (Fig. XII-6).

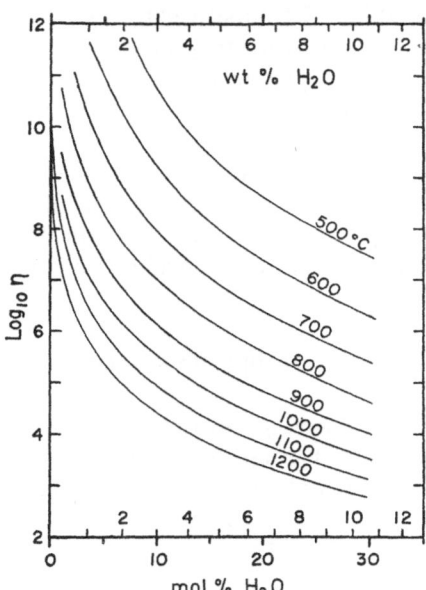

Fig. XII-6. Generalized graph of viscosity vs H_2O content for granitic liquids having bulk compositions approximating those in the low melting regions of the system $Ab–Or–SiO_2–H_2O$ (Shaw 1965). The effect of water is largest for water contents below 3 wt% H_2O

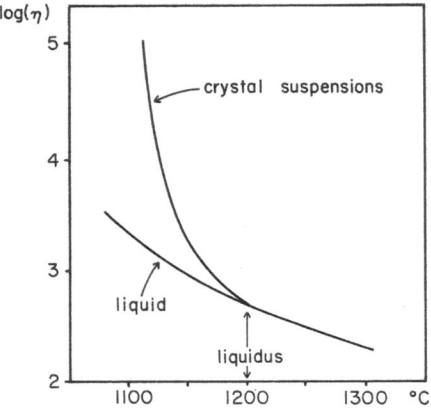

Fig. XII-7. The apparent viscosity vs temperature for a tholeiitic lava. The *lower curve* shows the viscosity of the liquid, while the *upper curve* shows the viscosity of liquid with suspended crystals (Shaw et al. 1968)

6 Non-Newtonian Behavior

The viscosity of a natural tholeiitic magma was estimated by the rotating spindle method by Shaw et al. (1968). A spindle was rotated in the lava beneath a 5 m thick crust at the Makaopuhi lava lake on Hawaii. It was observed that the 1150 °C hot magma displayed thixotropy, and that the yield stress was 700–1200 dyn/cm². As shown below this yield stress is far too high to allow settling of olivine phenocrysts, as was pointed out by Shaw et al. (1968), and the measured yield stress may not be generally representative. The estimation of phenocryst settling rates is further complicated by the fact that the yield stress may vary with time. Murase and McBirney (1973) estimated the variation in the yield stress of a tholeiitic melt with time. Using their data the yield stress after 10 h is calculated as 9707 dyn/cm² for a melt at 1195 °C. The initial yield stress is that of the melt at 1245 ° C which is zero, so there is a constant increase in yield stress with time, which no doubt is caused by an increased degree of polymerization with time. This result is in agreement with the increasing nucleation rates with increasing supercooling.

When the melts are cooled below the liquidus, the number of crystals in the melt increases, and the viscosity of the melt is, therefore, also increased (Fig. XII-7).

These results show first of all that the rheological behavior of silicate melts is complicated, and leaves us with the conclusion that it is rather difficult to calculate the settling rates of phenocrysts, using the experimental data.

7 Gravity Settling of Phenocrysts

If the viscosity and yield stress of a magma is known, then the settling rate may be calculated using the following equations (Shaw et al. 1968). The settling rate in a Newtonian liquid is given by Stokes' law:

$$v = \frac{2 g r^2 \Delta \varrho}{9 \eta} \tag{4}$$

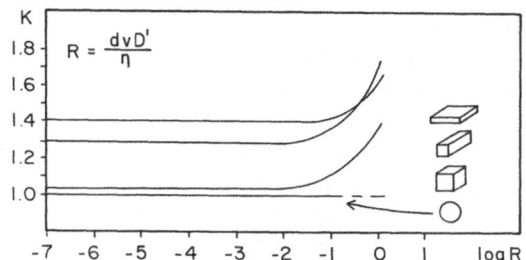

Fig. XII-8. The influence of crystal shape for the sinking velocity of crystals in liquids. R is Reynold's number for the liquid, for small numbers, that is, for R less than 2300 the flow is laminar. The constant K is the correction factor for the velocity, if v_s is the sinking velocity of a sphere, then the velocity for one of the shapes is obtained from v_s/K. The parameters defining R are, d: the density of the liquid; v: the velocity of the sphere; D': the nominal diameter of the crystals; and η: the viscosity of the liquid. The nominal diameter for the crystal is equal to the diameter for a sphere with the same volume as the crystal

where g is gravity acceleration, r is the radius of the spherical crystal, and $\Delta\varrho$ is the difference in density between the phenocryst and the magma. The buoyancy force is given by:

$$F = (\tfrac{4}{3})\pi r^3 g \Delta r \tag{5}$$

The shear stress, S, averaged over the surface of the sphere is given by the buoyancy force divided by the area of the sphere ($4\pi r^2$):

$$S = \frac{rg\Delta\varrho}{3} \tag{6}$$

If S is larger than the yield stress ψ, then the spherical crystal will sink in the liquid, otherwise it will just remain stationary. Let the diameter of an olivine phenocryst be 2 mm and $\Delta\varrho = 0.6$ g/cc, S is then calculated as 20 dyn/cm². This shear stress is small compared to the yield stress mentioned above, and the conclusion is that an olivine phenocryst of this size cannot settle, if the yield stress estimates are correct. How the laboratory measurements should be evaluated is not quite clear, however, it is quite clear that olivine phenocrysts do settle in nature. The variation in modes of olivine from drill cores of the Kilauea Iki lava lake, Hawaii show olivine settling (Richter and Moore 1966). Further, the presence of a bottom cumulate of olivine in many tholeiitic sills show that olivine settle by gravity. We must, therefore, consider the available rheological data as preliminary.

Stokes' law Eq. (4) is only valid for spherical bodies, and the obtained settling rates should be slightly modified if the law is applied to crystals which do not have a spherical shape. The influence of deviations from spherical shape has been estimated by McNown and Malaika (1950), and some of their results are shown in Fig. XII-8.

8 Mantle Rheology

The rheological properties of the mantle control the dynamic features of convection currents and plumes, and are, thus, of primary importance for the geophysical development of the Earth. We will here briefly consider some aspects of mantle rheology, more detailed accounts have been given by Carter (1976), Oxburgh and Turcotte (1978), Twiss (1976), Carter and Ave'Lallement (1970), Ave'Lallement et al. (1980).

The stress in a body may be described by two different types of stress, the normal stress σ, and the tangential stress τ. Consider that a stress component acts on a surface. The resultant stress in a direction perpendicular to the surface is the normal stress. The tangential stress is the resultant stress parallel to this surface. The units of both types of stress are forces per unit area, and are, thus, the same as for pressure, that is, bar or dyn/cm² (1 bar = 10^6 dyn/cm²).

The creep is defined as the relative deformation. Let the length of a bar be l, if the change in length at some stress is Δl, then the creep is:

$$\varepsilon = \frac{\Delta l}{l} \tag{7}$$

The creep is, thus, dimensionless (cm/cm). The creep rate is given by:

$$\dot{\varepsilon} = \frac{d\varepsilon}{dt} \tag{8}$$

and the unit is s^{-1}. By elastic deformation the relationship between the stress σ, and the deformation ε is given by:

$$\sigma = \varepsilon E \tag{9}$$

where E is Young's modulus. Poisson's ratio ν is the ratio of lateral expansion to longitudinal contraction. For solids we have $\nu = 0.25$, and for a Newtonian liquid $\nu = 0.5$. Poisson's ratio for the mantle is about 0.5.

Experimental investigations of the rheological properties of the mantle result is an estimation of the creep rate at a given stress. These results can be converted into a viscosity. It was shown by Griggs (1939) that for a pseudoplastic material:

$$\eta = \frac{\sigma}{3\dot{\varepsilon}} \tag{10}$$

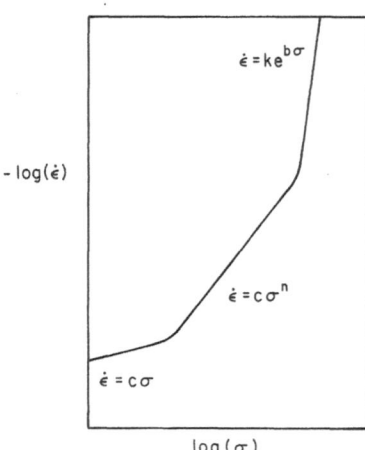

Fig. XII-9. The relationship between the creep rate and the strain at different strain levels (Carter 1976)

The rheological behavior of the mantle is not only dependent on temperature, but also on the stress. The creep rate depends on processes, like diffusion of lattice dislocations and deformation along glide planes, and these processes occur at different stress levels. Experimental as well as theoretical work suggest that the creep rate is related to the stress by the following function:

$$\dot{\varepsilon} = c\,\sigma^n \tag{11}$$

where c and n are material constants. The general relationship between the creep rate and the stress is shown in Fig. XII-9 (Weertman and Weertman 1970, Carter 1976). At small stress levels one observes the flow law:

$$\dot{\varepsilon} = c\,\sigma \tag{12}$$

Creep that is controlled by this relationship is called Nabarro-Herring creep. For a Newtonian material we have a similar relationship:

$$\sigma = \eta\dot{\varepsilon} \tag{13}$$

Nabarro-Herring creep can, thus, be related to Newtonian viscosity as c in Eq. (12) is equal to $1/\eta$.

At intermediate stress levels we have power law creep:

$$\dot{\varepsilon} = c\,\sigma^n \quad (1 < n < 8) \tag{14}$$

where the value of n will depend on the material and the temperature. At high stress levels we obtain:

$$\dot{\varepsilon} = k\,e^{b\sigma} \tag{15}$$

or:

$$\dot{\varepsilon} = k\,\sin h\,(B\sigma)^n \tag{16}$$

The temperature dependence of the creep rate at a given stress is given by:

$$\dot{\varepsilon} = k\,\exp\left(\frac{-Q_c}{RT}\right) \tag{17}$$

where Q_c is the activation energy for creep and T is the absolute temperature. At high temperature creep Q_c is nearly equal to the activation energy for self-diffusion, and the creep rate is then controlled by the diffusive movement of atoms.

An experimental determination of the flow rate for olivine is shown in Fig. XII-10 (Kohlstedt et al. 1976). The creep is power law creep between 10^2 and $5 \cdot 10^3$ bar and high stress creep occurs above $5 \cdot 10^3$ bar.

The strain rates of the lithosphere and asthenosphere has been estimated by various methods, including glacial rebounding (Peltier 1981). The estimated strain rates fall in the range 10^{-13}–10^{-15} s^{-1}. The viscosity of the asthenosphere has been estimated from these strain rates as between 10^{20} and 10^{22} poise. Using Eq. (13) the obtained stress in the mantle is in the range 1–100 bar (Carter 1976). This result suggests that the mantle is deformed by Nabarro-Herring creep, however, the lithosphere, and especially its upper parts may undergo deformation by power law creep and elastic deformation (Fig. XII-11).

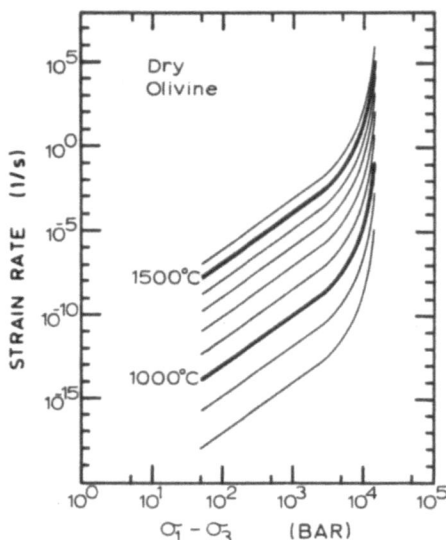

Fig. XII-10. The strain rate at different stress levels for olivine. The creep is power law creep between 10^2 and $5 \cdot 10^3$ bar (Kohlstedt et al. 1976)

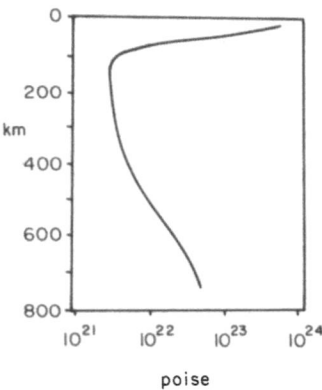

Fig. XII-11. The viscosity vs depth in the mantle as estimated by Houston and De Bremaecker (1975) and Weertman (1970)

9 Convection

One of the most fundamental ideas within geosciences was perceived by Wegener (1915) who suggested that the continents were drifting apart. Wegener's idea of the moving continents was, naturally enough, most difficult to accept by many geologists and geophysicists, since it was difficult to see how and why the continents should move. The continents do not move by themselves and Wegener did some speculation about the nature of the forces that could move the continents. Unfortunately, he proposed a force related to the Coriolis force. It was soon calculated by the geophysicists that the Coriolis force is far too small to cause continental drift, and many geologists consequently considered the drift theory as wrong. The basic mechanism for the continental drift was thought out by Bull (1921) and Holmes (1931), who suggested that there are convection currents in the Earth's mantle, and that these give origin to crustal deformation and continental drift. The convection was caused by the radioactive heating of the interior of the Earth and the cooling of the Earth's surface. These early ideas have later been fully confirmed, and the magmatic activity of the Earth is now explained on the basis of convection and plume activity.

The convective cooling does not only occur within the Earth's mantle, but also on a much smaller scale in the intrusive bodies, called layered intrusions. Some of the layered intrusions display textural and structural features which have a radial symmetry (Brothers 1964). This radial symmetry cannot be accidental, but must have some origin which can be related to a convective movement of the magma. The common occurence of cross-bedding also suggests current activity, and Hess (1960) and Wager and Deer (1939) suggested that the magma is in a convective state.

10 Convection Physics

The convective motion can arise when the upper part of a fluid body is cooled at a faster rate than its lower part. The volume of most liquids increases with increasing temperature, and the density will, therefore, decrease with increasing temperature. The density of the upper parts are, therefore, larger than that of the lower parts. This is an unstable situation as the high density material has a higher gravitative potential than the low density material. If the difference in density is large enough and the viscosity of the liquid is not too high, then the liquid may start to flow. Initially, numerous small eddies will form, and they will grow in size and decrease in number, and after a transient time, a large regular convection cell may form. The conditions for the onset of convection in incompressible media were estimated by Rayleigh (1916) who defined what now is called the Rayleigh number R:.

$$R = \frac{g \alpha \beta z^4}{\varkappa \nu} \tag{18}$$

where: g = the gravity acceleration, α = the volume coefficient of expansion, β = the thermal gradient minus the adiabatic gradient, z = the height of the fluid layer, $\varkappa$ = the thermal diffusivity, and ν = the kinematic viscosity. When R is larger than the critical Rayleigh number R_c, then convection will occur. The value of R_c depends on the boundary features of the system, as well as its shape, but its standard value is considered as 1700.

While the approximate determination of the physical possibility of convection is fairly straightforward, the calculations of the flow rates and the shape of the convection cells are highly complicated. The solutions to a convection system are estimated by solving three differential equations simultaneously, and solutions can only be obtained by using numerical methods (Jaluria 1980). We will not deal with the convection equations here, but instead consider some of the results obtained for convection in the upper mantle. For further reading, the textbooks by Jaluria (1980) and Turner (1973), and the papers by Oxburgh and Turcotte (1968, 1978), Moore and Weiss (1973), MacKenzie et al. (1974), Peltier (1980), and Velarde and Normand (1980) are recommended.

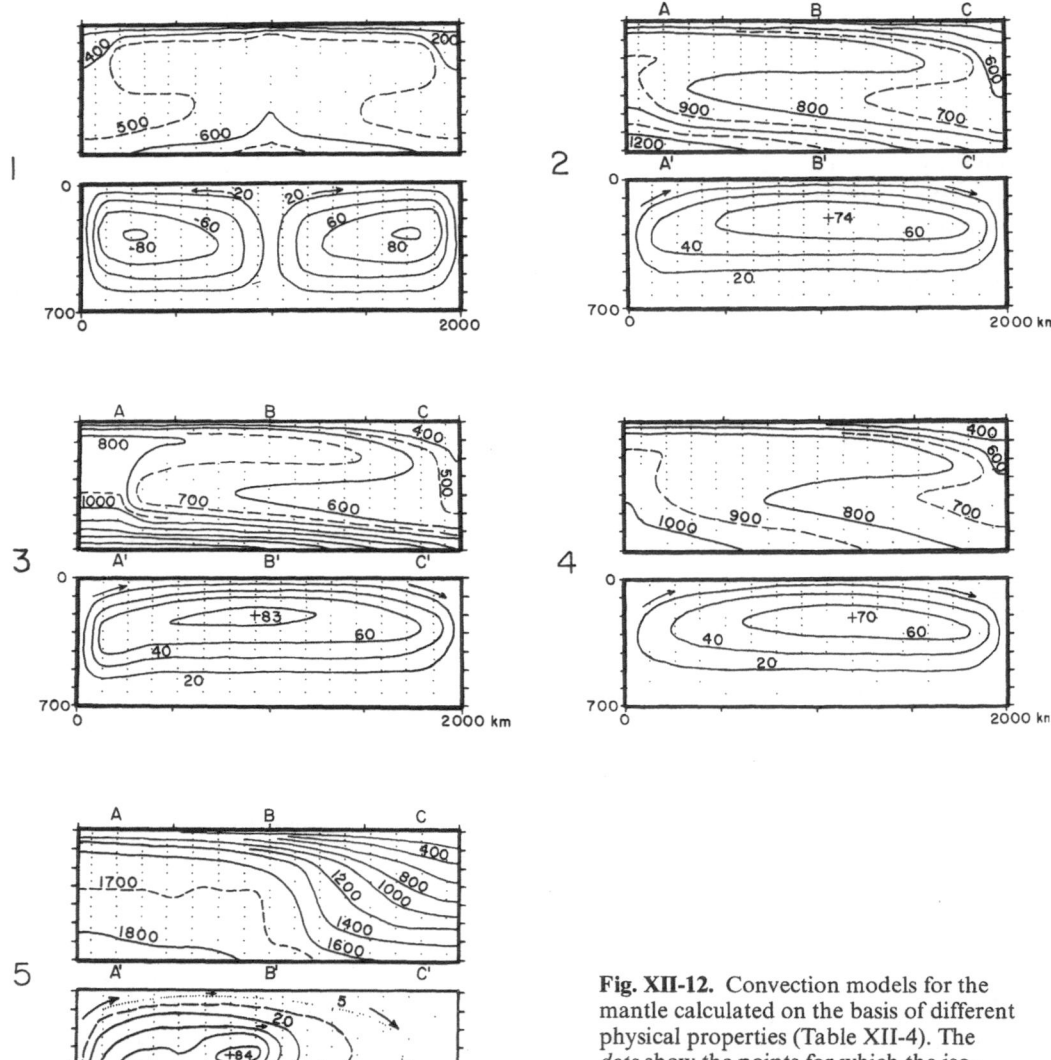

Fig. XII-12. Convection models for the mantle calculated on the basis of different physical properties (Table XII-4). The *dots* show the points for which the isotherms were calculated (*upper figures*). The *lower figures* show the stream lines

The numerical calculations of convection in the upper mantle have been carried out using widely different models. These differ in the manner in which the heat budget of the mantle is treated and in the manner in which the viscosity is dealt with. The models might assume that the temperature of the lower/upper mantle boundary is constant. Torrance and Turcotte (1971) assumed this temperature as 2100 °K, while Oxburgh and Turcotte (1968) considered 1900 °C as relevant. Other models assume a constant heat flow from the lower mantle into the upper one (Houston and De Bremaecker 1975). Another feature to be taken into account is the radioactive heat generation in the upper mantle, which is also included in some

Table XII-4. Parameters for the convection models 1 to 6

	Model number					
	1	2	3	4	5	6
Constant viscosity	×					
Depth dependent viscosity		×	×	×		
Temperature dependent viscosity					×	×
Heat flow at lower/upper mantle boundary	×	×	×		×	
Internal heat generation	×	×		×	×	
Constant temperature at lower/upper mantle boundary						×

models. A further complication is the phase transition of olivine at a depth of about 420 km which is endothermic for ascending mantle material (cf. Chap. I). This phase transition has not yet been included in the convection models.

The viscosity of the upper mantle was assumed constant in the first convection models (Oxburgh and Turcotte 1968), an approach which was justified since the convection models had to develop from a simple basic model. However, the viscosity of the mantle is not constant. Some models have assumed a constant variation in viscosity with depth (Foster 1969). This represents an improvement compared to a constant viscosity, but the viscosity is temperature dependent and this relationship is important because the viscosity will influence the flow rate, which again influences the temperature. Temperature dependent viscosity models have been worked out by Torrance and Turcotte (1971) and Houston and De Bremaecker (1975). These models assume Nabarro-Herring creep, i.e., a Newtonian viscosity. It is likely that a Bingham behavior is relevant for the upper mantle, but models based on Bingham properties have not yet been worked out.

There are, thus, several parameters involved in the determination of the convection patterns and the geothermal gradients of convection cells, and we will consider here some of the models calculated from numerical solutions of the convection equations. The different properties of the models are listed in Table XII-4, and the convection patterns are shown in Fig. XII-12.

Model 1: This model is the simplest one, it has been calculated assuming a constant kinematic viscosity of $5 \cdot 10^{-21}$ cm²/s. The ascending limbs are wide, and the descending ones narrow.

Model 2: By this model the viscosity is depth dependent, the variation in viscosity is given in Fig. XII-11. There is both bottom heat flux and internal heat generation for this model. The results suggest that a convection cell with a higher aspect ratio, i.e., a higher length/width ratio, is stable when the viscosity is depth dependent. The greater width results in a longer period of heating of the bottom and consequently in higher temperatures for the ascending limb.

Models 3 and 4: These models are similar to model 2, but model 3 has only bottom heat flux, and model 4 has only internal heat generation. The two models are largely similar, but the ascending limb of model 4 is wider than that of model 3.

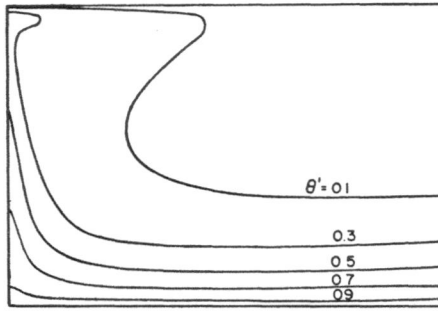

$\theta' = 0.1$

0.3

0.5
0.7
0.9

Fig. XII-13. The isotherms and stream lines for model 6, which assumes a constant temperature at the lower/upper mantle boundary (Torrance and Turcotte 1971). The *upper figure* shows the relative variation in temperature, and the *lower figure* shows the stream lines

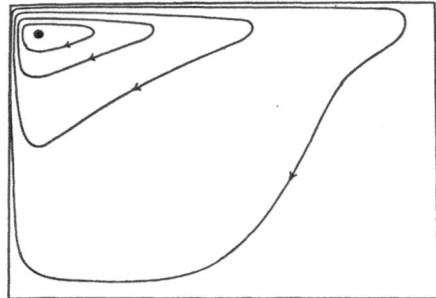

Model 5: This model has temperature dependent viscosity and the creep is Nabarro-Herring creep. There is both bottom heat flux and internal heat generation. This model is probably the most realistic one of the models considered here.

Model 6: This model assumes a constant temperature at the lower/upper mantle boundary (Fig. XII-13). The ascending limb is quite narrow and high temperatures are only attained in a small zone within the ascending limb.

It is not possible to point at one of these or other published models as the "final one", but there is no doubt that the progress within convection modeling has been substantial. One basic problem is that the models require an assumption either about the heat flow from the lower to the upper mantle, or a certain temperature at this boundary. The result is that the convection models are not fully constrained. The heat flux at the surface might constrain the heat flux parameters of the mantle to some degree, however, the heat flux from the mantle is difficult to estimate near the midoceanic ridge axes, due to the presence of magma chambers and hydrothermal circulation.

The different models will result in different temperature depth curves, an example is shown in Fig. XII-14 for model 5. These curves are of mandatory importance for the estimation of the depths of magma generation in the mantle. The pressure and temperature conditions for the generation of abyssal tholeiite have been estimated by Elthon and Scarfe (1980) and Maaløe and Jakobsson (1980). The pressure was in both cases estimated as 25 kbar, while the temperatures were somewhat different 1470 °C and 1580 °C, respectively. The difference in temperature is partly related to different calibration procedures. Since some heat is required for the generation of magma (cf. Chap. IX), then the temperature of the mantle that

Fig. XII-14. The temperature vs depth curves for the ascending limb (*ac*) and the descending limb (*dc*) of the convection current shown here as model 5. The temperature range required for the generation of abyssal tholeiite is shown by (*at*)

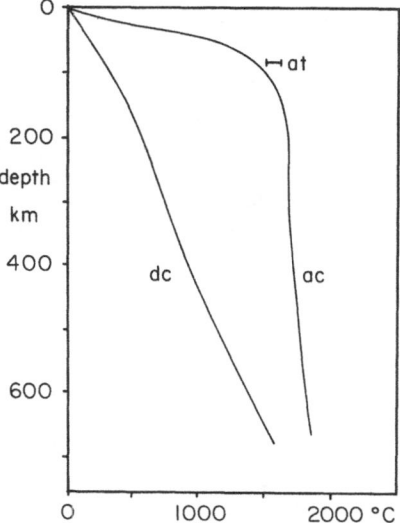

generates the primary abyssal tholeiite must be somewhat higher than about 1500° C. The experimental results show that the calculated convection models must imply a temperature of 1550°–1650°C at a depth of 80 km. The curve shown in Fig. XII-14 nearly fulfills this criterion, but the calculated temperature is too low. This is a general feature for most of the convection models, the temperatures calculated for the ascending convection limbs are too small to explain the generation of primary abyssal tholeiite.

It is, therefore, likely that the convection models need some revision. It should also be mentioned that the models considered here all have assumed the convection confined to the upper mantle. This assumption has been challenged by Peltier and Jarvis (1982) who suggest whole mantle convection on the basis of convection physics and the inferred properties of the mantle. However, this model appears in disconcert with the isotopic systematics of the noble gases (cf. Chap. XIII).

11 Plumes and Diapirs

Among the more important dynamic processes that control magma generation is the ascent of plumes and diapirs. The two terms are describing the same type of structure, but the plumes are generally used in connection with the mantle, while diapirs are used for salt and granite diapirs. Evidence for a diapiric motion was first observed for some granitic intrusions, which have a diapiric structure. A classic example investigated by Cloos (1936) is shown in Fig. XII-15. Cloos' observations and considerations about diapiric motions have later been confirmed from field evidence from many other places (Balk 1937) and the diapiric model is now well-established. Aspects of diapiric motion have been extensively studied by Ramberg (1967) using experimental methods, the results providing further support to the re-

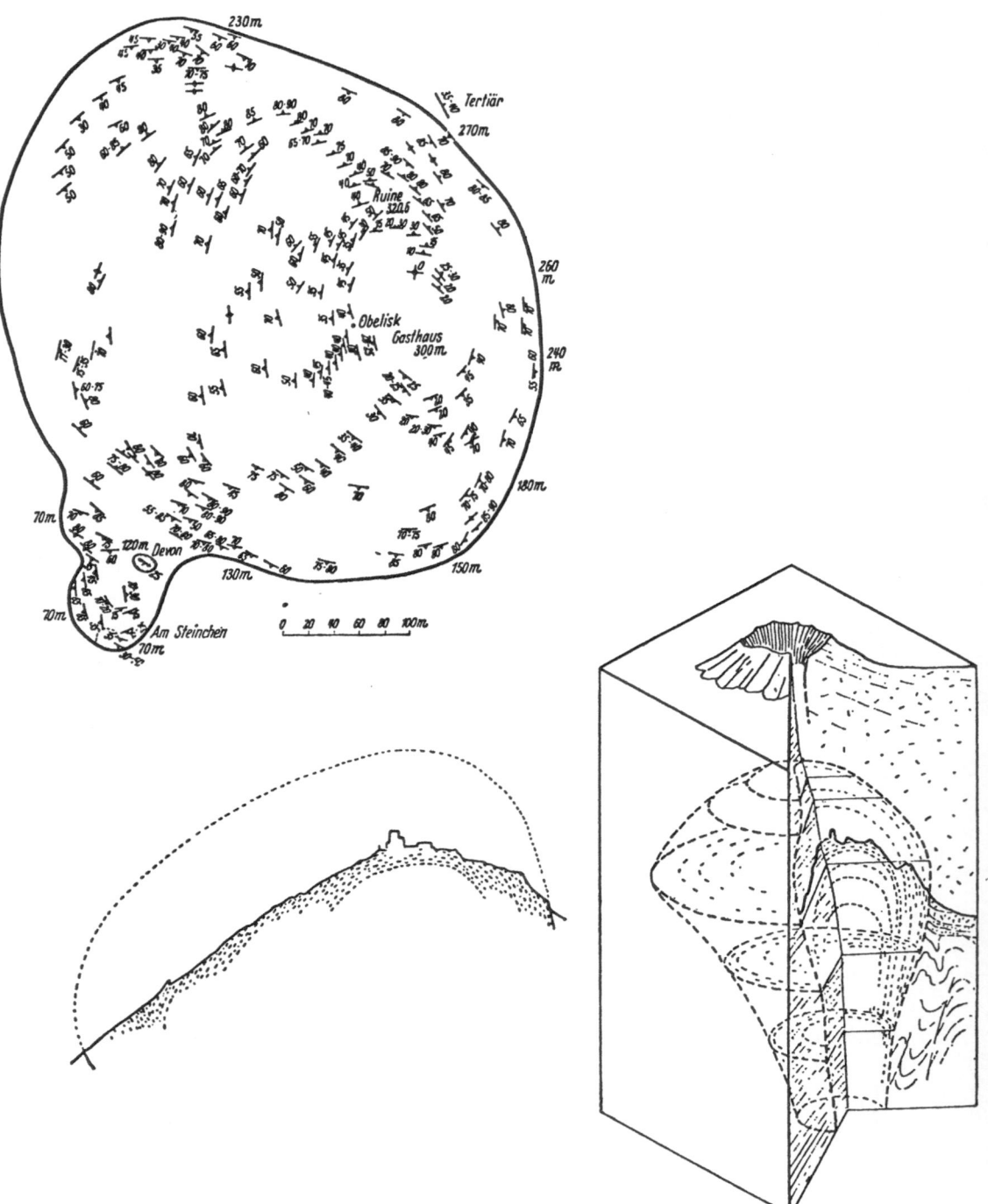

Fig. XII-15. Cloos' (1936) map section and block diagram of the trachytic intrusion of the Drachenfels, Germany. Showing the flow lines of the viscous magma during the intrusion, as derived from the orientation of tabloid sanidine phenocrysts

ality of diapiric motions in the crust and the mantle. Evidence for the processes that occur in the mantle cannot be observed as directly as the crustal ones, but several features suggest the presence of plumes in the mantle. The nature of the oceanic island chains, like the Hawaiian-Emperor chain, suggests that in independent dynamic source gives origin to these chains, and the only type of source consistent with the geophysical constraints is a plume (Morgan 1972). The differences in isotopic compositions between abyssal and oceanic island tholeiites also indicate two different source systems. The depth of origin of the plumes is still under dispute, but an origin from the lower mantle appears likely. The mantle plumes ascend through the upper mantle, and are first arrested by the lithosphere in oceanic regions, while they apparently ascend to the base of the continental crust in continental regions. The depth of origin of granite diapirs is not known, they may come all the way up from the Benioff zone, or they may initiate at the base of the continental crust. The granitic and trachytic domes associated with volcanoes have local character, and most likely stem from a subvolcanic magma chamber.

The theory which deals with plume motion is not yet fully developed (Whitehead and Luther 1975), but some of the important features can be dealt with quantitatively. The physics dealing with the dynamics of plumes and convection is very complicated and requires the solution of several differential equations using numerical methods (Chandrasekhar 1961; Berner et al. 1972). We will here briefly consider two aspects of plume motion, one being the initial instability of a horizontal low density layer, and the other the ascent of a spherical plume fed from below. More elaborate treatments are found in Danes (1964), Ramberg (1967), Marsh (1976, 1978, 1979, 1982, 1984) and Marsh and Kantha (1978).

12 Initial Instability

When a layer of low density is situated beneath a layer of higher density, an unstable situation arises because the high density layer will tend to sink down into the low density layer. The rate by which the high density layer moves downwards will depend on the viscosities of the layers as well as the difference in density between the layers. The boundary between the two layers will initially be planar, but after some time the shape will become sinusoidal, and the low density material will move towards the maxima, and the high density material will move towards the minima (Fig. XII-16). The amplitude of these maxima and minima will increase with time, and at some stage the shape becomes more plume-like, whereafter the low density material begins to ascend (Fig. XII-17). The initial shape of the boundary is given by the function:

$$z = z_0 \sin(kx) \qquad (19)$$

where:

$z_0 =$ the amplitude

k: wave number $\left(k = \dfrac{2\pi}{\lambda}\right)$

x: horizontal coordinate

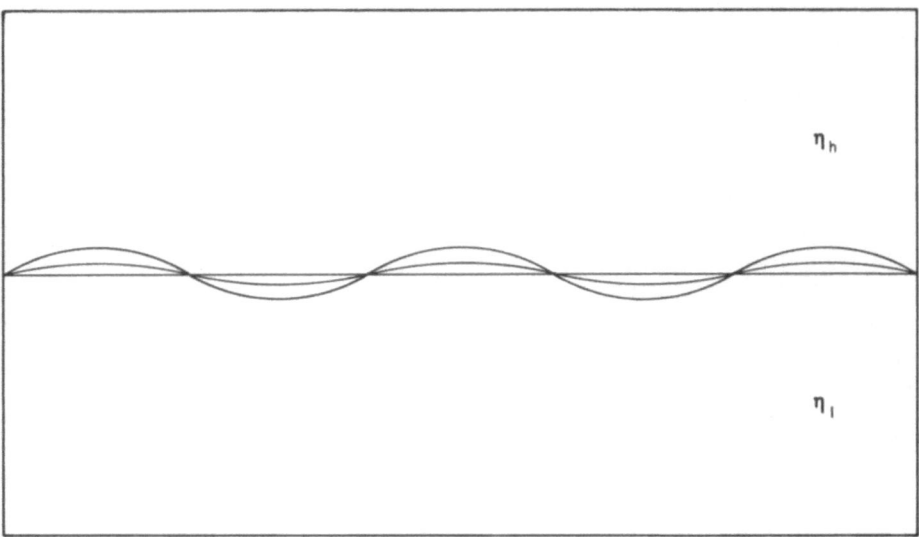

Fig. XII-16. The initial instability by plume formation. The horizontal layer between the layers of high density and low density become sinusoidal after some time. The plumes will ascend from the maxima and their spacing is, therefore, determined by the wavelength of the sinusoidal interface. High viscosity is indicated by η_h and low viscosity by η_l

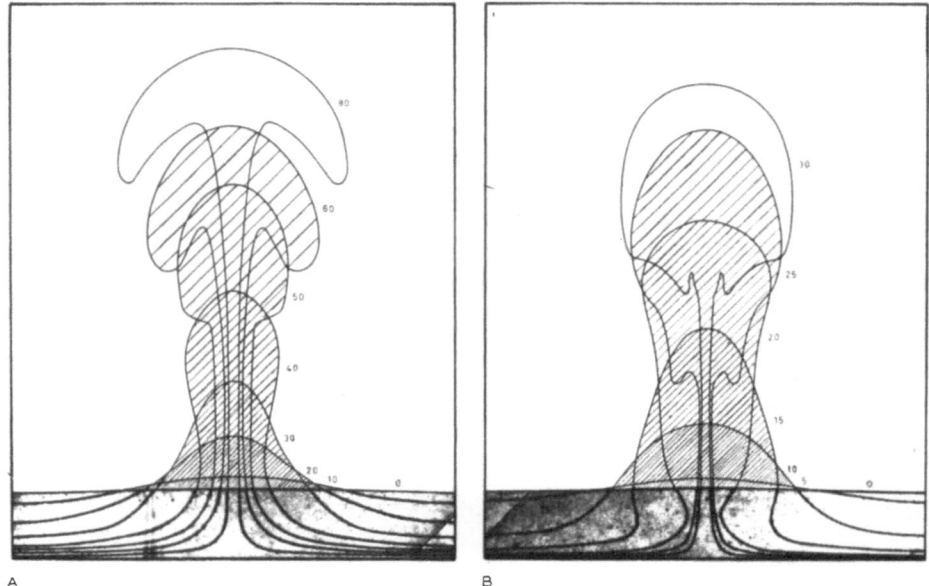

Fig. XII-17 A, B. Plumes are formed at the maxima of the interfacial layer, and their shape and ascent velocities depend on the differences in viscosity and density. Two examples are shown in the diagram (Berner et al. 1972)

The wave number is the reciprocal value of the wavelength, so the wavelength will be large for small values of k. The theory shows that the wavelength may attain different values along the boundary plane, however, it can be shown that one wavelength will grow with a faster rate than the other ones. This wavelength is given by the following equation:

$$\lambda = \frac{4\,\pi\,h\,(\varepsilon)^{\frac{1}{3}}}{2.88} \tag{20}$$

where:

$$\varepsilon = \frac{\eta_h}{\eta_l} \tag{21}$$

h = the height of the low density layer

The plumes will ascend where the maxima are situated, and one might, therefore, expect from Eq. (20) that the plumes will develop with a spacing which is characteristic for the properties of the material and the thickness of the low density layer. This regular spacing is observed both in nature and in experiments. Salt diapirs have a tendency to develop at a certain distance from each other, and the experiments show that an undulating boundary layer forms plumes one wavelength apart (Ramberg 1967).

13 Plume Formation and Ascent Rate

After the undulatory boundary has grown for some time in amplitude, the first plumes will be formed at the maxima. The shape and deformation rate of the boundary layer during the transition from the undulatory stage to the plume stage can be calculated using numerical methods, two examples are shown in Fig. XII-17. The shape of the plume can also be calculated in this manner, and it can be shown that the plume will attain different shapes depending on the values of viscosity and density of the two materials. The calculation of the plume dynamics is considerably simplified if the plume attains a spherical shape. The plume will be spherical when Reynolds' number for plume is less than 1.0 (Clift et al. 1978). Reasonable values of N_{Re} for mantle plumes and granite diapirs result in N_{Re} values much less than 1.0, so we will consider the ascent of a spherical plume, using the derivations of Whitehead and Luther (1975).

We will, thus, assume that the plume has a spherical shape and that its radius is a. Let the viscosity of the plume material be η, and that of the material surrounding the plume be μ. The density of the plume is ϱ, and the density of material outside the plume is $\varrho + \varDelta\varrho$. Further, let the acceleration of gravity be g, the ascent rate v, for the plume with radius a, v is then given by (Batchelor 1970):

$$v = \frac{a^2 \cdot g \cdot \varDelta\varrho}{\mu} \left(\frac{\mu + \eta}{\mu + \frac{3}{2}\eta} \right) \tag{22}$$

It is convenient to simplify this expression, by letting:

$$\xi = \frac{1}{\mu} \left(\frac{\mu + \eta}{\mu + \frac{3}{2}\eta} \right) \tag{23}$$

we obtain then:

$$v = a^2 \cdot g \cdot \Delta\varrho \cdot \xi \tag{24}$$

The ascent rate is proportional to a^2, and will increase with the size of the plume. The plume will develop in two stages when it is fed from below (Fig. XII-18). Initially it will grow in size, and after the size has increased to a critical value it will begin to ascend. The growth rate is given by da/dt, and the plume is in the state of growth as long as da/dt is less than v. When the growth rate becomes larger than v, the plume begins to ascend. Consider that the rate of flow of material to the plume is Q. The volume of the sphere is given by $V = \frac{4}{3}\pi a^3$, so that:

$$\frac{da}{dt} = \frac{Q}{4\pi a^2} \tag{25}$$

The plume will not start to ascend before v becomes larger than da/dt. The size of the plume at this critical stage can be calculated from the identity:

$$\frac{da}{dt} = v \tag{26}$$

so from Eqs. (24) and (26):

$$a_c = \left(\frac{3Q}{4\pi g \Delta\varrho \xi} \right)^{1/4} \tag{27}$$

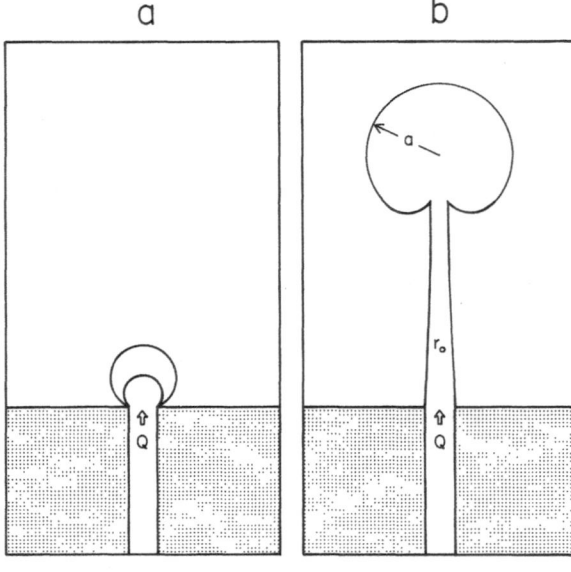

Fig. XII-18a, b. The development of plumes shown in two stages. Initially the head of the plume is formed (a) and when it has grown for some time its ascent rate becomes so large that it begins to ascend. The plume is thereafter fed from the pipe for some time (b) until its ascent rate becomes equal to the flow rate in the pipe and the time required is:

$$t_c = \frac{4\pi}{3} \cdot \frac{1}{Q} \left(\frac{3Q}{4\pi g \Delta\varrho\xi}\right)^{3/4} \tag{28}$$

We have now obtained expressions for the critical radius and the period of time required to reach this radius, and may now consider the dynamics of the ascending spherical plume. The plume is fed from below during its ascent by the feeder pipe (Fig. XII-18). The velocity distribution in the feeder pipe is given by Poiseuilles' law (see Eq. 62):

$$w = \frac{g \Delta\varrho}{4\eta} (r_0^2 - r^2) \tag{29}$$

where r_0 is the radius of the cylindrical feeder, and r the distance from the center line of the pipe. The mean flow rate in the pipe is obtained by integration of Eq. (29). The mean flow rate in the pipe must be equal to Q so that:

$$Q = \left(\frac{\pi}{8\eta}\right) g \Delta\varrho\, r_0 4 \tag{30}$$

or:

$$r_0 = \left(\frac{8\eta Q}{\pi g \Delta\varrho}\right)^{1/4} \tag{31}$$

During the ascent of the plume, the length of the feeder pipe must increase. Part of the flow Q is used for increasing the length and volume of the pipe; let us call this volume ΔV_f. Some of the material leaving the source reaches the plume as discussed above, and this volume must be given by:

$$\Delta V_p = Q - \Delta V_f \tag{32}$$

as:

$$\frac{dV}{dt} = 4\pi a^2 \frac{da}{dt} \tag{33}$$

One gets from Eqs. (32) and (33):

$$4\pi a^2 \frac{da}{dt} = Q - \pi r_0^2 \frac{1}{3} a^2 g \Delta\varrho\xi \tag{34}$$

Let $\beta = \pi r_0^2 \frac{1}{3} g \Delta\varrho\xi$, then:

$$\frac{da}{dt} = \frac{Q}{4\pi a^2} - \frac{\beta}{4\pi} \tag{35}$$

For $da/dt = 0$ the plume has attained its maximal size, which means that the flow rate into the plume is now equal to its own ascent rate, so that no material enters the plume. The maximal radius a_m, is given by:

$$a_m = \left(\frac{Q}{\beta}\right)^{1/2} \tag{36}$$

The differential Eq. (35) can be integrated by separating the variables, but the integral is so complex that Whitehead and Luther preferred to integrate using finite differences, an excellent account of this procedure is given by Carnahan et al. (1969). The theoretical results were compared with experimental ones by Whitehead and Luther (1975), and the experiments showed that the theory will afford satisfactory estimates of the various parameters.

Example

The equations derived by Whitehead and Luther (1975) can be used to calculate the size and rate of ascent of plumes in the mantle, and perhaps more importantly they allow an evaluation of the reality of the plume model. However, it will not be possible to obtain accurate results unless the various constants are carefully estimated, and we will here just assume an approximate constant in order to see if the plume model is reasonable. The following values will be used: $a = 100$ km, $g = 980$ cm/s^2, $\mu = 10^{22}$, $\eta = 10^{20}$, $\Delta\varrho = 0.1$ g/cm^3, and v is then obtained as $9.8 \cdot 10^{-7}$ cm/s, or as 30 cm/yr using Eq. (22). The ascent rate 30 cm/y appears acceptable, within the right order of magnitude.

14 Dynamic Aspects of Partial Melting

The partial melt generated in ascending plumes, in subduction zones, and in the anatectic zones of the continental crusts has to undergo an accumulation process before it can ascend through a feeder dyke towards the surface. The first magma is formed interstitially between the grains of the rocks and the magma must accumulate in some manner before the magmatic source region can surrender substantial volumes of magma to a feeder dyke. The observed eruption rates of magma are far too high to be explained by the direct ascent from a permeable source system.

We may distinguish between two types of partial melting as far as the accumulation process is concerned. The partial melting that results in the generation and extraction of the melt situated interstitially between the residual grains is called matrix melting. This type of melting may occur under sheared or unsheared conditions. If the accumulation occurs in the deep levels of ascending plumes or convection currents where the flow is parallel, then the residuum may remain essentially unsheared (Fig. XII-19). When the accumulation occurs in the topmost part of ascending plumes, then the residuum will undergo shear because the flow is highly divergent (Fig. XII-19).

The other possible type of partial melting is suggested by isotopic evidence. The Nd/Sm and Rb/Sr systematics of abyssal tholeiites may suggest that the present abyssal tholeiites have been generated from recycled abyssal crust (cf. Chap. XIII). The abyssal crust that is subducted may remain as kilometer-sized fragments in the upper mantle. The mantle that ascends beneath the midocean ridges may contain these fragments, and the new abyssal crust is in that case formed from magma generated by bulk or lump melting of the fragments. If this model is correct, then no magma accumulation is required for the generation of abyssal tholeiites, as the

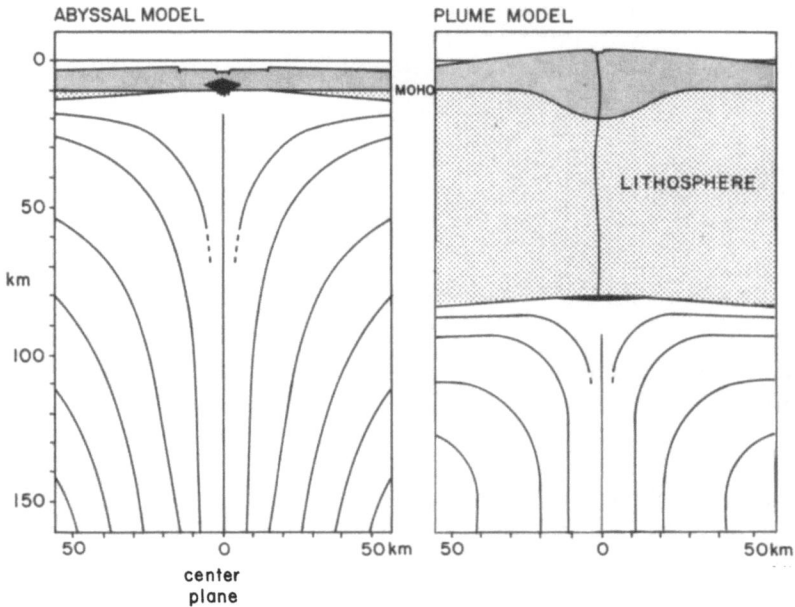

Fig. XII-19. (*Left*) The stream lines for an ascending convection current and the abyssal magma chamber. The stream lines are nearly parallel near the center plane at depths below 75 km (Oxburgh and Turcotte 1968). (*Right*) The stream lines for a plume (Parmentier et al. 1975). The flow becomes divergent beneath the lithospheric plate, and a magma chamber may form in the topmost part of the plume

lumps of tholeiitic eclogite are converted directly into magma chambers. It should, however, be mentioned that there is presently no consensus on what happens with the subducted abyssal crust.

We will subsequently deal with the permeability of partially molten rock and the flow in permeable media.

15 Permeability

Partial melting will result in the formation of melt where the solidus temperature is minimal, that is, where the maximal number of minerals are in contact with each other. The melt generated by partial melting may be distributed between the mineral grains in three different manners: (1) at the very corners of the grains, (2) along the edges of the grains as well as the corners; and (3) along the sides in addition to the corners and edges. The type of distribution is of prime importance for the permeability of mantle, as well as for its seismic attenuation. The present evidence suggests that the second type is most relevant for the partially molten rocks.

The minerals in a rock or any crystalline material under high-confining pressure will have polyhedral shapes. Various irregular polyhedrons can fill the space, but there is a definite relationship between the number of polyhedrons P, faces of poly-

Table XII-5. γ_{ss}/γ_{sl} ratios and θ values

γ_{ss}/γ_{sl}	θ
0	180°
1	120°
1.414	90°
1.730	60°
1.850	45°
1.930	30°
2.000	0°

hedrons F, edges between polyhedrons E, and corners C, which is given by Euler's formula:

$$P + F - E + C = 1 \tag{37}$$

The faces of three grains with the same surface energy will meet at an angle of 120°, and the angle of intersection at the corners is 109½° (cf. Chap. XI). These energy requirements are nearly met by the truncated octahedron, called the tetrakaidecahedron (Fig. XI-47).

The melt that is formed in the partially molten rock will be distributed in different manners depending on the wetability of the melt and the excess pressure of the melt. The wetability is described by the dihedral angle φ. If the interfacial energy between two solids is γ_{ss} and that between solid and liquid is γ_{sl}, then φ is given by (Kingery 1960):

$$\cos(\varphi/2) = \frac{\gamma_{ss}}{2\gamma_{sl}} \tag{38}$$

Some values of φ and γ_{ss}/γ_{sl} are shown in Table XII-5, and the shape of the melt pockets at the grain edges is shown in Fig. XII-20. For $120° < \varphi < 180°$ the melt will form isolated drops at the grain corners, and for $60° < \varphi < 120°$ the melt par-

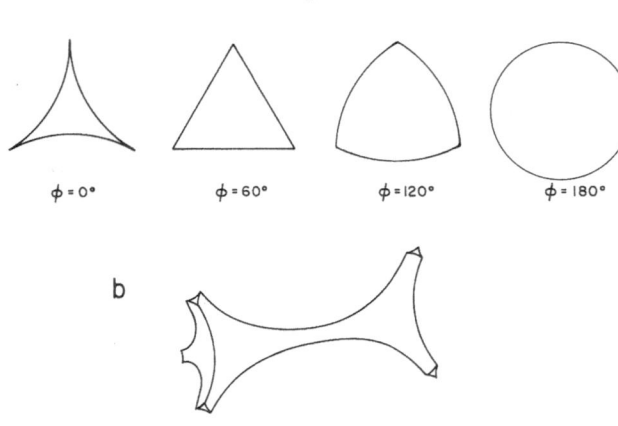

a

$\phi = 0°$ $\phi = 60°$ $\phi = 120°$ $\phi = 180°$

b

Fig. XII-20a, b. a The shape of the melt pockets along grain edges for different dihedral angles; **b** the shape of a melt pocket at two grain corners and one grain edge with negative curvature (Waff and Bulau 1979)

Fig. XII-21. Reflected light photo of the distribution of melt formed by partial melting of a lherzolite at 15 kbar and 1375 °C for 2½ mo (Maaløe 1981). The melt (*dark gray*) is situated at grain corners and along grain edges, and tends to concentrate in melt pockets. This feature is known as coarsening in metallurgy

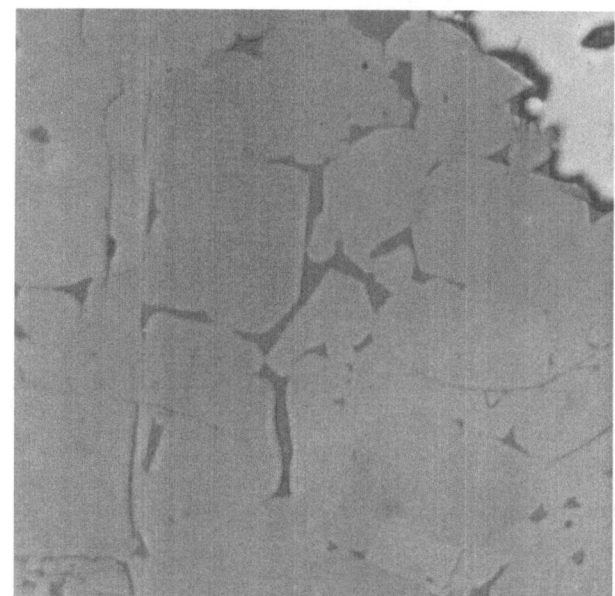

tially penetrates along grain edges. For $\varphi < 60°$ the melt partially penetrates along grain surfaces. The curvature of the melt/solid boundary is positive for $\varphi > 70°$ and negative for smaller values (Fig. XII-20). For $\varphi \leqq 60°$ the mantle will be permeable at small degrees of partial melting, and the flow of magma will be mainly controlled by pressure gradients and the permeability.

The φ values for partially molten rocks may be approximately estimated by melting at small degrees of partial melting. Such experiments have been made by Waff and Bulau (1979) and Maaløe (1981) using lherzolitic materials, and the results suggest a mean value of φ of about 50° (Fig. XII-21). The curvature of the melt pockets is negative, and the mantle will be permeable at small degrees of partial melting according to this result (Fig. XII-22). The value for φ may be expected to display some variation since the surface energies of the minerals must crystallographic orientation of the solid/liquid interfaces (Cooper and Kohlstedt 1982).

An increased pressure of the interstitial melt will change the shape and curvature of the melt pockets (Waff 1980). The excess pressure inside a soap bubble is given by:

$$\Delta P = 2\,\gamma_{lg}/r \tag{39}$$

where γ_{lg} is the surface tension, and r is the radius of the bubble. The shape of a melt pocket can be described by two radii, and the excess pressure is given by:

$$\Delta P = \gamma_{sl}\,(1/r_1 + 1/r_2) \tag{40}$$

where r_1 and r_2 are the two principal radii of the pocket. The radii are positive for a positive curvature and negative for a negative curvature. If ΔP is positive, then the pressure within the melt pocket is larger than the lithostatic pressure, and at least

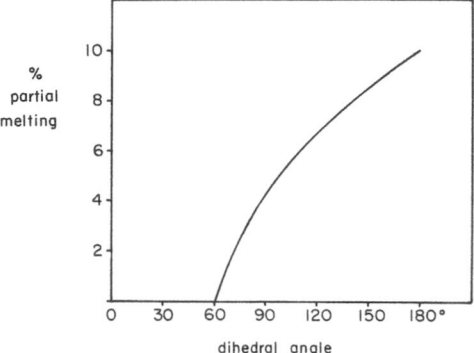

dihedral angle

Fig. XII-22. The minimum degree of partial melting required for permeability vs the dihedral angle. The textures of partially molten lherzolite suggest a dihedral angle of about 50°, and the minimum degree of partial melting is thus about 3–5%. The diagram is based on data from Beere (1981), who used iterative computer methods for the determination of the shape of the intergranular phase. The determination of the shape is not straightforward because the interstices change shape with the degree of partial melting

one radii must be positive and smaller than the negative one. Since the average dihedral angle for melt pockets in lherzolite is less than 60° the pressure inside the pockets must be slightly smaller than the confining pressure. The φ values mentioned here were obtained by experiments and apply for the experimental charges. The shape of the melt pockets in natural partial molten lherzolite might be somewhat different. The compressibility of silicate melts is small and the generation of partial melt will no doubt increase the pressure of the melt above that of the lithostatic pressure. This excess pressure has not yet been estimated. Its value will depend on the creep rate of the residuum, which is not known with the accuracy required for an evaluation of the excess pressure.

The present evidence suggests that the permeability threshold may be attained at small degrees of partial melting, perhaps about 3%, but the exact threshold is not known presently.

16 Flow in Permeable Media

By the partial melting of rocks some melt is formed which is situated interstitially between the grains. The new melt formed in the mantle is called primary magma, while the silicate melt generated in the crust is termed the anatectic magma, or the leucosome. Partial melting of the mantle and the crust may differ in detail, so there are some reasons to use different terms, but here it is convenient to use the same term, partial melting for both processes. At a certain degree of melting the small pockets of magma become interconnected and the magma may now flow through the rock if there is a pressure gradient. The flow may occur in various manners, the molten rock may only contain the pockets of magma, or numerous veins may form throughout the rock. Veined systems are known from the migmatites where the leucosome forms subparallel veins decimeters or meters apart. In the rock which only is permeable the flow will have to take place through the interstices and the rate of flow will be small. In the veined rock, the magma will only have to flow through the permeable residuum for a short distance before it enters a vein, where the resistance to flow is small. It is not known whether or not veins always form in a partially molten source, but the presence of the leucosome schlieren in migmatites suggests that veins may form in some rocks. The initial flow of the partial melt must occur through a permeable rock, and the relationships of permeable flow is, therefore, considered below.

The rate of flow through a permeable media will depend on the viscosity of the liquid, η (g/cm · s), the permeability of the rock, k (cm²), and the pressure gradient of the liquid, dP/dx (g/s² · cm²). The equation which relates these parameters assuming stationary flow was first estimated by Darcy, and is called Darcy's law. Detailed accounts and applications of this law are given by Hubbert (1969), Craft and Hawkins (1959), Matthews and Russel (1967) and Bear (1972). Let v (cm/s) be the rate of flow, then according to Darcy:

$$v = - \frac{k}{\eta} \frac{dP}{dx} \tag{41}$$

The rate of flow is proportional to k, and inversely proportional to the viscosity. The rate of flow v, is the apparent rate, that is, v is the rate of flow through a unit cross-section. The rate of flow through the interstitial channels is much faster. Consider a unit section with area 1 cm², and let the total cross-sectional area of the channels be 0.2 cm². If the flow rate in these channels is 10 cm/s, then the volume passing through the channels is 2 cc/s. The rate of flow through the unit cross-section is, thus, $2 \text{ cm}^3/\text{s}/\text{cm}^2 = 2 \text{ cm/s}$. The permeability k, will increase with increasing degree of partial melting, and is generally of the order of $10^{-7}–10^{-10}$ cm².

The first estimate of the permeability of partially molten rocks was made by Frank (1968) from theoretical considerations. Before we consider his model, a simpler, but similar model will be considered. Let the flow in a unit volume occur through parallel channels with a circular cross-section, and let the distance between the channels be the same and equal to d. The permeability of this unit volume may now be estimated as follows. The mean flow rate in circular channels is given by Poiseuilles' law:

$$v = \frac{8 \Delta P \, r^2}{64 \, L \, \eta} \tag{42}$$

where ΔP is the pressure difference between the two ends of the channel, r is the diameter, L is the length of the channel, and η is the viscosity. This equation may also be given in the following form:

$$v = \frac{r^2}{8\eta} \frac{dP}{dx} \tag{43}$$

The radius r will depend on the degree of partial melting f, and the number of channels which is $1/d^2$, so that we obtain:

$$f = \frac{1}{d^2} \pi r^2 \quad (cc) \tag{44}$$

The cross-sectional area of the channels is given by:

$$A = \frac{\pi \, r^2}{d^2} \quad (cm^2) \tag{45}$$

The flow rate in the $1/d^2$ channels is then given by:

$$v = \frac{r^2 \, \pi \, r^2}{8 \, d^2 \, \eta} \frac{dP}{dx} \tag{46}$$

and by using f expressed by r we get:

$$v = \frac{f^2 \, d^2}{8 \, \pi \, \eta} \frac{dP}{dx} \tag{47}$$

This is the flow in the channels, however, Darcy's law applies for the flow through a unit cross-section so that:

$$\frac{v}{f} = \frac{f \, d^2}{8 \, \pi \, \eta} \frac{dP}{dx} = \frac{k}{\eta} \frac{dP}{dx} \tag{48}$$

and we then get:

$$k = \frac{f \, d^2}{8 \, \pi} \tag{49}$$

The distance between the channels may here be considered equivalent to the grain size, and Eq. (49) suggests then that the grain size has an important influence on the permeability.

Frank (1968) developed a similar model for the tetrakaidecahedron and obtained the following expression for the permeability:

$$k = \frac{f^2 \, d^2}{144 \, \pi \, \sqrt{2}} \tag{50}$$

where d is the diameter of the grains. For $f = 0.05$ and $d = 0.1$ cm, k is calculated as $3.9 \cdot 10^{-8}$ cm^2. This model assumes that the cross-section of the channels is constant. However, the negative curvature of the melt pockets will cause a variation in the cross-section (Fig. XII-20). Since the flow rate in a channel varies with the second power of its radius we might expect that the flow rate in partially molten rock is smaller than that given by Eq. (50). The permeability of a flow system similar to that of the mantle was estimated by Maaløe and Scheie (1982) and the resulting relationship between the porosity or the degree of partial melting and the permeability is shown in Fig. XII-23. These measurements were made on compressed glass spheres which attain a shape similar to that of the grains of the partially molten lherzolite. These experiments suggest that realistic values for the permeability are given by:

$$k = \frac{f^2 \, d^2}{64,000} \tag{51}$$

In order to become familiar with the application of Darcy's law let us first consider the flow in a 1 m long bar with a cross-section of 1 cm^2, and a permeability of 10^{-7} cm^2. Let the pressure difference between its ends be 1 bar, so that dP/dx

Fig. XII-23. A comparison between the experimentally estimated permeabilities (*black dots*), and the permeability curve estimated using Frank's (1968) equation with a corrected constant (Eq. 51)

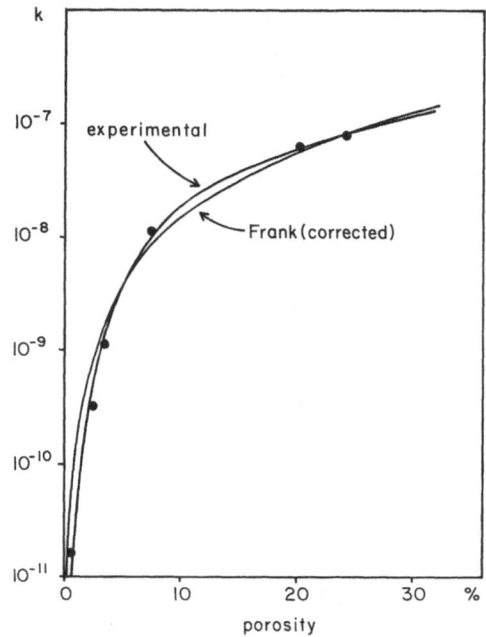

=0.01 bar/cm. Using the cgs system we should convert the bar unit to dyn/cm², and we have 1 bar = 10^6 dyn/cm². For $\eta = 100$ poise we then get:

$$v = -10^{-7}\ 10^{-2}\ (-10^6\ 10^{-2}) \tag{52}$$

$$v = 10^{-5}\ \text{cm/s}$$

It will, thus, take 10^5 s or 1.16 days before 1 cc of liquid has passed through the bar.

For model studies it might be convenient to consider some simple models, and we will here derive the equations for flow of spherical and cylindrical symmetry. The radius of a spherical magma chamber is r_c, and the radius of the spherical region that supplies the magma chamber is r_m. The total flow rate into the chamber is q, and the flow rate across a unit section of the spherical surface is v. The surface of the magma chamber is then $A = 4\pi r_c^2$, and v is given by:

$$v = \frac{q}{A} = \frac{q}{4\pi r_c^2} \tag{53}$$

The value of v should equal the unit flow rate obtained from Darcy's law, so that:

$$\frac{q}{4\pi r^2} = \frac{k}{\eta}\frac{dP}{dr} \tag{54}$$

The minus sign on the right side of Eq. (41) has been omitted as the absolute rate is warranted. Equation (54) is now integrated between r_m and r_c, and we obtain:

$$\frac{q}{4\pi}\left(\frac{1}{r_c} - \frac{1}{r_m}\right) = \frac{k}{\eta}(P_m - P_c) \tag{55}$$

For most problems considered r_m may be considered infinite.

The derivation for cylindrical flow is similar. Let the radius of the magma chamber be r_c, and the radius of the supply region be r_m. The length of the cylinder is h, and the surface area is given by:

$$A = 2\pi r h \tag{56}$$

and q is then estimated from:

$$\int_{r_m}^{r_c} \frac{q\,dr}{2rh} = \frac{k}{\eta} \int_{P_m}^{P_c} dP \tag{57}$$

and q is then given by:

$$q = \frac{2hk}{\eta}\left(\frac{P_c - P_m}{\ln(r_c/r_m)}\right) \tag{58}$$

17 Melt Accumulation by Matrix Melting

During its ascent the mantle will at first become partially melted, whereafter it becomes permeable at a certain degree of partial melting. The partially molten mantle will consist of a residuum with a high viscosity and an interstitial partial melt with a low viscosity. The physical behavior of the melt and its residuum may be understood by considering a column of residuum and partial melt. Let the amount of melt be the same at all levels of the column, and somewhat above that of the permeability threshold. The pressure in the residuum at depth h will be $gh\varrho_r$, and the pressure in the melt is $gh\varrho_m$, before any compaction occurs (Fig. XII-24). Just when the system becomes permeable the residuum will behave like a rigid solid, because it takes some time before creep in the residuum causes its deformation. However, the creep will start immediately, and the pressure on the residuum will be given by:

$$\Delta P = h(\varrho_r - \varrho_m)g \tag{59}$$

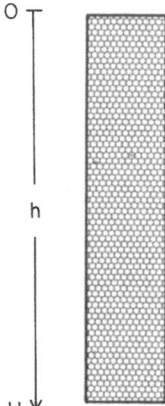

Fig. XII-24. A column of partially molten mantle with the height H. Compaction will start at the bottom and the interstitial melt will flow upwards towards the top of the column

The compaction of the residuum will start at the bottom of the column where ΔP is largest. When the residuum undergoes creep it will exert a pressure on the interstitial melt, and the melt will, therefore, begin to ascend through the interstitial channels towards the top of the column. The flow rate will be determined by the pressure gradient which is constant and equal to $(\varrho_r - \varrho_m)g$. The upward flow of melt will continue until the pressure gradient in the melt has disappeared, that is, until the overpressure in the melt has increased to $H(\varrho_r - \varrho_m)g$.

The pressure of the interstitial melt will exert a pressure on the grains of the residuum and if the pressure of the melt becomes large enough, a creep fracture might form in the residuum. The melt will then flow into the fracture, which will be horizontal, and form a horizontal layer of melt.

We have here considered a stationary situation. Since the partial molten mantle is ascending, the degree of partial melting will increase with decreasing depth, and the permeability will also increase with decreasing depth. The dynamic features of the accumulation process in the ascending mantle may be calculated, and the results suggest that the permeable mantle will consist of horizontal layers of melt, the vertical distance between the layers being about 50 m (Maaløe and Scheie 1982). Waff (1980) used a different approach, but his model also suggests the formation of horizontal layers of melt, the distance between the layers being a kilometer or more.

18 Dyke Propagation

The primary basaltic, kimberlitic, and andesitic magmas are formed at great depths, the entire depth range of the magma generation probably being from 60 km to 250 km. We are not presently able to give a general account on how the ascent takes place, but it is likely that the ascent occurs either by diapiric rise, or by dyke intrusion. The diapiric mode of intrusion of the viscous granitic and syenitic magmas is well-known from field relations. However, it is also possible that the initial ascent in the mantle occurs by diapiric motion. The mantle situated above a partially molten zone is likely to be ductile, and the initial ascent may, therefore, occur by diapiric ascent. However, seismic evidence indicates that dyke formation takes place at depths of 40–50 km, and it is possible that the major ascent mode is by dyke intrusion. Some aspects of dyke intrusion and propagation will be considered here, but it is emphasized that our present insight into dyke dynamics is still fairly limited.

Let us first consider a simple situation. A fracture filled with liquid is situated within an infinite medium. The pressure of the liquid is now raised by some means. The possible increase in pressure will be limited by the fracture thoughness of the medium, P_f. When the pressure of the liquid exceeds P_f, then the fracture will begin to propagate along its entire perimeter. Assume now that a dyke begins to form by hydraulic fracturing above a deep-seated magma chamber within the mantle. When the pressure within the magma chamber has increased to P_f, then a dyke may begin to ascend from the chamber. The distance the dyke can ascend will depend on the pressure in the magma chamber and its size. Let the compressibility of the magma be $\varkappa$, P the pressure, and V the volume, we then have:

$$\ln\left(\frac{P_2}{P_1}\right) = \frac{1}{\varkappa}\ln\left(\frac{V_1}{V_2}\right) \tag{60}$$

Hence, the flow of magma from the chamber will decrease the pressure in the chamber, and the dyke will only be able to propagate all the way to the surface if the volume of the chamber is large enough. If we now assume that the dyke can propagate to the surface, then the overpressure of the magma in the tip of the dyke will increase. The overpressure is here defined as the pressure difference between the pressure in the magma and the lithostatic pressure. A calculation will show that there not only will be an overpressure near the tip of the ascending dyke, but there will also be an overpressure in the magma situated in the deeper parts of the dyke. This overpressure will be sufficient to cause fracture propagation along almost the entire height of the dyke, and the dyke will, therefore, be able to propagate both in vertical and horizontal directions, and a calculation will show that the horizontal length of the dyke will be almost equal to its height when it reaches the surface. Thus, if the dyke is 100 km high, then it will be almost 100 km wide. This size of a feeder dyke is most unlikely and the simple model just considered must be considered unrealistic. In order that the dyke retain a limited dimension then there must be some mechanism that causes the dyke to propagate mainly in the vertical direction.

The possible nature of dyke propagation was first proposed by Weertman (1971, 1974) from theoretical considerations, who used modified dislocation theory, and the geophysical evidence for dyke and magma propagation was suggested by Eaton and Murata (1960) and evaluated quantitatively in detail by Aki et al. (1977) and Aki and Koyanagi (1981).

The pressure difference between the top and bottom parts of a vertical dyke will increase with the height of the dyke (Fig. XII-25). Let the density of the magma be d_m, and that of the mantle be d_M, the difference in pressure is then given by $(d_M - d_m) g z$, where z is the vertical extension of the dyke. If the overpressure in the magma is ΔP_0 and the difference in pressure between the top and bottom parts of the dyke is ΔP_z, then the overpressure near the upper tip of the dyke is $\Delta P_z + \Delta P_0 = \Delta P_d$. The dyke will propagate in the vertical direction as long as ΔP_d is larger than the pressure required for fracture. During its vertical movement the

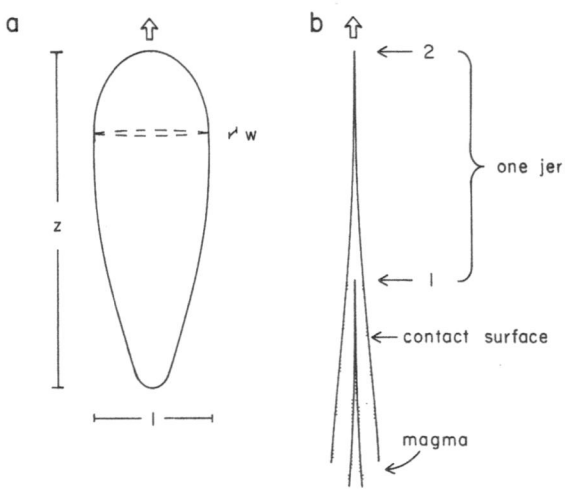

Fig. XII-25 a, b. a A hypothetical shape for an ascending body of magma that ascends while forming a dyke. The shape shown is entirely hypothetical, the actual shape has not yet been calculated. **b** The jerky movement of magma in an ascending dyke. The magma has reached stage 1 and has formed a new fracture to stage 2, and will move from 1 to 2 in a fraction of a second. This movement gives origin to the harmonic tremor. The height of the magma body is z, its lenght l, and its width is w

Fig. XII-26. The possible shape of a dyke extending from the mantle to the surface. The aspect ratio varies from a small one to a large one as the ductility of the rock decreases with increasing height. The diagram is not to scale because the vertical distance is of the order of 50–100 km

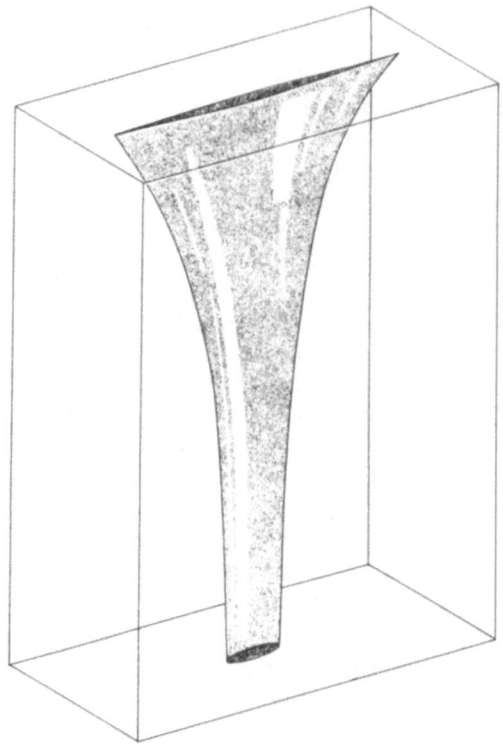

dyke will form new fractures at its upper end, and will close up at the bottom. The movement of the feeder dyke may be compared to that of a flat drop that moves in the vertical direction. The dimensions of the flat magma body that moves upward in this fashion is not yet known, but it is likely that the vertical dimension is a few kilometers or less.

In the deep parts of the mantle near the magmatic source, the dyke will propagate by creep fracture, since the mantle here is ductile, and the movement will probably be fairly slow. Near the surface the rocks become elastic and the velocity of fracture propagation will increase towards the surface. The fracture propagation velocity in elastic materials is of the order of a few km/s. During the ascent the cross-sectional shape of the dyke will change from a more or less ellipsoid shape to an elongated narrow fracture (Fig. XII-26).

Near the surface the elastic fracture propagation velocity is so high that the magma will lag behind. At first the pressure of the magma will cause a fracture to form. The fracture will then move for some distance with a high velocity. Thereafter the magma will flow into the new fracture, then the pressure again is built up near the tip of the dyke and a new fracture is formed (Fig. XII-25). The fracture propagation will, thus, occur in jerks and the frequency of these jerks is about 1–10 Hz. These jerks give origin to a characteristic sinusoidal seismic activity called harmonic tremor, which is observed during volcanic eruptions (Aki et al. 1977).

19 Pressure Relationships of Magmas

The forceful and violent nature of volcanic eruptions suggest that the erupting magmas possess a large overpressure when they reach the surface. Similarly the formation of dykes and dyke systems and the formation of magma chambers by stoping indicate that large pressures must be involved. A more peaceful and indirect evidence for the pressures involved is obtained from the height of volcanoes. A volcano of some height can only be formed if the pressure of the magma at the surface is higher than that of a column of magma with a height equivalent to that of the volcano. While this evidence is not as impressive as the activity during an eruption, it can afford some valuable quantitative evidence which we will consider subsequently. The overpressure of magma at the surface of the Earth has not yet been measured, but several sources of evidence suggest that this pressure is of the order of 500 to 3000 bar for basaltic magmas, while the overpressure of kimberlitic magmas may be even higher.

Geophysical investigations of the stress in the upper mantle and the continental crust indicate that the pressure at depths of more than 10 km is essentially lithostatic, i.e., the pressure at any given depth is mainly controlled by the weight of the overburden. A similar result is obtained from gravity anomalies and the isostatic recovery of the continents after the last ice age. This result suggests that the pressure of the magmas in deep-seated magma chambers will be similar to the lithostatic pressure in the surrounding rocks. We can evaluate this possibility by considering the pressure relationships of a volcano, and will here investigate the situation for Mauna Loa, the largest Hawaiian volcano.

The presence of an oceanic island like Hawaii will have some influence on the pressure in the mantle beneath the island, but the effect will be small at great depths. The pressure in the suboceanic mantle beneath Hawaii can be calculated using the average values for the thickness and densities of the different layers estimated by Watts (1976):

Layer:	Thickness	Density
Water	3.5 km	1.0 g/cm³
Basaltic layer	1.5 km	2.8 g/cm³
Gabbro layer	5.0 km	2.9 g/cm³
Mantle	Z km	3.3 g/cm³

From these figures the lithostatic pressure at the Moho discontinuity is calculated at 2.22 kbar, and the pressure within the mantle is $P_1 = (2.22 + 0.33\,Z)$ kbar, where Z is the depth beneath Moho in km. Assuming that the pressure within a magma chamber in the mantle is equal to the lithostatic pressure, the pressure of the magma at any depth is given by this equation. The magma that has formed Mauna Loa must have been able to ascend to the summit which is 4169 m high. The pressure of the magma at the bottom of the ocean must, therefore, be so high that it can carry a column of basaltic liquid being 7670 m high. With a density of 2.7 g/cm³ for the basaltic liquid, this pressure is calculated at 2.07 kbar. The magma can only attain this minimum pressure if it ascends from a certain minimum depth. The pressure exerted by a column of basaltic liquid is smaller than that exerted by a column of mantle material because the density of the liquid is smaller. This depth is, therefore, calculated from the following equation:

$$1.87 + 0.33Z = 1.87 + 2.07 + 0.27Z$$

so that

$$Z = 41.4 \text{ km}$$

Here Z is the depth beneath the Moho discontinuity, so the depth beneath the surface of the ocean is estimated at 51.4 km using the above figures. This depth is the minimum depth for the origin of the lavas of Mauna Loa, the actual depth must be somewhat larger, because the pressure of the source region will decrease as soon as some of the magma leaves this region. Seismic activity during the eruptions of Hawaiian volcanoes suggests a depth of origin of 60 km or more so the minimum depth calculated above is consistent with this result.

The position of andesitic volcanoes in relation to the Benioff zones suggests that the andesitic magmas ascend from great depths. It is not yet known if the andesitic magmas stem from the Benioff zone, or if they are formed from another type of magma generated in this zone, but let us consider the pressure relations of an andesitic magma generated in the Benioff zone. Let the thickness of the continent beneath an andesitic volcanoe be 35 km and the density of the andesitic magma be 2.5 g/cm³. The pressure of a 35 km high column of andesitic magma is then 8.75 kbar at the continental Moho. The difference in density between the magma and the mantle is 3.3 − 2.5 = 0.8 g/cm³, and the minimum depth of origin of the magma is, thus, estimated as 110 km beneath Moho or 145 km beneath the surface. Again this estimate is consistent with the geophysical data which suggests a depth of 100 to 150 km. Both these examples show that we are dealing with large pressures in magma generation, and that the initial pressures in the source regions can be assumed similar to the lithostatic pressure.

20 Laminar and Turbulent Flow

Fluids in motion may display two types of flow, laminar flow which is a regular type of flow, and turbulent flow being of an irregular nature due to the presence of transverse eddies. The two types of flow occur at different velocities, the laminar flow occurs at slow flow rates, while the turbulent flow occurs at relatively high flow rates.

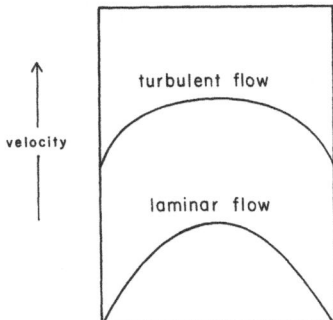

Fig. XII-27. The flow velocity profiles for turbulent and laminar flow in a circular channel

If the flow velocity of a liquid displaying laminar flow is increased continuously, then there will be a transition to turbulent flow at a certain flow velocity. The actual velocity of transition depends on the dimensions of the duct and the properties of the liquid. The transition velocity is calculated from Reynolds' number, and is about 70 km/h for basaltic liquids.

The characteristic features of laminar flow is shown in Fig. XII-27. All the particles of the liquid move parallel to each other, the stream lines of the liquid are parallel, and do not intersect each other. The velocity of the liquid at the walls is zero, and increases gradually towards the center of the duct, the velocity distribution having the shape of a parabola. The mean velocity v_m, of the liquid in a pipe is estimated from Poiseuilles' law (Kay 1968; Jacob 1949):

$$v_m = \frac{2 \Delta P D^2 10^6}{N L \eta} \tag{61}$$

where:

ΔP: pressure difference between ends of pipe (bar)
D: diameter of pipe (cm)
L: length of pipe (cm)
η: viscosity (poise)
N: a number (64 for circular cross-section)

The number N depends to some degree on the shape of the duct as shown in Fig. XII-28. The velocity distribution is given by:

$$v_r = \frac{16 \Delta P 10^6 (R^2 - r^2)}{N \eta L} \tag{62}$$

where R is the radius of the pipe. The maximal velocity of the liquid in the center of the pipe is given by:

$$v_{max} = \frac{4 \Delta P D^2}{N \eta L} 10^6 \tag{63}$$

These three equations are the basic equations for the liquid flow in ducts. They apply only for a pipe with circular cross-section, and cannot be used directly for the

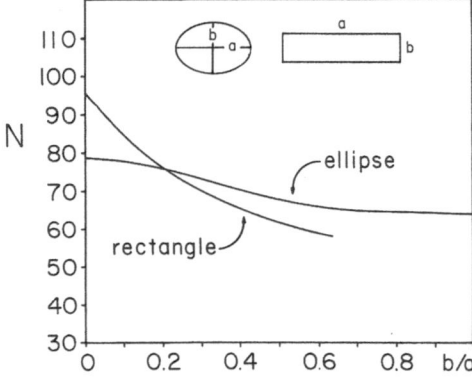

Fig. XII-28. The variation in the factor N as function of the aspect ratio b/a for a duct with parallel walls and with ellipsoidal shape. The ratio b/a is determined whereafter the value of N is determined from the diagram and used in Eq. (61) and (62)

flow in dykes and ducts of other shapes. However, it is possible to modify these equations so that they can be used for ducts of various shapes. It can be shown that whatever shape the cross-section of the duct, it is possible to calculate the diameter of a pipe with the same resistance to flow. This diameter is called the equivalent diameter D_e (Jacob 1949). Let the cross-sectional area of a duct be A, and its circumference C. The equivalent diameter is then given by:

$$D_e = \frac{4A}{C} \tag{64}$$

For a circular pipe the value of D_e is given by:

$$D_e = \frac{4\pi D^2}{4\pi D} = D \tag{65}$$

The equivalent diameter of the pipe is, thus, the diameter of the pipe as might be expected. Consider now that a dyke can be represented by a rectangular cross-section, the long side being a and the short one b, then:

$$A = a \cdot b$$
$$C = 2(a+b) \tag{66}$$

so that:

$$D_e = \frac{2a \cdot b}{a+b} \tag{67}$$

A more appropriate shape for the cross-section of a dyke is probably the ellipse. Let the semiaxes be a and b, then C is approximately given by:

$$C = 2\pi \sqrt{\frac{a^2+b^2}{2}} \tag{68}$$

and for A, one gets:

$$A = \pi a b \tag{69}$$

and D_e then becomes:

$$D_e = \frac{2ab}{\sqrt{\frac{a^2+b^2}{2}}} \tag{70}$$

The Eqs. (61) to (63) may now be used if D is replaced by D_e, and N is estimated from Fig. XII-28.

21 Reynolds' Number

Before the relationship between pressure and flow rate can be determined it is necessary to estimate the type of flow. This is done by calculating Reynolds' number which is given by:

$$N_{Re} = \frac{\varrho v_m D_e}{\eta} \tag{71}$$

If this dimensionless number exceeds the value 2300, the flow is turbulent, if on the other hand, N_{Re} is less than 2300, then the flow is laminar and the flow rate will be determined by the equations given above for the laminar flow. The value of the critical number 2300 has been obtained from experiments, and the actual value may differ slightly from 2300. One might except that the critical number would depend on the smoothness of the walls of the duct, but that is apparantly not the case (Jacob 1949), so that the critical value 2300 can be used with confidence for a variety of petrological problems.

22 Turbulent Flow

At high velocities the flow becomes turbulent, the motion of the liquid becomes irregular as eddies of all sizes are formed in the liquid. The resistance of the turbulent flow is much higher than for laminar flow, and the mean velocity of the liquid v_m will, therefore, only increase slowly with an increasing pressure gradient. The velocity distribution of the liquid is also different, the center of the flow has a nearly constant velocity, while the rate decreases rapidly at the margins of the duct (Fig. XII-27). Despite the irregularity of the flow, the mean flow velocity can be calculated and is given by (Jacob 1949):

$$v_m = \frac{1600 \, \Delta P^4 \, D_e^4 \, 10^{24}}{\eta \varrho L^4} \tag{72}$$

where ΔP again is given in bar. The application of these equations is demonstrated in the example given below.

Example:

Let us consider the ascent of magma from a subvolcanic magma chamber, the roof of which is situated at a depth of 3 km below the surface. Further, let the dyke be of 3 km length, and 3 m width, the aspect ratio of the dyke is then 10^{-3}, a fairly typical value. Let the viscosity of the magma be 1000 poise. We will first have to calculate the value of N_{Re} in order to estimate the critical flow velocity, and this calculation requires the value of D_e. Let the cross-section of the dyke be ellipsoid, then according to Eqs. (69), (68), and (70):

$$4A = 28.27 \cdot 10^7 \, cm^2$$
$$C = 6.66 \cdot 10^5 \, cm \tag{73}$$
$$D_e = 424 \, cm$$

Having obtained the value for D_e we can now calculate the critical value of v_m using Eq. (17):

$$N_{Re} = \frac{2.7 \, v_m \, 424}{1000} = 2300 \tag{74}$$

From this eqation $v_{critical}$ is estimated as $2 \cdot 10^3$ cm/s, or 72 km/h. If the magma is ascending with velocities lower than this value, the flow is laminar, otherwise it is turbulent. The critical pressure for the flow is now estimated from Eq. (61), the result being 132 bar. This is a relatively small pressure in connection with eruptions, and we might expect turbulent flow rather than laminar flow by the eruption of magma from a subvolcanic magma chamber. For flow rates above the critical one, the flow rate is estimated from Eq. (72), and the result is shown in Fig. XII-29. For pressure differences being smaller than 132 bar the flow is laminar, and the flow rate increases linearly with pressure. Above this pressure the character of the flow changes and it becomes turbulent. In the turbular regime the increase in flow rate with increasing pressure is small, and large pressure gradients are now required to increase the rate of flow.

23 Ascent Rate and Eruption Rate

The rate of ascent is the average flow rate in the feeder dyke. The rate of flow will be different in different parts of the dyke because its width varies, and the rate of flow increases towards the center of the dyke. However, the flow rate for the dyke as a whole will be represented by an average flow rate. The eruption rate, on the other hand, is the rate by which the lava is pouring out at the surface of the Earth. In general, the eruption rate must be larger than the ascent rate as the lava only erupts from selected parts of the surface fissure of the feeder dyke. It is a general observation that a long fissure is formed at the beginning of an eruption, but the lava is only pouring out from a small part of this fissure. A lava fountain might form along some length of the fissure, but the length of the fountain will only be a fraction of the entire length of the fissure. The eruption rate and the ascent rate is related to each other by a simple relationship. Let the area of the dyke be A_d and that of the lava fountain be A_f. If the ascent rate is R_a and the eruption rate is R_e, then according to Pascals' law we have:

$$\frac{A_d}{A_f} = \frac{R_f}{R_d} \tag{75}$$

The velocities for the lava fountains can be estimated from their height s. Typical heights of Hawaiian lava fountains are about 200 m, while the maximum height has

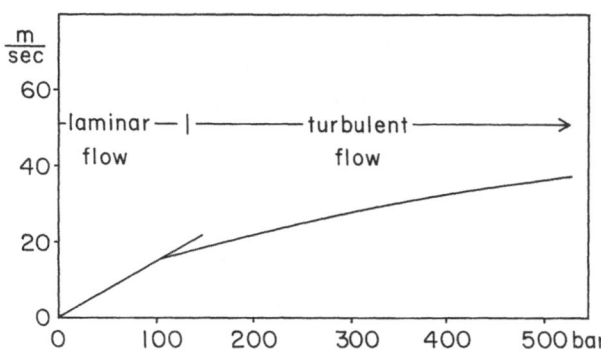

Fig. XII-29. The transition from laminar to turbulent flow for a dyke of 3 m width. The transition occurs at about 132 bar

been estimated as 600 m. Let g be the acceleration of gravity (980.7 cm/s²) and t the time, then:

$$s \; = \frac{1}{2} g t^2 \tag{76}$$

and

$$R_f = gt \tag{77}$$

so far s = 200 m we get R_f = 226 km/h. For s = 600 m, R_f = 391 km/h. These velocities are quite impressive, and indicate that large pressures must be involved.

However, it is unlikely that the magma ascend with these velocities in the feeder dyke. If the cross-sectional area of the fountain is ten times smaller than that of the dyke, we obtain ascent rates of the order of about 30 km/h, which appears as a more reasonable value.

The ascent rate of magma in dykes is difficult to estimate, but there are two methods which might afford some evidence. The presence of mantle-derived nodules in a lava flow shows that the ascent rate must have been large enough for the nodules to be carried to the surface. The size of such nodules varies in diameter from 1 cm to about 30 cm, a typical value being 5 cm. The nodule with the largest diameter will indicate the minimal flow rate in the feeder dyke, so let us calculate the flow rate required to transport a nodule of this size to the surface. According to Stokes' law one gets:

$$R_d = \frac{2 g r^2 \Delta \varrho}{9 \eta} \tag{78}$$

where:

   r: radius of sphere
   $\Delta \varrho$: difference in density between sphere and liquid
   η: the viscosity of the liquid

For g = 980.7 cm/s², r = 15 cm, $\Delta \varrho$ = 0.5 g/cm³, and η = 100 poise, we get R_d = 245 cm/s, or R_d = 8.8 km/h. The value for R_d calculated in this manner is a minimum value for the ascent rate, but it is clear that the ascent rate must be much smaller than the eruption rate as suggested above.

The second method was proposed by Komar (1976) who considered the variation in the amounts of phenocrysts in dykes. In some dykes the phenocrysts have been concentrated in the central part of the dyke, and it was shown by Komar (1976) that the concentration is a function of the ascent rate. However, the concentration is also dependent on other partly unknown parameters, so it has not yet been possible to obtain values for the ascent rate using this method.

XIII Isotope Geology

1 Introduction

The initial significance of isotopes was their potential for precise age determinations, and isotopic systematics still provide the most accurate age determinations. After isotopic data had been accumulated for some years it became evident that the source of igneous rocks also may be estimated from the isotopic ratios. The reason for this additional advantage is that isotopes with atomic numbers higher than 40 are not fractionated during partial melting and fractional crystallization, so that any uncontaminated magma will obtain exactly the same isotopic ratios as its source. As an example, the granite controversy may be mentioned, which has lasted several decades. Two major types of origin have been considered for the granites. The granitic magmas may either have originated by anatexis from a gneiss or sediment, or by fractional crystallization from an andesitic magma. Both these theories have proven correct, as granites may form in both manners. The combined evidence from the $^{87}Sr/^{86}Sr$, $^{143}Nd/^{144}Nd$ ratios and the lead isotopes will now provide the fundamental information about the origin of a granite.

In addition, the isotopes have given information about the origin of the solar system and the development of the Earth. The isotopes formed by rapid decay like ^{129}Xe give information about the condensation rate of the solar nebulae, and isotopic anomalies of the Mg and Al isotopes have shown that the solar system formed from at least two different sources. The isotopic data for recent and tertiary basalts show that the basalts must originate from different mantle sources. It is apparent that the isotopic systematics of basalts may define the geochemical processes within the mantle, and provide evidence about the convection pattern within the mantle. By studying the variation in isotopic ratios with time it might also be possible to estimate the variation in convection patterns within the mantle. The isotopes may not only be used for geochemical applications, their systematics may also be used as a geophysical tool.

There are now many different isotopic pairs in use, each of which may serve a specific purpose. The isotopes of uranium first aroused interest as the decay of uranium accounts for some of the heat generation within the Earth. In 1907 Bootwood published age determinations based on uranium, and subsequently Holmes (1913) published a book which emphasized the potential impact of isotopes on geology, but his ideas did not attract much attention at that time. Age determinations based on Rb/Sr isotopes were first considered by Hahn and Walling (1938), while the U-Pb, Th-Pb, and K-Ar methods became established around 1950. In 1949 Arnold and Libby applied the ^{14}C method for dating. The use of the important Sm-Nd isotopic pair was pioneered by Lugmair (1974), and the high precision methods required for this isotopic pair and modern high precision dating were developed by Wasserburg et al. (1969).

We will here consider the Rb-Sr, Sm-Nd, and K-Ar methods, and the relationships of the O and He isotopes. The general aspects of the methods and applications of isotopes have been given in an excellent textbook by Faure (1977), and the Rb-Sr isotopic pair have been dealt with in detail by Faure and Powell (1972). Reviews of the petrological significance of isotopes have been given by DePaolo (1981), O'Nions et al. (1979), and Allegre (1982) as well as by Hofmann and Hart (1978). The significance of isotopes for the evaluation of the development of the solar system have been considered by Wetherill (1975), Lee (1979) and Wasserburg et al. (1980).

2 Basic Equations

From their work with isotopes Rutherford and Soddy (1902) were led to the conclusion that the radioactivity of a sample is proportional to the number of atoms in the sample. Hence, the number of decays per time unit is constant for a given amount of sample. This relationship shows, that the statistical chance for decay is constant for each of the atoms. The intensity of the radioactivity will be dependent on the isotopic composition of the sample and its age. The number of decays per time unit is given by:

$$-\frac{dN}{dt} = \lambda N \tag{1}$$

where λ is a proportionality constant related to the isotopic composition of the sample. The negative sign is required as the number of radioactive atoms is decreasing with increasing time. The differential equation is solved by isolating the variables:

$$\int_{N_0}^{N} \frac{-dN}{N} = \int_0^t \lambda t \tag{2}$$

and by integration we obtain:

$$\ln (N/N_0) = -\lambda t \tag{3}$$

By exponentiation we have:

$$\frac{N}{N_0} = e^{-\lambda t} \tag{4}$$

This equation is the fundamental equation for all age determinations. N_0 is the number of radioactive atoms present at $t = 0$, and N is the number of radioactive atoms left at time t. The practical application of this equation requires some additional treatment as shown below.

The decay constant for ^{87}Rb is $1.39 \cdot 10^{-11}$. Let the initial number of ^{87}Rb atoms be 1000, the number of radiogenic atoms is then 986.2 after 10^9 yr.

The atoms represented by N_0 are called the parent atoms, and the atoms generated by the decay are called the daughter atoms or the radiogenic isotopes. When ^{87}Rb decays to ^{87}Sr, then ^{87}Rb is the parent, and ^{87}Sr the daughter atom.

A rock will contain some ^{87}Rb and some ^{87}Sr, but not all the ^{87}Sr present are radiogenic. In fact the major part of the ^{87}Sr isotopes present will generally have been formed before the Earth was formed. Some of the ^{87}Sr was present already when the Earth was formed, and we should, therefore, distinguish between the radiogenic daughter atoms D^*, and the daughter atoms that were present when the radioactive clock started, D_0. Equation (4) may be changed so that we can distinguish between D^* and D_0. The number of radiogenic daughter atoms will be given by:

$$D^* = N_0 - N \tag{5}$$

By representing N_0 by the $N e^{\lambda t}$ we get:

$$D^* = N e^{\lambda t} - N = N (e^{\lambda t} - 1) \tag{6}$$

The total amount of daughter atoms is given by:

$$D = D_0 + D^* \tag{7}$$

so that:

$$D = D_0 + N (e^{\lambda t} - 1) \tag{8}$$

Both D and N are the recent values and can be measured using mass spectrometry, while D_0 and t have to be estimated. The value of D_0 can sometimes be assumed, however, D_0 can be estimated accurately by performing several estimates of D and N on minerals or rocks with the same age. By estimating D and N for at least two different samples the two unknowns in Eq. (8) can be determined.

3 Rubidium–Strontium Dating

Rubidium has two naturally occurring isotopes, $^{87}_{37}$Rb and $^{85}_{37}$Rb the first one being radioactive and decaying to $^{87}_{38}$Sr by the emission of a beta particle:

$$^{87}_{37}\text{Rb} = ^{87}_{38}\text{Sr} + \beta + v + Q \tag{9}$$

where v is a neutrino and Q is the decay energy. As ^{87}Sr is the radiogenic daughter isotope we get according to Eq. (8):

$$^{87}\text{Sr} = ^{87}\text{Sr}_0 + ^{87}\text{Rb} (e^{\lambda t} - 1) \tag{10}$$

The atomic abundances of both ^{87}Sr and ^{87}Rb are those of the sample today, and they can be measured by various methods including mass spectrometry. When using mass spectometry one obtains isotopic ratios rather than the absolute abundances, and it is, therefore, more accurate to work with isotopic ratios. Equation (10) is,

Table XIII-1. Rubidium–strontium data

	%	amu
$^{85}_{37}\text{Rb}$	72.1654	84.9117
$^{87}_{37}\text{Rb}$	27.8346	86.9094
Average		854678
$^{88}_{38}\text{Sr}$	82.53	87.9056
$^{87}_{38}\text{Sr}$	7.04	86.9088
$^{86}_{38}\text{Sr}$	9.87	85.9092
$^{85}_{38}\text{Sr}$	0.56	83.9134

Decay constants:
$\lambda = 1.39 \cdot 10^{-11} \text{ y}^{-1}$
$\lambda = 1.41 \cdot 10^{-11} \text{ y}^{-1}$

therefore, applied in a different form. The equation is divided with the abundance of one of the nonradiogenic, stable isotopes, and as ^{86}Sr is stable and occurs in about the same amount as ^{87}Sr (Table XIII-1), this isotope is used as denominator:

$$\frac{^{87}\text{Sr}}{^{86}\text{Sr}} = \left(\frac{^{87}\text{Sr}}{^{86}\text{Sr}}\right)_0 + \frac{^{87}\text{Rb}}{^{86}\text{Sr}}(e^{\lambda t} - 1) \tag{11}$$

This equation is the generally applied equation for Rb–Sr dating. The terms $^{87}\text{Sr}/^{86}\text{Sr}$ and $^{87}\text{Rb}/^{86}\text{Sr}$ are the recent ratios, and these are measured and estimated from the sample. The $(^{87}\text{Sr}/^{86}\text{Sr})_0$ ratio is the initial ratio for the sample when it was generated, and this ratio is unknown. As the time t is also unknown, there are two unknowns in the equation.

The isotopic ratio $^{86}\text{Sr}/^{87}\text{Sr}$ is estimated directly from mass spectrometry using a pure strontium salt obtained by acid solution and exchange chromatography. The abundance of Rb and Sr is estimated by XRF or isotopic dilution. The ratio $^{87}\text{Rb}/^{86}\text{Sr}$ is determined from the absolute abundances of Rb and Sr, and the estimate of this ratio is more elaborate.

Let Rb and Sr be the amounts of these elements in a rock, normally given in ppm. Using mass spectrometry the fractions of the Sr isotopes are determined, and let the fraction of ^{87}Sr be $F(\text{Sr})$. In order to obtain the atomic abundances we also introduce the at. wts. $W(\text{Rb})$ and $W(\text{Sr})$. The atomic abundance of ^{87}Rb is given by:

$$^{87}\text{Rb} = 0.278346\,(\text{Rb ppm}/85.4678)$$

The atomic abundance of ^{87}Sr is given by:

$$^{87}\text{Sr} = F(\text{Sr})\,(\text{Sr ppm}/W(\text{Sr}))$$

The fraction of ^{87}Rb isotopes is the same in all terrestrial and meteoritic material. The constant ratio shows that the initial fraction of ^{87}Rb was the same in the different parts of the solar nebula when the planets were formed. The constant ratio between ^{87}Rb and ^{85}Rb also results in a constant at. wt. for Rb. The at. wt. of Sr will depend on the sample since ^{87}Sr has been added in various amounts to the sample by the decay of ^{87}Rb. The at. wt. of Sr is estimated from the isotopic ratios obtained from mass spectrometry. Let us assume that we have obtained the following ratios (Faure 1977):

	Ratio	Abundance
87/88	0.0846	0.0698
86/88	0.1194	0.0986
84/88	0.0068	0.0056
88/88	1.0000	0.8259

Using the at. wts. in Table XIII-1, the at. wt. of Sr is 87.6079 amu. Having estimated $W(Sr)$ we can then estimate the atomic abundance of ^{87}Sr using the above equation.

We can now return to Eq. (11), however, we are still not able to estimate the age as the $(^{87}Sr/^{86}Sr)_0$ ratio is unknown. This ratio can be estimated if we perform at least two estimates of the isotopic ratios for two different samples with different $^{87}Rb/^{86}Sr$ ratios. In general, this requirement does not constitute any limitation as one can estimate the ratios for two different minerals or rocks of the same age. Equation (11) is linear and of the form:

$$y = \alpha x + \beta \tag{12}$$

where:

$$\alpha = (e^{\lambda t} - 1)$$
$$\beta = (^{87}Sr/^{86}Sr)_0$$
$$x = (^{87}Rb/^{86}Sr)$$

Both α and β will be the same for the different minerals of a rock, because they have formed from the same source at the same time. The different minerals of a rock will have different $^{87}Rb/^{86}Sr$ ratios and these ratios will result in different $^{87}Sr/^{86}Sr$ ratios. The two ratios will plot along a line in a diagram where $^{87}Sr/^{86}Sr$ is plotted as a function of $^{87}Rb/^{86}Sr$ (Fig. XIII-1). At the time where the rock is formed all minerals will have the same $^{87}Sr/^{86}Sr$ ratio, and these ratios will plot along a horizontal line in the diagram. As the rock ages the amount of ^{87}Sr increases, while the amount

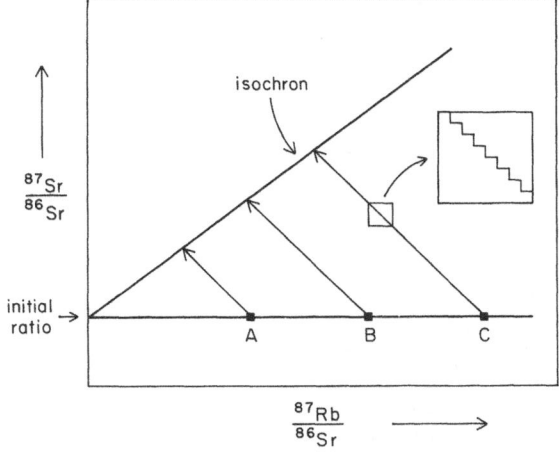

Fig. XIII-1. The change in $^{87}Sr/^{86}Sr$ ratios with time for three different compositions, *A, B,* and *C.* The ratios will increase with time from the initial ratio, and the increase *is* proportional to the $^{87}Rb/^{86}Sr$ ratio of the compositions, which may be minerals or rocks. The $^{87}Sr/^{86}Sr$ ratios will at any given time fall along a straight line called the isochron. The *insert* shows how the ratios vary with time

of ^{87}Rb decreases. Each time an ^{87}Rb atom decays to an ^{87}Sr atom one of the compositions in Fig. XIII-1 will move one step to the left and one step up. The isotopic ratios will, therefore, move along lines with the slope −1.0. Further, they will all be situated along a line, called the isochron, which has a positive slope. The slope α thus defines the age of the sample, while the initial ^{87}Sr/^{86}Sr ratio is given by the intersection between the line and ^{87}Rb/^{86}Sr = 0.

The mineral dating is performed by estimating the ^{87}Rb/^{86}Sr and ^{87}Sr/^{87}Sr ratios of different minerals, which must be separated from the rock. The larger the difference of the ^{87}Rb/^{86}Sr ratios, the more accurate the dating will be.

A combined mineral and whole rock isochron is shown in Fig. XIII-2 for the Algoman granites of the Ontario province in Canada. The two sets of data lie nearly on the same isochron, so that they define the same age. The coincidence shows that the granite must have been left unmetamorphosed since its intrusion, otherwise the minerals would have defined a different isochron.

The effect of metamorphism is demonstrated in Fig. XIII-3. A granite intrudes at time t_i, and the metamorphism starts at time t_s. After the intrusion took place each of the minerals change their ^{87}Sr/^{86}Sr ratios because their ^{87}Rb/^{86}Sr ratios are different. During the metamorphism the temperature increases and the diffusion rate also increases. The ions of the ^{87}Sr isotope will be able to move freely around between the different minerals, and the minerals will, therefore, all attain the same ^{87}Sr/^{86}Sr ratio if the temperature is high enough. When the metamorphism reaches amphibolite facies the rock will be completely homogenized with respect to isotopes. As the metamorphism goes on all the minerals will retain constant ^{87}Sr/^{86}Sr ratios. The ^{87}Sr/^{86}Sr ratios will not begin to differ before the rock undergoes cooling, and the minerals again become closed systems with respect to diffusion. This will happen at different temperatures for the different minerals, the closing temperatures being between 300 °C and 600 °C. If all the minerals have closed at t_e, then the mineral age of the rock will be t_e, and it will not be possible to estimate the

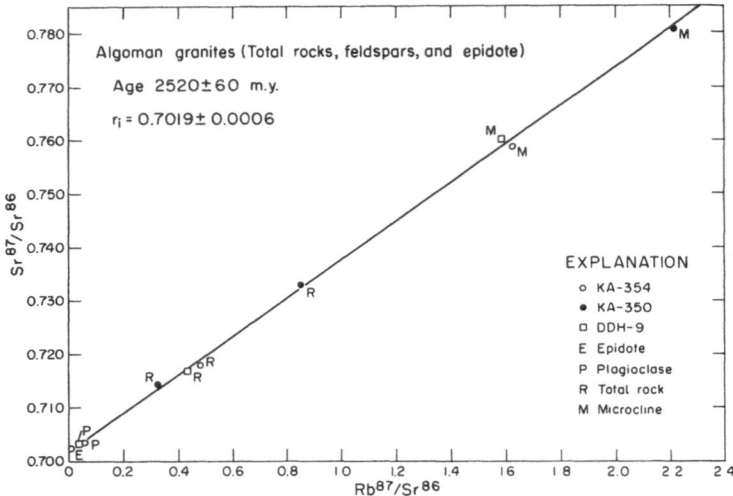

Fig. XIII-2. Rb-Sr isochron plot for total rocks and constituent minerals, feldspars, and epidote for the Algoman granites, Rainy lake (Peterman et al. 1975)

Fig. XIII-3. The variation in $^{87}Sr/^{86}Sr$ ratios by metamorphism. The rock is intruded at time t_i, and begins to undergo metamorphism at time t_s. This results in a homogenization of the ratios for the different minerals, which lasts until the metamorphism ends at time t_e. Thereafter the minerals develop their individual ratios again

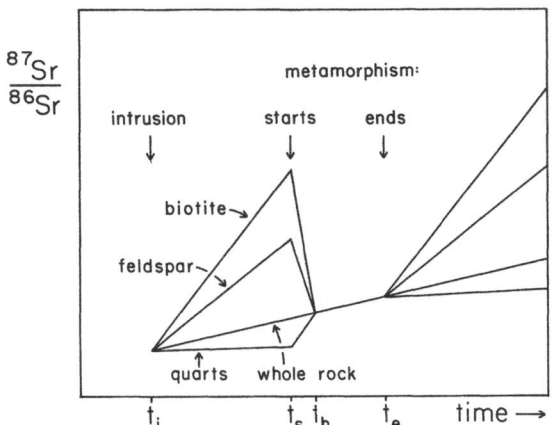

original age of the rock from the minerals. The original age may, however, still be determined from the dating of whole rocks with different $^{87}Rb/^{86}Sr$ ratios. An example of the metamorphic resetting is shown in Fig. XIII-4. The dating was carried out on the Cairn Chuninneag granite complex in Scotland. The whole rock dating yields an age of 560 ± 10 m.y., while the mineral isochron yields a younger age of 412 m.y. This age is the closing age, and we cannot say anything about when the metamorphism started, only that it ceased 412 m.y ago. The K–Ar dating will also yield the closing age of the minerals, and hence, the end of the metamorphism.

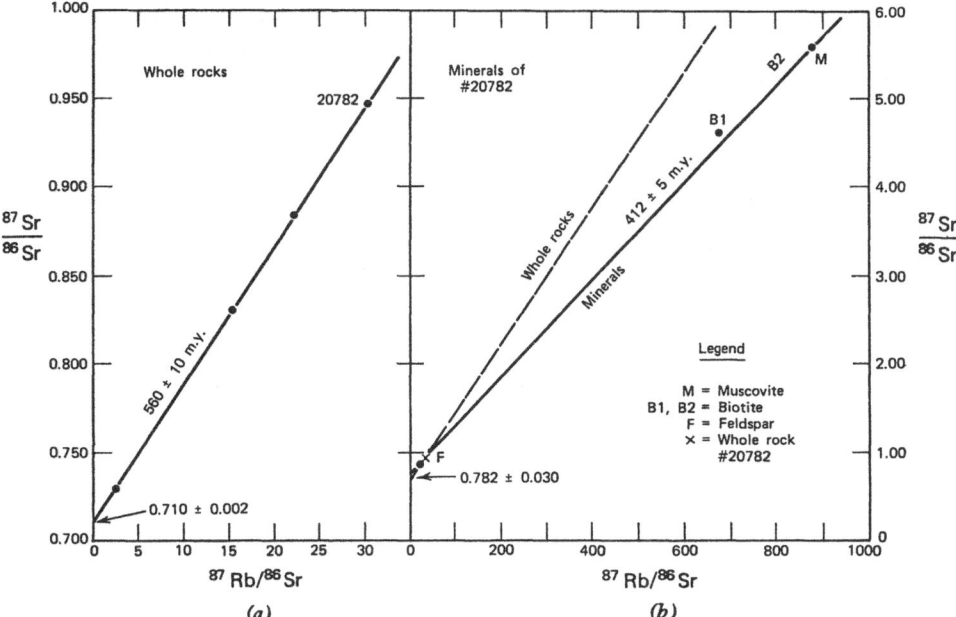

Fig. XIII-4a, b. a Whole rock isochron of granitic rocks from the Carn Chuinneag complex in the northern Scottish Highlands; **b** Mineral isochron for the same rocks formed by the minerals. The granite crystallized 560 m.y. ago, and was later affected by the Caledonian metamorphism 412 m.y. ago (Long 1964; Faure 1977)

4 Meteorites

The isotopic ratios of the meteorites form the basis for the study of the evolution of
the isotopic ratios of the Earth, because the initial ratios of the Earth are the same as
those of the meteorites. Both the meteorites and the planets formed from the con-
densing nebulae that were hurled out from the sun about 4.6 aeons ago. The sun is
classified as a normal star, and the atomic chain reactions within the sun can only
form elements with at. wts. up to 12. The conclusion is, therefore, that elements like
Si, Fe, Ca, Ti, and Al that form the major constituents of rocks, must have been
formed before the sun was formed. The nuclear synthesis of the heavy elements is
believed to occur in supernovas, a type of star that is quite common in our galaxy. It
has been proposed that the elements of the solar system stem from an exploding
supernova, an explanation which at first sight might appear a bit exotic. However,
the interstellar shock wave from an exploding supernova may concentrate the interstellar dust, which thereafter may become concentrated due to the gravitational
attraction – and form a star. During the gravitative collapse the dust begins to rotate
around the gravity center, and this rotation caused the rotation of the sun. If this
theory is correct, then the age of the sun is only slightly larger than the age of the
planets and the meteorites.

The present photosphere of the sun has a Rb/Sr ratio which has been estimated
as 0.5 and 0.65, and it is considered that the primordial nebuale had the same ratio,
as neither Rb nor Sr can be formed within the sun. This ratio is much higher than
that of common rocks, and is equivalent to an $^{87}Rb/^{86}Sr$ ratio of 1.4 and 1.8, respectively. Most meteorites have lower $^{87}Rb/^{86}Sr$ ratios as evident from Table XIII-2.
Apparently a large fraction of the rubidium in the nebulae must have been lost from
the material that formed the Earth and the meteorites. Due to the relatively high
$^{87}Rb/^{86}Sr$ ratio the initial increase of the $^{87}Sr/^{86}Sr$ ratio must have been quite high,

Table XIII-2. Chondrite data

Type	$^{87}Rb/^{86}Sr$ Atomic	$(^{87}Sr/^{86}Sr)_0$ Atomic	Age aeons
Enstatite chondrites	0.304–4.60	0.6993	4.54
Bronzite chondrites	0.516–3.937	0.6983	4.69
Amphoterites	0.102–1.359	0.7005	4.56
Carbonaceous chondrites	0.154–0.997	0.6983	4.69 (?)
BABI	0.00234–0.024	0.69899	4.39
Allende meteorite	–	0.69877	4.62
Juvinas basaltic achondrite	–	0.698976	4.60
Bulk Earth	0.084396	0.69899	4.55
Anorthosite Moon	–	0.69884	4.60

and the variation in this ratio of the meteorites might give evidence about the condensation history of the solar system. Both the $^{87}Sr/^{86}Sr$ ratio and other isotopic evidence suggests that the condensation took place within about 50 m.y.

The meteorites form a complex group of rocks, but their initial $^{87}Sr/^{86}Sr$ ratios display a narrow range of values between 0.6883 and 0.7000 (Table XIII-2), and their ages vary between 4.5 and 4.6 aeons. Some of the initial ratios have been estimated from the minerals of single meteorites and these ratios are called internal ratios, and are considered the most reliable estimates. The ages and initial ratios have also been estimated from groups of samples with the same petrographic compositions. One of these groups, the basaltic achondrites, contains a very small amount of Rb, and their initial ratios are, therefore, significant as their ratio is virtually independent of age. The isochron estimated for the basaltic achondrites is shown in Fig. XIII-5. The basaltic achondrites were formed from a silicate melt, and are in this respect similar to the material that formed the Earth. The similarity suggests that the initial $^{87}Sr/^{86}Sr$ ratio of the basaltic achondrites may be used as a reference value for the initial ratio of the Earth. The initial $^{87}Sr/^{86}Sr$ ratio of the basaltic achondrites is 0.69899, and is called the BABI value (Basaltic Achondrites Best Initial value). The most primitive ratio estimated so far is that of the refractory chondrules of the Allende meteorite, i.e., 0.68977. The refractory chondrules are 1–10 mm in diameter and consist of Ca–Al silicates like spinel, corundum, fassaite, and anorthite, and these chondrules are considered the first condensates of the nebulae (Grossman 1980).

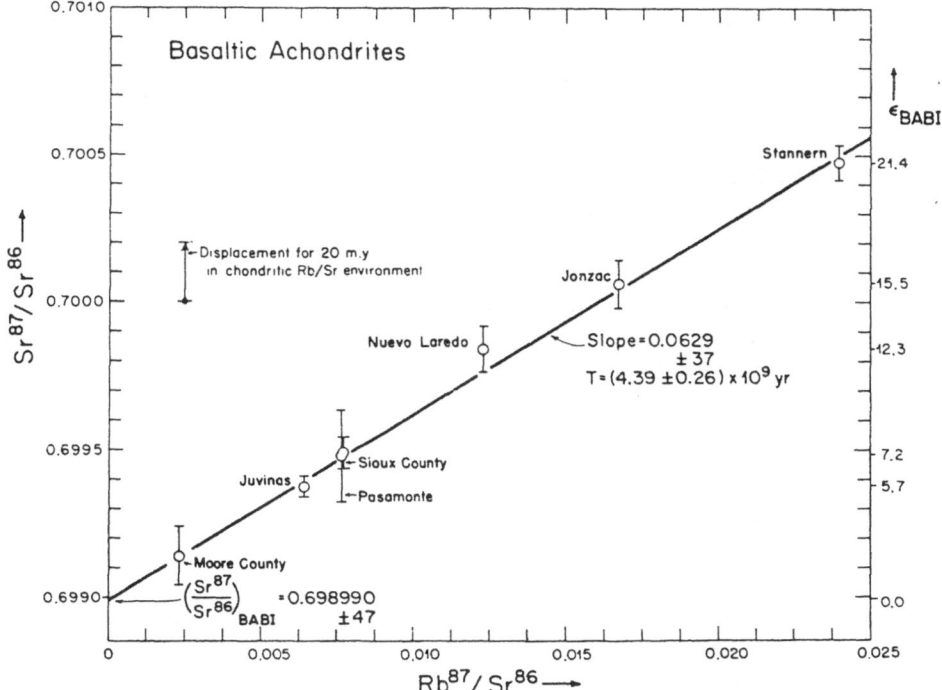

Fig. XIII-5. The isochron plot for the basaltic achondrites. The slope suggest an age of 4.39 m.y. and the initial ratio is estimated as 0.698990 (Papanastassiou and Wasserburg 1969)

Assuming the Rb/Sr ratio of the sun is 0.65, a difference in the $^{87}Sr/^{86}Sr$ ratio of $5 \cdot 10^{-5}$ will be equivalent to $2 \cdot 10^{6}$ yr. Thus, the difference in the initial ratio of the Allende meteorite and the BABI value suggest a time difference of about $10 \cdot 10^{6}$ yr.

The moon is smaller than the Earth and has, therefore, cooled much faster than the Earth. The planet mars has also cooled quite rapidly, and both the moon and mars provide valuable information about the early formation of the solar system and the initial formation of the planets. The rocks on the moon are generally older than on the Earth. The earliest formed rocks were the anorthosites that form the high lands on the moon, and they were generated from 4.65 to 4.1 aeons ago. The oldest date for the anorthosites is about 4.6 aeons, and this rock, anorthosite 60015 had an initial $^{87}Sr/^{86}Sr$ ratio of 0.69884. The basaltic mares are the youngest rocks with ages from 3.8 to 3.2 aeons. The isotopic systematics appear to indicate that the magmatic activity was especially high on Earth in this time interval, the major part of the continents is considered to have formed within this period. The Pb and Rb–Sr dating of the lunar rocks suggests that the moon was formed 4.65 aeons ago. The age of the Earth is probably the same, but the consolidation of the Earth's crust may have occurred later, as the cooling rate of the Earth was smaller than that of the moon. The oldest terrestrial rocks are the tonalitic gneisses from West Greenland with ages of about 3.8 aeons. The oldest age estimated so far for any terrestrial material is 4.2 aeons. This age was obtained from grains of zirkon in an archaic sediment using the ion microprobe. This instrument allows the analysis of isotopic compositions of micron-sized spots. As zirkon is a typical accessory mineral of granites the result suggests that at least some granite was formed very early in the archaic, and it is very likely that some granite magma was formed when the upper mantle consolidated. These zirkons were situated in quartzites in the Mt. Narryer region of Western Australia nearly 3630 m. y. old orthogneiss (Froude et al. 1983).

5 $^{87}Sr/^{86}Sr$ Ratios of Basalts

As the Rb–Sr isotopic data began to accumulate it became evident that abyssal tholeiites (MORB) and plume basalts (OIB) have different $^{87}Sr/^{86}Sr$ ratios (Tatsumoto et al. 1965; Hofmann and Hart 1978). The distribution of the measured ratios is shown in Fig. XIII-6, the averages are 0.7028 and 0.7034, respectively. The ratios have been measured on tertiary and recent basalts, and the ratios may be considered the initial ones, as the change in ratio is small in $60 \cdot 10^{6}$ yr. Although there is some overlap in the distributions it is evident that the ratios for the abyssal tholeiites are lower than the ratios for the plume basalts.

In dealing with these isotopic data we should remember that the magmas formed by partial melting from a given source have the same isotopic ratio as the source. Consider that the source consists of lherzolite. The different minerals of the lherzolite will have different $^{87}Rb/^{86}Sr$ ratios, but the $^{87}Sr/^{86}Sr$ ratios will nevertheless be the same for all the minerals at high temperatures. This is due to the rapid diffusion at high temperatures which allows the free diffusion of Sr between the minerals. At low temperatures, less than about 1100 °C the diffusion rate becomes too small for homogenization and the minerals get different $^{87}Sr/^{86}Sr$ ratios. As the partial melting of the mantle occurs at high temperatures of about 1500 °C, the iso-

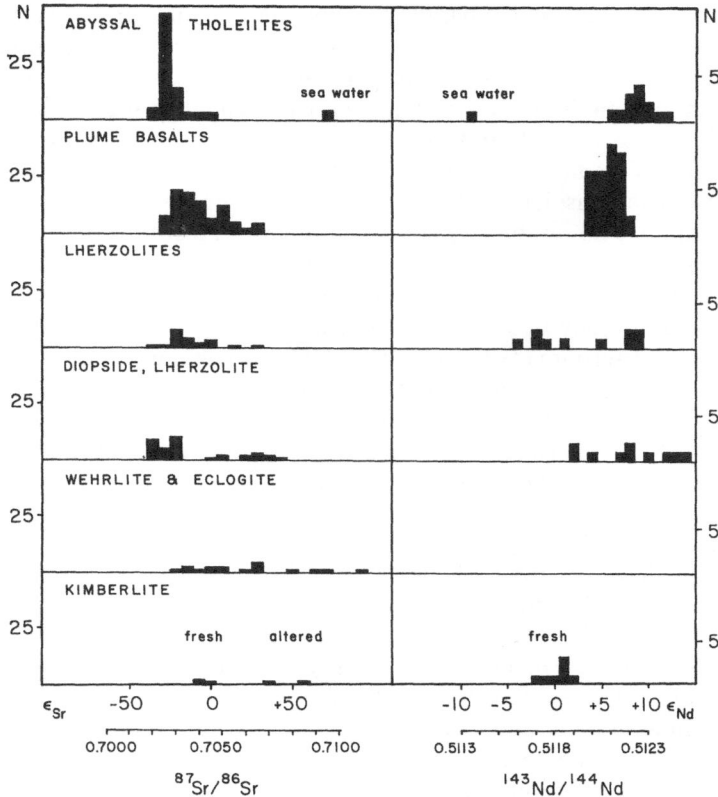

Fig. XIII-6. Histograms for the ε_{Sr} and ε_{Nd} values of different mantle derived rocks and seawater. The values were compiled from the literature, and the histograms may change as additional analyses become available

topic ratios of the primary magmas will be identical to the average ratio of the source. Apart from the diffusion there will also be phase changes in the ascending mantle material which ensure similar ratios for the different minerals.

The difference in isotopic ratios of the plume basalts and the abyssal tholeiites, thus, shows that their sources must have different ratios, and the sources must, therefore, be of different nature. Another explanation could have been that the abyssal tholeiites have been contaminated with seawater, but the ^{87}Sr/^{86}Sr ratio for seawater is quite high, about 0.7090 (Fig. XIII-6).

The two types of basalt do not only differ in their isotopic compositions, but also in their content of incompatible trace elements. The abyssal tholeiites are relatively depleted in Rb, K_2O, TiO_2, and LREE. Hence, the source of the abyssal tholeiites must be depleted compared to the source of the plume basalts.

In addition, the plume basalts and the abyssal tholeiites differ in their geophysical setting. The plume basalts occur in continental rift zones and on oceanic islands that may form chains, like the Hawaiian–Emperor chain. The presence of these chains suggests that the source for the plume basalts belongs to a dynamic system which is different from that of the abyssal tholeiites. Apparently, there is one system that generates plume basalts and another one that forms abyssal tholeiites. It is presently considered that the abyssal tholeiites are formed from the upper mantle, while the plume basalts are generated from plumes that ascend from the lower mantle, but there is as yet no general agreement on this model.

Let us evaluate this model on the basis of the Sr isotopic evidence. Further aspects of the mantle evolution will be evident from the Sm–Nd evidence, but it is relevant to consider the Rb–Sr evidence separately. The initial $^{87}Sr/^{86}Sr$ ratio of the mantle is no doubt very similar to that of BABI. In order to calculate the $^{87}Sr/^{86}Sr$ variation of the pristine mantle, we should also know the $^{87}Rb/^{86}Sr$ ratio of the mantle. This ratio cannot be estimated from the Rb–Sr evidence alone. However, this ratio may be estimated from the Sm–Nd data as was first shown by DePaolo and Wasserburg (1976). The $^{87}Rb/^{86}Sr$ ratio obtained on this basis is 0.084396.

Using these values we can calculate the $^{87}Sr/^{86}Sr$ evolution curve for the unmodified mantle material, and the curve, which is nearly a straight line is shown in Fig. XIII-7, is called the bulk Earth evolution curve. The $^{87}Sr/^{86}Sr$ ratios of the recent abyssal tholeiites plot below this curve, and the major part of the plume basalts also plot below the curve although some of these basalts have values above the curve (Fig. XIII-7). It is evident from the data in Fig. XIII-7 that the last depletion must have occurred 1.5 aeons ago. The abyssal tholeiites do contain some Rb and the $^{87}Sr/^{86}Sr$ ratio can only have increased with time. Since both abyssal tholeiites and plume basalts have ratios below the bulk Earth curve, the sources of both types of basalt must be depleted, although to various degrees. Various models have been proposed to explain these data. It has been suggested by Allegre (1982) that the mantle initially underwent whole mantle convection with the result that the whole mantle was depleted. Approximately 2–3 aeons ago the convective system changed, and an upper and lower mantle were formed. The upper mantle was thereafter further depleted by the formation of the continental crust, while only the lower

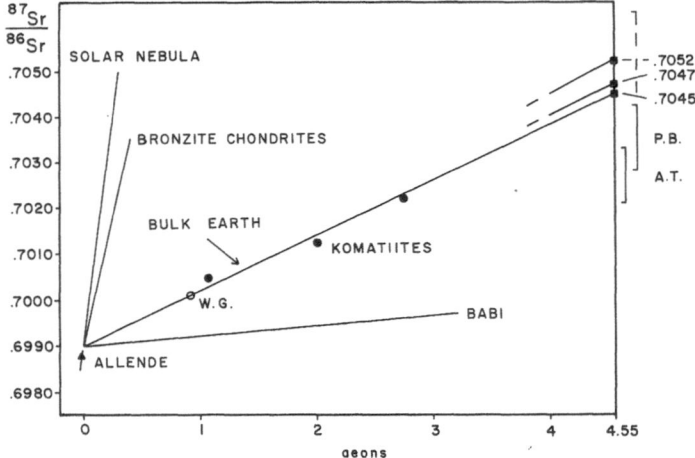

Fig. XIII-7. The bulk Earth evolution curve and the initial $^{87}Sr/^{86}Sr$ ratios for komatites. The evolution curves for BABI, bronzite chondrites, and the solar nebula are also shown for comparison. *P.B.:* plume basalts; *A.T.:* abyssal tholeiite; *W.G.:* the Isortoq rocks from West Greenland (Moorbath et al. 1972)

mantle was further depleted by the generation of mantle plumes, which generate the oceanic islands. It was also suggested that the oceanic crust may be recycled so that the present abyssal tholeiites are generated partly from material that at an earlier state formed oceanic crust. Another model favored by DePaolo (1981) assumes that the lower mantle may be essentially undepleted. The low $^{87}Sr/^{86}Sr$ ratios of the plume basalts are then due to a mixing which occurs during ascent of the plume which occurs between the plume and the surrounding mantle. These models show that the interpretation of the isotopic systematics is tightly connected with the dynamic processes that occur in the mantle, and Allegre (1982) proposed the term geochemical dynamics in order to emphasize this relationship.

6 Samarium–Neodynium Dating

The radioactive isotope ^{147}Sm decays to ^{143}Nd by emitting an α particle:

$$^{147}Sm = {}^{143}Nd + \alpha$$

The abundance of ^{143}Nd is then given by:

$$^{143}Nd = (^{143}Nd)_0 + {}^{147}Sm\,(e^{\lambda t} - 1) \tag{13}$$

The element Nd has 7 isotopes, and ^{143}Nd constitutes about 12.3% of the total amount of isotopes (Table XIII-3). In order to deal with isotopic ratios rather than the absolute amount of isotopes, Eq. (13) is divided by the amount of a stable non-radiogenic isotope, ^{144}Nd. Actually ^{144}Nd is radiogenic, but the half-life is rather large, $5 \cdot 10^{15}$ yr, so that the decay has no significance for the calculations. By division we obtain:

$$\frac{^{143}Nd}{^{144}Nd} = \frac{^{143}Nd}{^{144}Nd_0} + \frac{^{147}Sm}{^{144}Nd}\,(e^{\lambda t} - 1) \tag{14}$$

Table XIII-3. Nd-Sm data

	Abundance %	amu
^{144}Sm	3.09	143.9117
^{147}Sm	14.97	146.9146
^{148}Sm	11.24	147.9146
^{149}Sm	13.83	148.9169
^{150}Sm	7.44	149.9170
^{152}Sm	26.72	151.919
^{154}Sm	22.71	153.9220
^{142}Nd	27.11	141.9075
^{143}Nd	12.17	142.9096
^{144}Nd	23.85	143.9099
^{145}Nd	8.30	144.9122
^{146}Nd	17.22	145.9127
^{148}Nd	5.73	147.9165
^{150}Nd	5.62	149.9207

$\lambda\,(^{147}Sm)$: $6.54 \cdot 10^{-12}$ yrs^{-1}

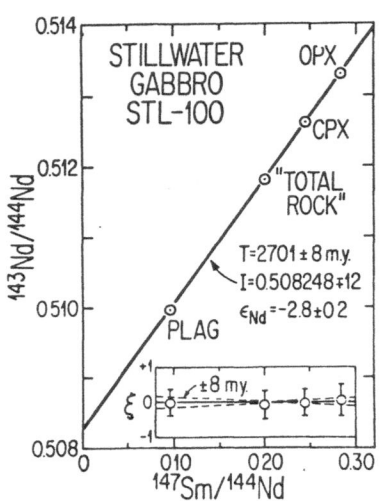

Fig. XIII-8. Sm–Nd mineral isochron for a gabbro from the Stillwater intrusion, southwestern Montana (DePaolo and Wasserburg 1979)

This equation allows age determinations in exactly the same manner as the equation for Rb–Sr, and an example is shown in Fig. XIII-8.

The chemical properties of the rare earth elements are very similar and the determination of the abundances of Sm and Nd, therefore, require the separation of these elements from the other rare earth elements, a separation which is quite difficult to perform. Using mass spectrometry the relative intensities of the different isotopes are measured, and these measurements define the relative proportions between the isotopes. Some of the Nd isotopes are nonradiogenic, like ^{142}Nd, ^{143}Nd, and ^{145}Nd. As these isotopes are not fractionated by geological processes, the proportions between these isotopes will be constant. The ratios between these isotopes could be measured by each sample, but the ratios will then display some variance due to the error of analysis. It is, therefore, more convenient to use some standard values for the ratios. A similar definition of the standard ratios has been made for the Sr isotopes which is used by all laboratories. Unfortunately, a general agreement on the ratios for Nd has not been made, and the different laboratories use different ratios, with the result that the reported ^{143}Nd/^{144}Nd ratios are different for the same sample. One group of investigators (A: DePaolo, Wasserburg, Allegre) use:

$$^{146}\text{Nd}/^{142}\text{Nd} = 0.63615$$

$$^{146}\text{Nd}/^{144}\text{Nd} = 0.724134$$

and these ratios have been used here. Another group (B: O'Nions, Hawkes, Menzies, Lugmair, Hofmann) use the following ratios:

$$^{146}\text{Nd}/^{144}\text{Nd} = 0.72190$$

$$^{148}\text{Nd}/^{144}\text{Nd} = 0.241572$$

To convert the $^{143}/^{144}$Nd ratio from one group to the other, one can use the following equation:

$$1.001563 \,(^{143}\text{Nd}/^{144}\text{Nd})_A = (^{143}\text{Nd}/^{144}\text{Nd})_B$$

The Sm–Nd pair differs in one important respect from the Rb–Sr pair. By the Rb–Sr pair the radioactive isotope, ^{87}Rb is the most incompatible element, while by the Sm–Nd pair the radioactive isotope ^{147}Sm is the most refractory one. A small degree of partial melting and a high degree of fractional crystallization will, therefore, enrich the magma in Rb and Nd.

The decay constant is quite small for ^{147}Sm and the variations in the $^{143}Nd/^{144}Nd$ ratios will consequently be small. For an average meteorite the variation will be from 0.50599 to 0.511836 in a period of 4.55 aeons. The ratios are estimated with six decimals and since it is inconvenient to deal with a small difference between long numbers, a new representation of the Sm–Nd data was proposed by DePaolo and Wasserburg (1976) and O'Nions et al. (1978). These new parameters are different and both definitions are given below.

The initial values of the $^{143}Nd/^{144}Nd$ and $^{147}Sm/^{144}Nd$ ratios for the Earth are the basis for both definitions, and this ratio is obtained for meteorite data. Investigations have shown that the basaltic achondrites and the chondrites have similar Sm/Nd ratios, and the value of this ratio has been reported as 0.308 and 0.311. By comparison, the Rb/Sr ratio displays a large variation for the meteorites. The basaltic achondrites defined the initial $^{87}Sr/^{86}Sr$ value for the Earth, the BABI value. Hence, it is relevant to use a similar value for the Nd ratio. One of the basaltic achondrites, the Juvinas achondrite has been chosen as the standard, and its initial $^{143}Nd/^{144}Nd$ value is 0.50599 (DePaolo 1983, pers. com.). However, this ratio was originally reported as 0.50677 (Lugmair et al. 1975) and this value is used by some other investigators instead. DePaolo (1979) defined the ε_{Nd} value as follows:

$$\varepsilon_{Nd} = \left[\frac{I_s}{I_c} - 1 \right] 10^4 \tag{15}$$

where:

I_s: is the initial $^{143}Nd/^{144}Nd$ ratio at the time of formation of the sample.
I_c: is the $^{143}Nd/^{144}Nd$ ratio of the bulk Earth at the time of formation of the sample.

The essence of this definition is that we compare the Nd ratio of a sample with the same ratio for the average Earth, and the bulk Earth or average Earth value is estimated from the isotopic ratios of meteorites. If the source of a magma has been depleted in Nd relative to Sm, then its I_s value will be larger than I_c, and the ε_{Nd} value will be positive for both the source and the magma, as the magma attains the same isotopic ratio as the source. A tholeiite which has formed from a depleted mantle will, therefore, have a positive ε_{Nd} value, and abyssal tholeiites have ε_{Nd} values of about + 10. The value of I_c is calculated from the following equation:

$$I_c := 0.50599 + 0.19355 \, (e^{\lambda t} - 1) \tag{16}$$
$$t = 4.55 \text{ aeons} - T$$

where T is the age of the sample, i.e., the period of time since the formation of the sample.

O'Nions et al. (1978) used the $^{147}Sm/^{144}Nd$ ratio instead and proposed the following definition:

$$\Delta Nd = \left[\frac{(^{147}Sm/^{144}Nd)_{ss} - (^{147}Sm/^{144}Nd)_{BE}}{(^{147}Sm/^{144}Nd)_{BE}} \right] 10^2 \tag{17}$$

where:

$$(^{147}Sm/^{144}Nd)_{BE} = 0.192027$$

$$(^{147}Sm/^{144}Nd)_{ss} = \frac{(^{143}Nd/^{144}Nd)_M - 0.50682}{(e^{\lambda T} - 1)}$$

and T is the age of the sample, and $(^{143}Nd/^{144}Nd)_M$ is the value of this ratio measured on the sample today. With the introduction of these parameters for the Sm–Nd pair, it is convenient to introduce the same parameters for the Rb–Sr pair, and we get for ε_{Sr}:

$$\varepsilon_{Sr} = \left(\frac{I_s}{I_c} - 1\right) 10^4 \tag{18}$$

where:

I_s: is the initial $^{87}Sr/^{86}Sr$ ratio at the time of formation of the sample.

I_c: is the $^{87}Sr/^{86}Sr$ ratio of a chondritic source or the bulk Earth at the time of formation of the sample.

$$I_c = 0.69899 + 0.084396 \, (e^{\lambda t} - 1)$$

For the ΔSr parameter we have the following definitions:

$$\Delta Sr = \left[\frac{(^{87}Rb/^{86}Sr)_{SS} - (^{87}Rb/^{86}Sr)_{BE}}{(^{87}Rb/^{86}Sr)_{BE}}\right] 10^3 \tag{19}$$

$$(^{87}Rb/^{86}Sr)_{SS} = \frac{(^{87}Sr/^{86}Sr)_M - 0.69899}{e^{\lambda t} - 1} \tag{20}$$

$$(^{87}Sr/^{86}Sr)_{BE} = 0.69899 + 0.0846 \, (e^{\lambda t} - 1) \tag{21}$$

A magma that generated from a depleted source will have negative ε_{Sr} and ΔSr values.

7 The Nd–Sr Correlation

The constant Sm/Nd ratio of the meteorites of about 0.308 or 0.311 suggest that the Sm/Nd ratio of the Earth must be that of the meteorites. Since the ratio is known for the average Earth we can calculate the variation in the $^{143}Nd/^{144}Nd$ ratio for the bulk Earth accurately as the initial $^{143}Nd/^{144}Nd$ ratio is also known from the meteorite values. By plotting the ε_{Nd} values against the ε_{Sr} values, it becomes evident that the ε_{Nd} and ε_{Sr} values are correlated (Fig. XIII-9). Rocks that have $\varepsilon_{Nd} = 0$, like the kimberlites, have the same $^{143}Nd/^{144}Nd$ ratio as the bulk Earth, and must have formed from a mantle that is pristine and undepleted, if contamination can be excluded. As the two ε values are correlated, then it is reasonable to assume that the rock must also be undepleted with respect to Rb, and the $^{87}Sr/^{86}Sr$ ratio of the bulk Earth today must be at the intersection between the mantle array line and the $\varepsilon_{Nd} = 0$ line (Fig. XIII-9). This intersection occurs at $^{87}Sr/^{86}Sr = 0.7045$ according to DePaolo and Wasserburg (1976 b). Later estimates are slightly higher, this ratio has been estimated at 0.7047 by O'Nions et al. (1977), and as 0.7052 by Zindler et al. (1982). Since the recent bulk Earth value of the ratio is known, then we can also

Fig. XIII-9. The mantle array line as defined by the ε_{Nd} and ε_{Sr} values (DePaolo 1979)

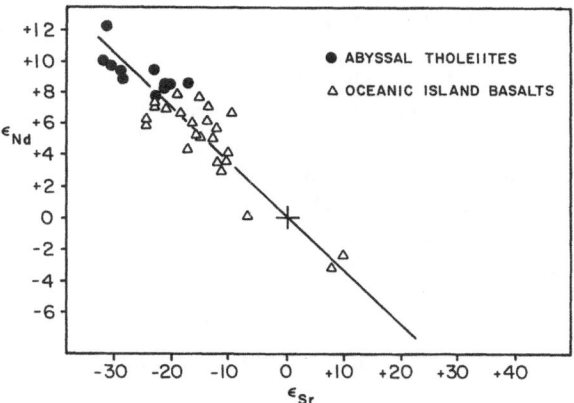

calculate the initial $^{87}Rb/^{86}Sr$ ratio of the Earth using the BABI value and Eq. (11), and using DePaolo's estimate we obtain $^{87}Rb/^{86}Sr = 0.084396$. This estimate may be controlled by comparing the bulk Earth evolution curve for $^{87}Sr/^{86}Sr$ with the initial $^{87}Sr/^{86}Sr$ ratios of komatiites, and the curve is nearly coincident with these ratios (Fig. XIII-7). The comprehensive investigation by White and Hofmann (1982) of the Nd–Sr correlation has shown that the correlation apparently is less well-defined than originally thought, which could be due to contamination. The data shown in Fig. XIII-7 and Fig. XIII-10 suggest that the correlation is real. The origin of the correlation was investigated by Allegre et al. (1979) and DePaolo (1979), and it was shown that the slope of the correlation is related only to the partition coefficients of Sm, Nd, Rb, and Sr, and is independent of time. However, the detailed explanation for the slope of the mantle array has not yet been found. It will depend both on the partition coefficients and the type of partial melting, but no model has yet been able to account for the mantle array.

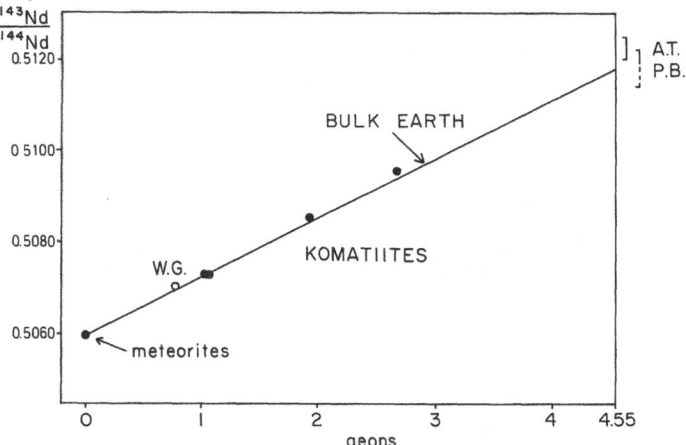

Fig. XIII-10. The bulk Earth evolution curve for the $^{143}Nd/^{144}Nd$ ratio. The initial ratios for komatiites and the old Isortoq rocks from West Greenland (*W.G.*) are situated near the curve. Abyssal tholeiites (*A.T.*) have values above the curve, while plume basalts (*P.B.*) have values on either side of the curve. The komatiite data are from Zindler (1982)

8 Andesites and Granites

The ε_{Sr} and ε_{Nd} values have provided fundamental evidence about the origin of andesites and granites, because the ε values indicate the source material for these rocks. We must here distinguish between two types of andesites, the intraoceanic andesites and the continental andesites. The intraoceanic andesitic volcanoes are situated along island chains within the oceans, far from the continents. Examples are the Marianas in the western Pacific, and the South Sandwich islands of the southern Atlantic. The continental andesites are formed in the subduction zones beneath the continents. A considerable amount of sediments may accumulate in the trenches near the continents and these sediments may become subducted together with the oceanic crust, so that the continental andesites might form from a mixture of oceanic crust and sediments.

The ε_{Sr} values of both types of andesites are higher than the ε_{Sr} values of the abyssal tholeiites. If we consider that the andesites have formed by partial melting of the subducted oceanic crust, then there is apparently an inconsistency in their ε_{Sr} values. This difference in isotopic compositions originally caused some concern, and led some petrologists to the conclusion that andesites are not formed by partial melting of the abyssal crust. Instead, it was considered that the andesites were formed by partial melting, in the presence of water, of the mantle situated above the subducting crust. Considering that this part of the mantle once must have formed abyssal tholeiites, then this explanation was hardly likely. The reason for the difference in the ε_{Sr} values became clear after the hydrothermal acticity of the midocean ridges were discovered, and after the ε_{Nd} values became available.

The $^{87}Sr/^{86}Sr$ ratios of the abyssal tholeiites are changed by hydrothermal activity which occurs above the magma chambers at the midocean ridges, while the $^{143}Nd/^{144}Nd$ ratio remains unchanged. Both the ε_{Sr} and ε_{Nd} values of seawater differ from those of the abyssal crust (Fig. XIII-6), but it is only the ε_{Sr} value that is changed by the hydrothermal activity. The Nd concentration of seawater is extremely small, and Nd is more resistant to alteration that Sr. The result is that the hydrothermal circulation of the seawater changes the $^{87}Sr/^{86}Sr$ ratio, but leaves the $^{143}Nd/^{144}Nd$ ratio unchanged. The intraoceanic andesites, therefore, attain the isotopic characteristics of the hydrothermally altered oceanic crust. In order to avoid confusion, it should be mentioned here that the abyssal tholeiites that are dredged stem from the surface of the oceanic crust. These dredged samples have undergone fast cooling, and even the glass of the pillow lavas have remained unchanged in composition. The hydrothermal circulation occurs beneath the surface in the lower part of the sequence of the pillow lavas and in the dyke complex.

The continental andesites, on the other hand, have smaller ε_{Nd} values than those of abyssal tholeiite, and larger ε_{Sr} values than the intraoceanic andesites (Fig. XIII-11). These differences are due to crustal contamination. The continental andesitic magmas may receive material from the continental crust in two different manners. The crustal material may stem from the subducted sediments which accumulate in the deep trenches. The contamination could also occur while the andesitic magmas ascend up through the continental crust. The $\delta^{18}O$ values and the U-Th-Pb isotopic systematics show that a component from the continental crusts must be involved in the genesis of the continental andesites (Harmon et al. 1981; Barreiro and Tilton

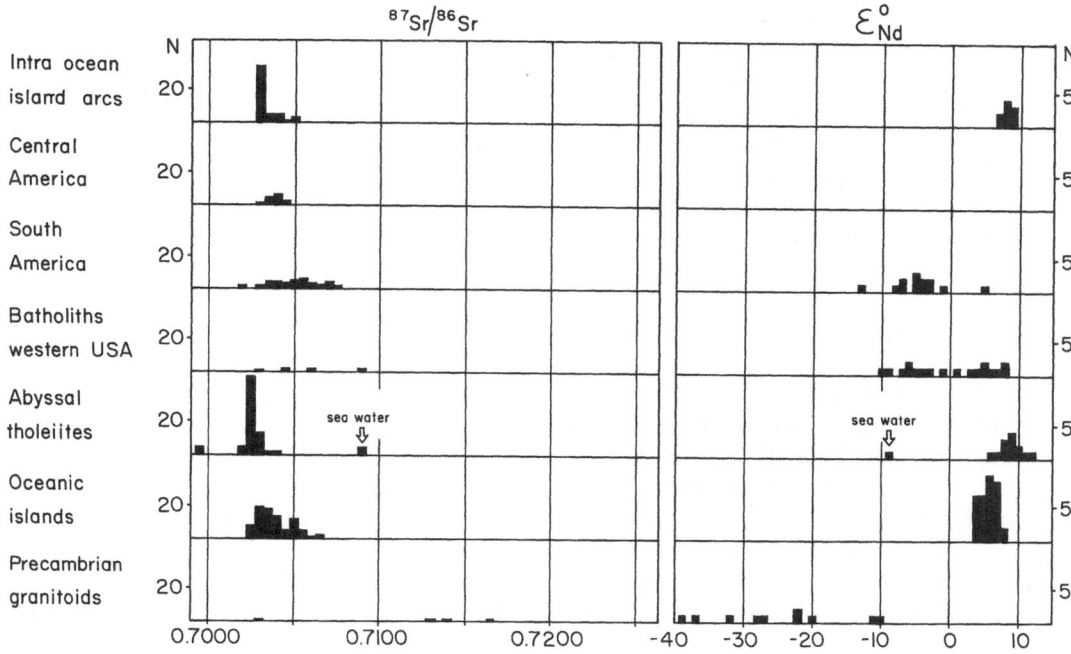

Fig. XIII-11. Isotopic data for andesites and granites from different regions compared with those of abyssal tholeiite (Maaløe and Petersen 1981)

1980). Thorpe et al. (1979) presents several arguments in support of contamination during ascent through the continental crust, while Margaritz et al. (1978) favor contamination by the subduction of sediments. The major element trends of the continental and intraoceanic andesites are different, even at low SiO_2 contents, which suggests that their initial primary magmas differ in composition (Maaløe and Petersen 1981). The difference in primary composition could be related to the incorporation of sediments in the continental andesites. However, it is also likely that the andesites may become contaminated during their ascent through the continental crusts. The SiO_2 and K_2O contents and the $^{87}Sr/^{86}Sr$ ratios of the recent and Tertiary andesites of the Andes increase with increasing distance to the subduction zone (Harmon et al. 1981). Since detrital sediments may contain more granitic components than tholeiites, one would expect that SiO_2-rich magmas may form at small depth rather than at large depths. The relationship between depth of the subduction zone and the composition of the andesitic magmas must be related to some other process than contamination by sediments. One possibility is that the temperature of the generated andesites may increase with increasing depth of generation, with the result that the potential for assimilation increases. Another factor which may be taken into account is the increased thickness of the continental crust which occurs beneath some parts of the Andes. The crust may be up to 70 km thick beneath Peru, and the deep parts of the sialic crust may undergo substantial heating, and it is possible that the crust may even undergo anatexis and form granitic magmas. The $^{87}Sr/$ ^{86}Sr ratios of some of the andesitic and dacitic rocks are so high, attaining values of about 0.7100, that the fraction of material from the abyssal tholeiites must be very small or virtually nil. The magmas formed by anatexis may either erupt without any mixing with andesitic magmas or they may become mixed with the andesitic magmas when they ascend through the continental crust.

In conclusion, the present evidence suggests that the continental andesites may be contaminated with subducted sediments, but the evidence probably favors assimilation during ascent as the most important process. In addition, it is possible that some of the andesitic magmas are generated by anatexis alone of material from the continental crust.

9 Noble Gas Isotopes

The systematics of the noble gas isotopes have shown that the mantle is not completely out gassed, so that the mantle must still retain some of its original amount of incompatible elements. The noble gas isotopes have also provided important information about the early condensation history of the solar system. It has been shown by Staudacher and Allegre (1982) that the aggregation of the Earth took place within about 60 m.y. after the meteorites were formed.

Helium has two naturally occurring isotopes, ^3He and ^4He. Both isotopes are present in the atmosphere and in seawater. The He atoms are so light that they es-

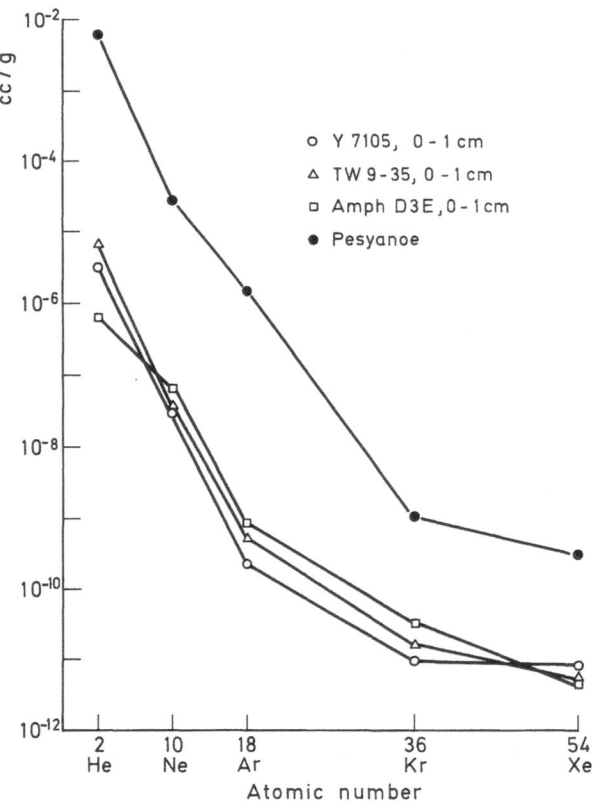

Fig. XIII-12. Rare gas abundance pattern of three glassy basalts compared with the pattern of a meteorite (Pesyanoe). The rare gas content of the meteorite is higher, but the patterns are similar

Fig. XIII-13. Rare gas content
of two holocrystalline basalts
compared to seawater and air
abundances. The patterns are
similar and show that the holocrystalline parts of the pillow
lavas have been contaminated,
while the data of Fig. XIII-12
show that the glass has retained
its original content of rare
gases

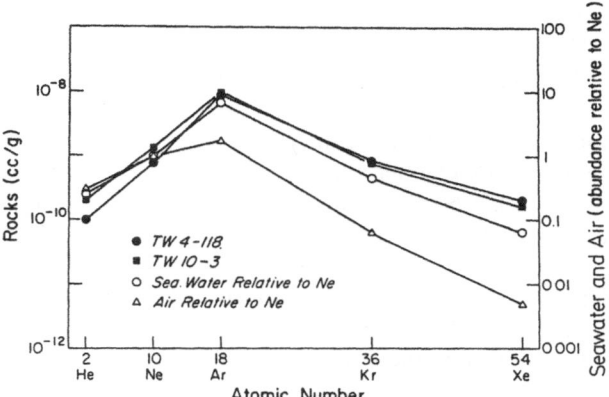

cape from the Earth's atmosphere and the escape rate has been estimated at 3–6 atoms/cm² s. This escape rate is about ten times higher than the production rate from cosmic ray interaction, and some ^{3}He must be produced from the Earth itself. The flux of ^{3}He from the continental crust is low, about 0.1 atom/cm². The crustal ^{3}He is accounted for by the decay of ^{6}Li, according to the reaction: ^{6}Li + n = α + ^{3}He. As the Li content of the crust is low, the production of ^{3}He is low. As the Li content of the mantle is even lower, there is virtually no recent production of ^{3}He in the mantle. Since the continental crust cannot provide the required ^{3}He, the ^{3}He must nevertheless stem from the mantle. It was shown by Clarke et al. (1969) that the ^{3}He content of the oceans is higher than that of the atmosphere, and subsequent analyses of the glass rims of abyssal pillow lavas showed conclusively that the ^{3}He/^{4}He ratio is uniformly about $1.1 \cdot 10^{-6}$ (Kurtz and Jenkins 1981). There is no radioactive process that can generate ^{3}He in the mantle, while ^{4}He is generated by the decay of U and Th. Hence, it has to be concluded that the ^{3}He present in the abyssal tholeiites must be primordial, i.e., it must have been present when the Earth was formed, and must stem from the solar nebulae. This result is substantiated by the evidence of other noble gas isotopic ratios like the ^{20}Ne/^{36}Ar and ^{4}He/^{40}Ar ratios (Craig and Lupton 1976).

The noble gas contents in abyssal tholeiite is compared with the content in a meteorite in Fig. XIII-12 and with seawater in Fig. XIII-13 (Dymond and Hogan 1973). The noble gas abundances of the glassy rims has the same pattern as the meteorite, while the crystalline center of the pillows has the same abundance as seawater. The crystalline parts of the pillows have been contaminated with seawater, while the glassy rims have remained unchanged, probably due to the rapid cooling of the pillow margins. The similarity between the pattern for the glassy margins and that of the meteorite suggest a primordial origin of the noble gas content in agreement with the ^{3}He evidence. The abyssal tholeiites are more depleted than the plume basalts, and this difference is also valid for the noble gases. The gas emanations of the Kilauea volcano of Hawaii has a ^{3}He/^{4}He ratio of 20.9, twice that of abyssal tholeiites.

The xenon isotope ^{129}Xe provides similar evidence for the presence of primordial gas in the mantle. The isotope ^{129}Xe is formed by the decay of ^{129}I, and the half-life is very small, and has been reported as 17 m.y. and 15.7 m.y. Virtually no ^{129}Xe is produced today and the presence of this isotope in the mantle-derived rocks again demonstrates the presence of primordial gas in the mantle. Due to the short half-life the abundance of ^{129}Xe will provide evidence about the rate of condensation of the solar nebulae and the aggregation of the Earth (Crabb et al. 1982). It was shown by Staudacher and Allegre (1982) that the aggregation of the Earth was completed within 60 m.y. after the formation of the meteorites. This estimate is in agreement with the dynamic model for the aggregation developed by Wetherill (1980).

The ^{129}Xe/^{130}Xe ratios as well as the ^{4}He/^{3}He and ^{40}Ar/^{36}Ar ratios provide evidence about the convection pattern within the mantle. Xenon has nine isotopes with the numbers, 124, 126, 128, 129, 130, 131, 132, 134, and 136. Of these 131, 132, 134, and 136 form by the decay of ^{244}Pu, and 131, 132, 134, and 136 form by the decay of ^{238}U, while ^{130}Xe is a stable nonradiogenic isotope. The degassing occurs without any isotopic fractionation so that all the xenon isotopes are degassed in the same proportions as they occur within the source. Changes in the isotopic ratios will, therefore, be related to the decay of ^{244}Pu, ^{238}U, and ^{129}I. As ^{130}Xe is a nonradiogenic isotope, one uses the ratio ^{129}Xe/^{130}Xe for an indication of changes in the abundance in ^{129}Xe. The degassing of the Earth occurred very early in its history, and some xenon was lost, while iodine and plutonium remained within the mantle. Since ^{129}Xe was formed from ^{129}I, a high ^{129}Xe/^{130}Xe ratio implies loss of xenon. The abyssal tholeiites have ^{129}Xe/^{130}Xe ratios within the range 6.6 to 7.0, while the basalts of the oceanic islands have ratios from 6.4 to 6.6 (Allegre et al. 1983). The lower ratios for the plume-derived ocean island basalts suggest that their source was degassed to a lesser extent than the source of the abyssal tholeiites. As the abyssal tholeiites stem from the upper mantle, and the plume basalts are believed originating from the lower mantle, the difference in the ratios suggests that the upper mantle was degassed to a larger extent than the lower mantle. A difference in the ^{129}Xe/^{130}Xe ratios would not be observed if all the ^{129}I had decayed before the degassing occurred, and the obtained ratios show that the degassing must have occurred at least 4.4 aeons ago (Allegre et al. 1983). Further, the fact that the two mantle sources differ in their ^{129}Xe/^{130}Xe ratios show that they must have remained separated and largely unmixed since the degassing occurred. This is only possible if the convection within the mantle has occurred exclusively within the upper mantle. The xenon isotopic data, thus, apparently exclude whole mantle convection since 4.4 aeons ago. This result supports the presence of a pristine unmodified lower mantle as suggested by DePaolo (1981). The Nd/Sr data for kimberlites also suggest a pristine character of the lower mantle (Fig. XIII-6). The isotopic data obtained so far indicate that the lower mantle may have remained largely unmodified since it was formed. The major element composition of the lower mantle is still unknown, but it is probably of lherzolitic composition with a MgO content between 35% and 42% MgO, according to the compositions of meteorites and lherzolites.

10 K–Ar Dating

The K–Ar dating method has proven particularly suitable for the dating of tertiary and even quaternary rocks, which is an advantage as the Rb–Sr and Sm–Nd

Table XIII-4. The abundances of isotopes
of potassium, argon, and calcium

Isotope	%
^{39}K	93.08
^{40}K	0.0119
^{41}K	6.91
^{36}Ar	0.337
^{38}Ar	0.063
^{40}Ar	99.60
^{40}Ca	96.94
^{42}Ca	0.65
^{43}Ca	0.14
^{44}Ca	2.08
^{46}Ca	0.003
^{48}Ca	0.19

methods become inaccurate for rocks of young age. In addition, the K–Ar method allows dating on one sample, while the two other methods just mentioned require at least two samples with different isotopic ratios. The disadvantage with the K–Ar method is that radiogenic argon may be lost due to heating, with the result that the estimated ages become too young.

Potassium has three isotopes, and one of them, ^{40}K is radioactive and decays to either ^{40}Ca or ^{40}Ar (Table XIII-4). About 88.8% of the ^{40}K isotopes decay to ^{40}Ca, and 11.2% decay to ^{40}Ar (Dalrymple and Lanphere 1969; Faure 1977). The decay to ^{40}Ca does not have any extensive application as ^{40}Ca consitutes 96.94% of Ca. The K–Ca method is only used for minerals with a high K content and a small Ca content.

Let λ be the total decay constant for ^{40}K, the increase in ^{40}Ca and ^{40}Ar is then given by:

$$^{40}\text{Ca} + {}^{40}\text{Ar} = {}^{40}\text{K}(e^{\lambda t} - 1) \tag{22}$$

The decay constants for ^{40}Ca and ^{40}Ar are different, the most widely used decay constants are:

$$\lambda_e = 0.585 \cdot 10^{-10} \quad (\text{to } {}^{40}\text{Ar})$$
$$\lambda_b = 4.720 \cdot 10^{-10} \quad (\text{to } {}^{40}\text{Ca})$$

The total decay constant λ is then given by the sum of these two constants:

$$\lambda = \lambda_e + \lambda_b = 5.305 \cdot 10^{-10} \tag{23}$$

The amount of ^{40}Ar is then calculated from:

$$^{40}\text{Ar} = \frac{\lambda_e}{\lambda} \, {}^{40}\text{K}(e^{\lambda t} - 1) \tag{24}$$

It is generally assumed that there is no ^{40}Ar present in the minerals when they are formed and Eq. (24) is, therefore, the basic dating equation for the K–Ar method. In some cases one has observed an amount of ^{40}Ar which cannot be radiogenic, and

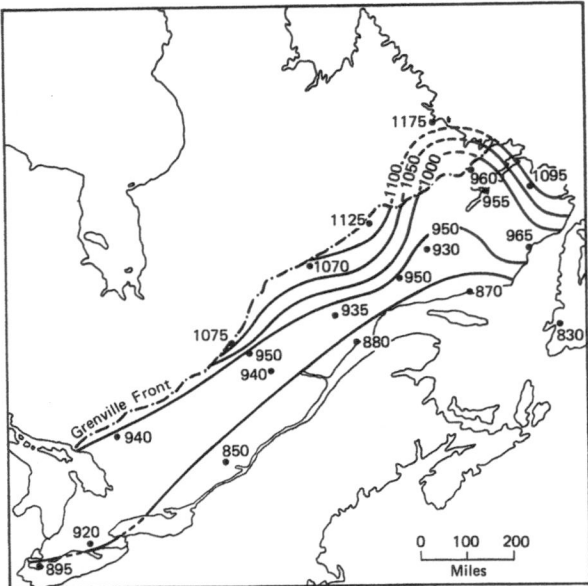

Fig. XIII-14. The metamorphic veil of the Grenville province of Canada based on K-Ar dates of biotite from granitic gneisses (Harper 1967)

this ^{40}Ar is called the excess argon. Some of the excess argon may be primordial as discussed above, or the excess argon may be introduced by the decay of ^{40}K from other sources.

Minerals like the feldspars, hornblende, biotite, and muscovite retain their radiogenic ^{40}Ar and may be used for dating if they have remained at low temperatures since their formation. Volcanic glasses also tend to keep their radiogenic ^{40}Ar. The K–Ar method may be used on granites, gabbros, and lavas, and have frequently been used for the dating of abyssal tholeiites. The geomagnetic reversals during the last 5 m. y. have been dated using the K–Ar method on abyssal tholeiites. The decay constant is so small that the method can also be used on Precambrian rocks. However, the method can only be used under the condition that the rocks have remained unmetamorphosed, as an increase in temperature leads to argon loss. The temperature-induced argon loss which occurs during metamorphism may, on the other hand, be used for a dating of the terminal stage of metamorphism, and the Grenville province in Canada has been K–Ar dated. The results show a systematic variation in the obtained dates related to metamorphism (Harper 1967), and Armstrong (1966) called this apparent age variation the metamorphic veil (Fig. XIII-14).

11 Oxygen Isotopes

The atomic mass of oxygen is so small that the isotopes of oxygen may be fractionated during the crystallization of magmas. The fractionation factor for the oxygen isotopes differs for different minerals and is dependent on temperature. The temperature of crystallization may, therefore, be estimated from the oxygen isotopic composition of minerals if their fractionation factors differ.

Oxygen isotopes have provided valuable information about the origin of volcanic gases. Isotopic analyses of volcanic gases have shown that the major part of the water vapor stem from meteoric water and that juvenile water only forms a small fraction of the vapors generated during eruptions. This shows definitively that the gas composition of juvenile gases cannot be estimated by gas sampling during eruptions.

It has been suggested that oxygen isotopes could indicate the presence of subducted sediments in the source of andesites, but that is probably not the case. Seawater has a different oxygen isotopic composition from that of igneous rocks. However, sediments only contain a small fraction of seawater, and their oxygen isotopic composition is essentially that of their sources.

While oxygen isotopic systematics have had some interesting applications to igneous petrology, the major impact has so far been within paleontology, as these isotopes can be used for an estimation of paleotemperatures in sediments and fossils. The oxygen isotopic composition of fossils depends on the temperature of their environment, and variations in their oxygen isotopic composition may be used for an estimation of changes in climate. The oxygen isotopic composition of ice and snow is also dependent on temperature, and the temperature variations during Pleistocene have been estimated from variations in the oxygen isotopic ratios (Dansgaard et al. 1969).

Oxygen has three stable isotopes and their approximate abundances are:

^{16}O: 99.756%

^{17}O: 0.039%

^{18}O: 0.205%

The oxygen abundances in ocean water and the atmosphere is related to the isotopic composition of water, which also depends on the three isotopes of hydrogen, ^{1}H, ^{2}H, and ^{3}H. The isotopic variant with the smallest mass is $^{1}H_2^{16}O$, and the heaviest one is $^{3}H_2^{18}O$. The water molecules with the smallest mass will have the highest vapor pressure and the light isotopes will be enriched in the vapor phase by evaporation and in the liquid phase by crystallization.

The isotopic composition of oxygen is reported in terms of the $^{18}O/^{16}O$ ratio relative to an international standard called SMOW (Standard Mean Ocean Water) (Craig 1961). The isotopic composition of oxygen is expressed relatively to SMOW by this equation which is similar to other equations defining relative isotopic abundances:

$$\delta^{18}O\permil = \left[\frac{(^{18}O/^{16}O)_{sample} - (^{18}O/^{16}O)_{SMOW}}{(^{18}O/^{16}O)_{SMOW}} \right] 10^3 \qquad (25)$$

As the values of $\delta^{18}O$ are given relative to seawater, the values of $\delta^{18}O$ are about zero for seawater.

Most igneous rocks have $\delta^{18}O$ values in the range from +4 to +14. Ultramafic rocks have the lowest values in the range from +5.4 to 6.6, and the range for basalts and gabbros is from +5.5 to 7.4. Granites and pegmatites have the highest ratios, the range being from 7 to 13.

Sediments also have positive $\delta^{18}O$ values in the range from +5 to +25. Metamorphic rocks have values in a similar range. There is, thus, no marked difference in the $\delta^{18}O$ values between the different rock types, except that values above $+15$ may indicate a sedimentary origin.

12 Fractionation of Oxygen Isotopes

The isotopes of oxygen are fractionated in nature, and the degree of fractionation is expressed by the fractionation factor α:

$$\alpha = \frac{R_a}{R_b} = \frac{(^{18}O/^{16}O)_{phase\,A}}{(^{18}O/^{16}O)_{phase\,B}} \tag{26}$$

The temperature dependence of α is with some application given by:

$$\ln(\alpha) \propto 1/T^2 \tag{27}$$

The fractionation factor will, thus, decrease with increasing temperature, and not all minerals can be used as temperature indicators at magmatic temperatures.

It is convenient to obtain an expression that relates $\delta^{18}O$ and α, and this is done as follows. Using the Eqs. (25) and (26) that define $\delta^{18}O$, we obtain by rearrangement:

$$\delta^{18}O = \frac{R_A - R_{SMOW}}{R_{SMOW}} 10^3 \tag{28}$$

So that:

$$R_A = \frac{\delta^{18}O\ R_{SMOW}}{1000} + R_{SMOW} \tag{29}$$

and similarly we get for R_B:

$$R_B = \frac{\delta^{18}O\ R_{SMOW}}{1000} + R_{SMOW} \tag{30}$$

From the last two equations and Eq. (26) we obtain:

$$\alpha = \frac{\delta^{18}O_A + 1000}{\delta^{18}O_B + 1000} \tag{31}$$

We will thereafter derive an equation that gives α as a function of temperature. The fractionation factor α has values very near 1.000, the value is rarely greater than 1.004. One has, therefore, applied the approximation:

$$1000 \ln(1.00X) \sim X \tag{32}$$

where X is an arbitrary digit between 0 and 9. The temperature dependence is expressed with good approximation by the following equation:

$$\alpha = A\,(10^6\,T^{-2}) + B \tag{33}$$

so that:

$$\alpha = 1000 \ln(\alpha) = A\,(10^6\,T^{-2}) + B \tag{34}$$

Fig. XIII-15. Temperature variation of the isotopic fractionation factor for oxygen between minerals and water at different temperatures (Faure 1977)

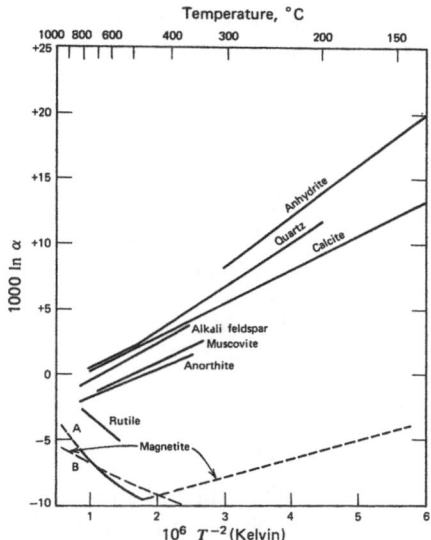

The relationship between α and T is generally shown in coordinates of $(1000 \ln \alpha)$ and $(10^6 T^{-2})$ (Fig. XIII-15). Some values of A and B for mineral/water are listed in Table XIII-5.

The $\delta^{18}O$ values of minerals are obtained from the $^{18}O/^{16}O$ ratios that are measured directly by mass spectrometry. It has, therefore, been found convenient to derive an equation that gives the difference in $\delta^{18}O$ values between minerals as a function of temperature, and we will finally consider this equation.

Let again α be given by:

$$\alpha = \frac{R_A}{R_B} \tag{35}$$

Table XIII-5. Fractionation factors for oxygen in mineral/water systems (Faure 1977)

Mineral	A	B	Temperature range °C
Quartz	3.38	−3.40	200–500
Quartz	4.10	−3.70	500–800
Muscovite	1.90	−3.10	500–800
Plagioclase	2.91 −0.76X[a]	−3.41 −0.41X[a]	350–800
Anorthite	2.15	−3.82	350–800
Magnetite	−1.59	−3.60	700–800
Calcite	2.78	−3.4l0	0–800

[a] X is the anorthite content of plagioclase

By rearrangement we obtain:

$$\alpha - 1 = \frac{R_A - R_B}{R_B} \tag{36}$$

By using Eq. (31) we obtain:

$$\alpha - 1 = \frac{\delta_A - \delta_B}{\delta_B + 1000} \tag{37}$$

As $\delta + 1000 \sim 1000$ we get:

$$\alpha - 1 = \frac{\delta_A - \delta_B}{1000} \tag{38}$$

By using the approximation given by Eq. (32) we obtain with some approximation:

$$1000 \ln(\alpha) = \delta_A - \delta_B = A \, (10^6 \, T^{-2}) + B \tag{39}$$

The values of the fractionation factor are estimated and given by mineral/water systems, but the obtained values can easily be applied for mineral/mineral systems. As an example let us consider the quartz/magnetite system in the temperature range 700° to 800 °C. Using Table XIII-5, we obtain these equations:

$$1000 \ln(\alpha_{QW}) = 4.1 \, (10^6 \, T^{-2}) - 3.70 \tag{40}$$

$$1000 \ln(\alpha_{MW}) = -1.59 \, (10^6 \, T^{-2}) - 3.60 \tag{41}$$

By subtraction we get the equation for α_{QM}:

$$1000 \ln(\alpha_{QM}) = \delta_Q - \delta_M = 5.69 \, (10^6 \, T^{-2}) - 0.10 \tag{42}$$

The oxygen isotopic geotherms have been applied to various types of basalts using plagioclase and magnetite, and the results are in excellent agreement with independent estimates (Faure 1977).

References

Adams LH (1924) Temperatures at moderate depths within the Earth. J Wash Acad Sci 14, no. 20, pp 459–472

Aki K, Koyanagi R (1981) Deep volcanic tremor and magma ascent mechanism under Kilanea, Hawai. J Geophys Res 86:7095–9109

Aki K, Fehler M, Oas S (1977) Source mechanism of volcanic tremor: Fluid driven crack models and their application to the 1963 Kilauea eruption. J. Volcanol. Geother. Res. 2:259–287

Akimoto S, Matsui Y, Syono Y (1976) High pressure chemistry of other silicates and the formation of the mantle transition zone. In: Strens RG (ed) The physics and chemistry of minerals and rocks. Wiley, London, pp 327–363

Albarede F, Provost A (1977) Petrological and geochemical mass-balance equations: An algoritm for least square fitting and general error analysis. Comput Geosci 3:309–326

Allegre CJ (1982) Chemical geodynamics. Tectonophysics 81:109–132

Allegre CJ, Minster JF (1978) Quantitative models of trace element behaviour in magmatic processes. Earth Planet Sci Lett 38:1–25

Allegre CJ, Othman DB, Potvé D, Richard P (1979) The Ned-Sr isotopic correlation in mantle materials and geodynamic consequences. Phys Earth Planet Int 19:293–306

Allegre CJ, Standacher T, Sarda P, Kurz M (1983) Constraints on evolution of Earth's mantle from rare gas systematics. Nature 303:762–771

Allsopp HL, Nicolaysen LO, Hahn-Weinheimer P (1969) Rb/K values and Sr-isotopic composition of minerals in eclogitic and peridotitic rocks. Earth Planet Sci Lett 5:231–244

Anderson O (1915) The system anorthite-forsterite-silica. Am J Sci Lett ser. 39:407–454

Anderson AT, Greenland LP (1968) Phospherous fractionation diagrams as a quantitative indicator of crystallization differentiation of basaltic liquids. Geochim Cosmochim Acta 33:493–505

Armstrong RL (1966) K-Ar dating of plutonic and volcanic rocks in orogenic belts. In: Schaeffer OA and Zähringer J (eds), Potassium-argon dating. Springer, Heidelberg, Berlin New York Tokyo, 117–131

Arnold JR, Libby WF (1949) Age determinations by radiocarbon content: Checks with samples of known age. Science 110:678–680

Ave'Lallement HG, Mercier JCC, Carter NL, Ross JV (1980) Rheology of the upper mantle: inferences from peridotite xenoliths. Tectonophysics 70:85–113

Baker EH (1962) The calcium oxide-carbon dioxide system in the pressure range 1–300 atmospheres. J Ceram Soc 87:464–470

Balk R (1937) Structural behaviour of igneous rocks. US Geol Soc Am Mem 5, 177 pp

Barreiro B, Tilton GR (1980) U-Th-Pb studies on quaternary volcanics and Precambrian basement rocks from southern Peru (abstract). Geol Soc Am Ann 76th Meet 12, p 95

Batchelor GK (1970) An introduction to fluid dynamics. Cambridge Univ Press pp 336–338

Bear J (1972) Dynamics of fluids in porous media. Elsevier, New York, 764 pp

Becker R, Døring W (1935) Kinetische Behandlung der Keimbildung in übersättigten Dämpfern. Ann Physik 24:719–752

Beere W (1981) The internal morphology of continuously interconnected two phase bodies. Trans J Br Ceram Soc 80:133–138

Bernal JD (1936) Discussion. Observatory 58:267–268

Berner H, Ramberg H, Stephansen O (1972) Diapirism in theory and experiment. Tectonophysics 15: 197–218

Biggar GM (1974) Phase equilibrian studies of the chilled margins of some layered intrusions. Contrib Mineral Petrol 46: 159–167

Bottwood BB (1907) On the ultimate disintegration products of the radioactive elements. Am J Sci Ser 4 vol 23, pp 77–88

Bowen NL (1912) The binary system: $Na_2Al_2Si_2O_8$ (nephelinite-carnegieite) – $CaAl_2Si_2O_8$ (anorthite). Am J Sci 33: 551–573

Bowen NL (1913) The melting phenomena of the plagioclase feldspar. Am J Sci 4th ser vol 35, pp 577–599

Bowen NL (1915) The crystallization of haplobasaltic, haplodioritic, and related magmas. Am J Sci 4th ser vol 40, pp 161–185

Bowen NL (1928) The evolution of igneous rocks. Princeton University Press, Princeton, 334 pp

Bowen NL, Schairer JF (1935) The system MgO-FeO-SiO_2. Am J Sci 4th ser 29: 151–217

Bowen NL, Schairer JF (1947) The system anorthite-leucite-silica. Bull Soc Geol Finlande 20: 67–87

Boyd FR (1973) A pyroxene geotherm. Geochim Cosmochim Acta 37: 2533–2546

Boyd FR and England JL (1959) Quartz coesite transition. Carnegie Inst Wash Yb 58, 87–89

Boyd FR, England JL (1960) Apparatus for phase-equilibrium measurements at pressures up to 50 kilobars and temperatures up to 1750 °C. J Geophys Res 65: 741–748

Boyd FR, England JL (1962) Mantle Minerals. Carnegie Inst Wash Year book 61: 107–112

Boyd FR, Nixon PM (1973) Structure of the upper mantle beneath Lesotho. Carnegie Inst Wash Year book 72: 431–445

Braun G, Stout JH (1975) Some chemigraphic relationships in n-component systems. Geochim Cosmochim Acta 39: 1259–1267

Brey G, Green DH (1977) Systematic study of liquidus phase relations in olivine melilitite $+ H_2O + CO_2$ at high pressures and petrogenesis of an olivine melilitite magma. Contrib Mineral Petrol 61: 141–162

Brothers RN (1964) Petrofabric analysis of Rhum and Skaergaard layered rocks. J Petrol 5: 255–274

Bryan WB, Finger LW, Chayes F (1969) Estimating proportions in petrographic mixing equations by least squares approximation. Science 163: 926–927

Bulau JR, Waff HS, Tybuvezy JA (1979) Mechanical and thermodynamic constraints on fluid distribution in partial melts. J Geophys Res 84: 6102–6108

Bull AJ (1921) A hypothesis of mountain building. Geol Mag 58: 364–367

Burnham CW (1975) Thermodynamics of melting in experimental silicate-volatile systems. Fortschr Miner 52: 101–118

Burnham CW (1979) The importance of volatile constituents. In: Yoder HS (ed) The evolution of the igneous rocks. Princeton Univ Press Princeton pp 439–482

Burnham CV and Jahns RH (1958). Experimental studies of pegmatite genesis: the solubility of water in granitic magmas (Abstract). Bull Geol Soc Amer 69: 1544

Burnham CW, Holloway JR, Davis NF (1969): Thermodynamic properties of water to 1000 °C and 10.000 bars. Geol Soc Am Spec Pap 132: 96 pp

Burton WK, Cabrera N, Frank FC (1950) The growth of crystals and the equilibrium structure of their surfaces. Philos Trans Soc A 243: 299–358

Carnahan B, Luther HA, Wilkes JO (1969) Applied numerical methods. Wiley, New York, 604 pp

Carter NL (1976) Steady stall flow of rocks. Rev Geophys Space Phys 14: 301–360

Carter NL, Ave'Lallement HG (1970) High temperature flow of dunite and peridotite. Geol Soc Am Bull 81: 2181–2202

Chalmers B (1964) Principles of solidification. Wiley, New York, 319 pp

Chandrasekhar S (1961) Hydrodynamics and hydromagnetic stability. Oxford Univ Press, New York, 652 pp

Chayes F (1962) Numerical correlation and petrographic variation. J Geol 70: 440–452

Chayes F (1968) A last squares approximation for estimating the amounts of petrographic partition products. Mineralog Petrog Acta 14: 111–114

Chayes F (1969) The chemical composition of Cenozoic andesites. Oreg Dept Geol Mineral Indust Bull 65:1–11

Chayes F (1971) Ratio correlation. University of Chicago Press, Chicago, 99 pp

Christian JW (1965) The theory of transformations in metals and alloys. Pergamon, New York, 973 pp

Clague OA, Frey FA (1982) Petrology and trace element geochemistry of the Honolulu volcanics, Oahu. Implications for the oceanic mantle below Hawaii. J Petrol 23, pp 447–504

Clark SP (1966) Handbook of physical constants. US Geol Soc AM Mem 97:587 pp

Clark SP, Ringwood AE (1964) Density distribution and constitution of the mantle. Rev Geophys 2:35–88

Clarke WB, Beg MA, Craig H (1969) Excess ^{3}He in the sea: evidence for terrestial primordial helium. Earth Planet Sci Lett 6:213–220

Clift R, Grace JR, Weber ME (1978) Bubbles, drops and particles. Academic, New York, 380 p

Cloos H (1936) Einführung in die Geologie. Borntraeger, Berlin, 503 p

Cohen RS, Ito K, Kennedy GC (1967) Melting and phase relations in an anhydrous basalt to 40 kilobars. Am J Sci 265:475–518

Cooper RF, Kohlstedt OL (1982) Interfacial energies in the olivine-basalt system. Adv Earth Planet Sci 12:217–228

Crabb J, Lewis RS, Anders E (1982) Extinct I^{129} in C3 chondrites. Geochim Cosmochim Acta 46:2511–2525

Craft BC, Hawkins MF (1959) Applied petroleum reservoir engineering. Englewood Cliffs New York, pp 437

Craig H (1961) Standard for reporting concentrations of deuterium and oxygen-18 in natural waters. Science 133:1833–1934

Craig H, Lupton JE (1976) Primordial neon, helium, and hydrogen in oceanic basalts. Earth Planet Sci Let 31:369–385

Crank J (1975) The mathematics of diffusion. Claredon Press, Oxford, 414 pp

Dahlquist G, Bjørck Å (197) Numerical methods. Englewood Cliffs, N J Prentice Hall, 573 pp

Dalrymple GB, Lanphere MA (1969) Potassium-argon dating. WM Freeman, San Francisco, 258 pp

Danes ZF (1964) Mathematical formulation of salt dome dynamics. Geophys 29:414–424

Dansgaard W, Johnson SJ, Møller J, Longway CC (1969) One thousand centuries of dimatic record from Camp Century on the Greenland ice sheet. Science 166:377–381

Darken LS (1948) Diffusion of carbon in austenite with a discontinuity in composition. Trans Am Inst Min Metall Engrs 2443:1–9

Darken LS, Gurry RW (1953) Physical chemistry of metals. McGraw-Hill, New York, 535 pp

Daubrée A (1860) Etudes et expériences synthetiques sur le metamorphisme et sur la formation des roches cristallines. Min Acad Sci Paris 17

Davis BTC (1965) The system diopside-forsterite-pyrope at 40 kilobars. Carnegie Inst Wash Year book 63:165–171

Dawson JB (1966) Oldoinyo Lengai – an active volcano with sodium carbonatite lava flows. In: Tuttle OF, Gittins J (eds.) Carbonatites Interscience Pub New York, pp 155–168

Deer WA, Howie RA, Zussman J (1965) Rock-forming minerals, I–V. Longmans, Green, London

Deines P, Nafriger RM, Ulmer GC, Woermann E (1974) Temperature-oxygen fugacity tables for selected gas mixtures in the system C-H-O at one atmosphere total pressure. Bull Earth Miner Sci Penn State Univ. 88, 129 pp

Delaney JR, Muenow OW, Graham DG (1978) Abundance and distribution of water, carbon and glassy rims of submarine pillow basalts. Geochim Cosmochim Acta 42:581–594

Denbigh K (1971) The principles of chemical equilibrium. Cambridge Univ Press, Cambridge, 494 pp

DePaolo, DJ (1979) Implications of correlated Nd and Sr isotopic variations for the chemical evolution of the crust and mantle. Earth Planet Sci Lett 43:201–211

DePaolo DJ (1981) Nd isotopic studies: Some new perspectives on Earth structure and evolution. EOS 62:137–140

DePaolo DJ, Wasserburg GJ (1976a) Nd isotopic variations and petrogenetic models. Geophys Res Utt 3:249–252

DePaolo DJ, Wasserburg GJ (1976b) Inferences about magma sources and mantle structure from variations of $^{143}Nd/^{149}Nd$. Geophys Res Lett 3:743–746

DePaolo DJ, Wasserburg GJ (1977) The sources of island arcs as indicated by Nd and Sr isotopic studies. Geophys Res Lett 4:465–468

DePaolo DJ, Wasserburg GJ (1979) Sm-Nd age of the stillwater complex and the mantle evolution curve for neodynium. Geochim Cosmochim Acta 43:999–1008

Dickinson SK (1970). Investigation of the synthesis of diamonds. US Air Force Cambridge Research Lab Phys Rep Pap 434:101 pp

Donaldson CH (1976) An experimental investigation of olivine morphology. Contrib Mineral Petrol 57:187–213

Donaldson CH (1979) An experimental investigation of the delay in nucleation of olivine in mafic magmas. Contrib Mineral Petro 69:21–32

Drake MJ (1976) Plagioclase-melt equilibria. Geochim Cosmochim Acta 40:457–465

Draper NR, Smith H (1966) Applied regression analysis. Wiley, New York, 407 pp

Dziewonsky AM, Anderson DL (1981) Prelaminary reference Earth model. Phys Earth Planet Int 25:297–356

Dziewonski AM, Hales AL, Lapwood E-R (1975) Parametrically simple Earth models consistent with geophysical data. Phys Earth Planet Int 10:12–48

Dymond J, Hogan L (1973) Noble gas abundance patterns in deep-sea basalts-primodial gases from the mantle. Earth Planet Sci Lett 20:131–139

Eaton JP, Murata KJ (1960) How volcanoes grow. Science 132:925–938

Edgar AD (1973) Experimental petrology. Claredon Press, Oxford, 217 pp

Eggler DH (1972) Water-saturated and undersaturated melting relations in a Paricutin andesite and an estimate of water content in the natural magma. Contrib Mineral Petrol 34:261–271

Eggler DH (1973) Role of CO_2 in melting processes in the mantle. Carnegie Inst Wash Year book 72:457–467

Eggler DH (1974) Effect of CO_2 on the melting of peridotite. Carnegie Inst Wash Year book 73:215–224

Eggler DH (1978): The effect of CO_2 upon partial melting of peridotite in the system $Na_2O - CaO - Al_2O_3 - MgO\text{-}SiO_2 - CO_2$ at 35 kb., with an analysis of melting in a peridotite – $H_2O - CO_2$ system. Am J Sci 278:305–343

Eggler DH, Baker N (1982) Reduced volatiles in the system C-O-H: Implications to mantle melting, fluid formation, and diamond genesis. Adv Earth Planet Sci 12:237–250

Eggler DH, Burnham CW (1973) Crystallization and fractionation trends in the system andesite – $H_2O - CO_2 - O_2$ at pressures to 10 kb. Geol Soc Am Bull 84:2517–2532

Eggler DH, Mysen BO (1976) The role of CO_2 in the genesis of olivine melilitite: Discussion. Contrib Mineral Petrol 55:231–236

Ehlers EG (1972) The interpretation of geological phase diagrams. WH Freeman, San Francisco, 280 p

Einstein A (1921) Progradation of sound in partly dissociated gases. Preuss Akad Wiss Berlin 19:380–385

Elthon D (1979) High magnesia liquids as the parental magma for ocean floor basalts. Nature 278:514–518

Elthon D, Scarfe CM (1980) High pressure phase equilibria of a high-magnesia basalt: implications for the origin of mid-ocean ridge basalts. Carnegie Inst Wash Year book pp 276–281

Eugster HP, Wones DR (1962) Stability relations of the ferruginous biotite, annite. J Petrol 3:82–125

Fairbain HW (1951) A cooperative investigation of precision and accuracy in chemical, spectrochemical, and modal analysis of silicate rocks. US Geol Surv Bull 980, 71 pp

Faure G (1977) Principles of isotope geology. Wiley, New York, 464 pp

Faure G, Powell JL (1972) Strontium isotope geology. Springer, Berlin Heidelberg New York Tokyo 188 pp

Fenner W (1926) The Katmai magmatic province. J Geol 34:673–772

Ferrara G, Treuil M (1973) Petrological implications of trace element and Sr isotope distributions in basalt-pantellerite series. Bull Vol 38:548–574

Flannagan FJ (1969) US Geological survey standards II. First compilation of data for the new U.S.G.S. rocks. Geochim Cosmochim Acta 33:81–120

Forsyth DW (1977) The evolution of the upper mantle beneath mid-ocean ridges. Tectonophysics 38:89–118

Foster TD (1969) Convection in a variable viscosity fluid heated from within. J Geophys Res 74:685–693

Franco JJ and Schairer JF (1951) Liquidus temperatures in mixtures of the feldspars of soda, potash and lime. J Geol 59:259–267

Frank FC (1949) The influence of dislocations on crystal growth. Disc Faraday Soc 5:48–54

Frank FC (1968) Two component flow models for convection in the Earth's upper mantle. Nature 220:350–352

French SM, Eugster HP (1965) Experimental control of oxygen fugacities by graphite-gas equilibrium. J Geophys Res 70:1529–1539

Frischat GH (1975) Ionic diffusion in oxide glasses. Transtech, Bay Village, Ohio, 91 pp

Froude DO, Ireland TR, Kinney PO, Williams IS, Compston W (1983) Ion microprobe identification of 4100–4200 Myr old terrestial zircons. Nature 304:616–618

Fudali RF (1965) Oxygen fugacities of basaltic and andesitic magmas. Geochim Cosmochim Acta 29:1063–1075

Garcia HO, Liu NWK, Muenow OW (1979) Volatiles in submarine volcanic rocks from the Mariana island arc and through. Geochim Cosmochim Acta 43:305–312

Gast PW (1968) Trace element fractionation and the origin of tholeiitic and alkaline magma types. Geochim Cosmochim Acta 32:1057–1086

Gibb FGF (1974) Supercooling and the crystallization of plagioclase from a basaltic magma. Mineral Mag 39:641–653

Gibbs JW (1873) Graphical methods in the thermodynamics of fluids. Trans Conn Acad 2:309–342

Gibbs JW (1876) On the equilibrium of heterogeneous substances. Trans Conn Acad 3:108–248

Gibbs JW (1878) On the equilibrium of heterogeneous substances. Trans Conn Acad 3:343–524

Gordon P (1968) Principles of phase diagrams in materials systems. Mc-Graw-Hill, New York, 232 pp

Gradshteyn IS, Ryzhik IM (1965) Table of integrals, series and products. Academic, New York, 1086 pp

Gray CM, Papanastassiou DA, Wasserburg GJ (1973) The identification of early condensates from the solar nebula. Icarus 20:213–239

Green DH (1970) The origin of basaltic and nephelinitic magmas. Trans Leicester Lit Philos Soc 64:28–54

Green DH, Ringwood AE (1967) The genesis of basaltic magmas. Contrib Mineral Petrol 15:103–190

Greenland LP (1970) An equation for trace element distribution during magmatic crystallization. Am Mineral 55:455–465

Greig JW, Barth TFW (1938) The system $Na_2 - Al_2O_3 - 2 SiO_2$ (nepheline-carnegieite) $Na_2O - Al_2O_3 - 6 SiO_2$ (albite). Am J Sci 35A:93–112

Griggs J (1939) Creep of rocks. J Geol 47:225–251

Grossman L (1975) Petrography and mineral chemistry of Ca-rich inclusions in the Allende meteorite. Geochim Cosmochim Acta 39:433–454

Grossman L (1980) Refractory inclusions in the Allende meteorite. Ann Rew Earth Planet Sci 8:559–608

Hahn O, Walling E (1938) Über die Möglichkeit geologischer Altersbestimmungen rubidiumhaltiger Mineralen und Gesteine. Z Anorg Allgem Chem 236:78–82

Hall J (1805) Experiments on whin stone and lava. Trans Soc Edinburgh 5:43–75

Hanson GN (1980) Rare earth elements in petrogenetic studies of igneous systems. Ann Rev Earth Planet Sci 8:371–406

Harmon RS, Thorpe RS, Francis PW (1981) Petrogenesis of andean andesites from combined O–Sr isotope relationships. Nature 290:396–399

Harker A (1900) Igneous rock series and mixed igneous rocks. J Geol 8:389–399

Harker A (1909) The natural history of igneous rocks. Methuen, London, 384 p

Harper CT (1967) On the interpretation of potassium-argon ages from precambrian shields and phanerozoic orogens. Earth Planet Sci Lett 3:128–132

Harris DM (1979) Geobarometry and geothermometry of individual crystals using H_2O, CO_2, S_1 and major element concentrations in silicate melt inclusions: The 1959 eruption of Kilanea volcano, Hawaii (Abstract). Geol Soc Am 1979 Ann Mtg

Harrison WJ (1979) Partitioning of REE between garnet peridotite minerals and coexisting melts during partial melting. Carnegie Inst Wash Year book 78:562–568

Helz RT (1976) Phase relations of basalts in their melting ranges at $^PH_2O = 5$ kb. Part II. Melt compositions. J Petrol 17:139–193

Henderson P (1970) The significance of the mesostasis of basic layered igneous rocks. J Petrol 11:463–473

Henderson P (1975) Geochemical indicator of the efficiency of fractionation of the Skaergaard intrusion, East Greenland. Mineral Mag 40:285–291

Hertogen J, Gijbels R (1976) Calculation of trace element fractionation during partial melting. Geochim Cosmochim Acta 40:313–322

Hess HH (1960) Stillwater igneous complex, Montana: a quantitative mineralogical study Geol Soc Am Mem 80:230 pp

Hess GB (1972) Heat and mass transport during crystallization of the Stillwater igneous complex. Geol Soc Am Mem 132:503–520

Hofmann AW (1980) Diffusion in natural silicate melts: A critical review. In: Hargraves RB (ed) Physics of magmatic processes. Princeton Univ Press, Princeton, pp 385–417

Hofmann AW, Hart S (1978) An assessment of local and regional isotopic equilibrium in the mantle. Earth Planet Sci Lett 38:44–62

Holloway JR (1973) The system pargasite–H_2O–CO_2: a model for melting of a hydrous mineral with a mixed-volatile fluid-I. Experimental results to 8 kbar. Geochim Cosmochim Acta 37:651–666

Holloway JR, Burnham CW (1972) Melting relations of basalt with equilibrium water pressure less than total pressure. J Petrol 13:1–29

Holloway JR, Eggler DM (1976) Fluid-absent melting of peridotite containing phlogopite and dolomite. Carnegie Inst Wash Year book 75:636–639

Holmes A (1913) The age of the Earth. Harper and Brothers, London, 194

Holmes A (1931) Radioactivity and earth movements. Trans Geol Soc Glasgow 18:559–606

Houston MH, De Bremaecher JC (1975) Numerical models of convection in the upper mantle. J Geophys Res 80:742–751

Huang WL and Wyllie PJ (1975) Melting relations in the system $NaAlSi_3O_8$–$KAlSi_3O_8$–SiO_2 to 35 kilobars, dry and with excess water. J Geol 83, pp 737–748

Hubbert MK (1969) The theory of ground-water motion and related papers. Hafner, 311 pp

Hytønen K, Schairer JF (1961) The plane enstatite-anorthite-diopside and its relation to basalts. Carnegie Inst Wash Year book 60:125–141

Iddings J (1892) The origin of igneous rocks. Bull Phil Soc Wash 12:89–214

Irvine TN (1977) Definition of primitive liquid compositions for basic magmas. Carnegie Inst Wash Year book 76:454–461

Irving AJ, Wyllie PJ (1973) Melting relationships in CaO–CO_2 and MgO–CO_2 to 36 kilobars with comments on CO_2 in the mantle. Earth Planet Sci Lett 20:220–225

Irving AJ, Wyllie PJ (1975) Subsolidus and melting relationships for calcite, magnesite and the join $CaCO_2$–$MgCO_3$ to 36 kb. Geochim Cosmochim Acta 39:35–53

Ito E, Yamada H (1982) Stability relations of silicate spinels, ilmenites and perovskites. Adv Earth Planet Sci 12:405–419

Ito K, Kennedy GC (1967) Melting and phase relations in a natural peridotite to 40 kilobars. Am J Sci 265:519–538

Jackson EO (1961) Primary textures and mineral associations in the ultramafic zone of the Stillwater complex, Montana. US Geol Surv Prof Paper 358:1–106

Jackson KA (1958) Liquid metals and solidification. Am Soc Metals, Cleveland, Ohio, pp 174–186

Jaeger JC (1968) Cooling and solidification of igneous rocks. In: Hess HH, Poldesvaart A (eds) Basalts Interscience, New York pp 503–536

Jacob H (1949) Heat transfer. Wiley, New York, 758 pp

Jaluria Y (1980) Natural convection. Pergamon, Oxford, 326 pp

Jaques AL, Green DH (1980) Anhydrous melting of peridotite at 0–15 kb pressure and the genesis of tholeiitic basalts. Contrib Mineral Petrol 73:287–310

Jeffreys M (1962) The Earth. Cambridge Univ Press 438 pp

Jost W (1960) Diffusion in solids, liquids, gases. Academic, New York, 558 pp

Judd JW (1881) Volcanoes. C Kegan Pall, London, 381 pp

Kawada K (1977) The system Mg_2SiO_4 – Fe_2SiO_4 at high pressures and temperatures and the Earth's interior. PhD Thesis, University of Tokyo

Kay JM (1968) An introduction to fluid mechanics and heat transfer. Cambridge Univ Press, 327 pp

Kelly KK (1960) Contribution to the data on theoretical metallurgy XIII. High temperature heat capacity, and entropy data for the elements and inorganic compounds. US Bur Mines Bull 584:232 pp

Killingley JS, Muenow DW (1975) Volatiles from Hawaiian submarine basalts determined by dynamic high temperature mass spectrometry. Geochim Cosmochim Acta 39:1467–1473

Kingery WO (1960) Introduction to ceramics. New York, 781 pp

Kirkpatrick RJ (1974) The kinetics of crystal growth in the system $CaMgSi_2O_6$ – $CaAl_2SiO_6$. Am J Sci 273:215–242

Kirkpatrick RJ (1975) Crystal growth from the melt: A review. Am Mineral 60:798–814

Kirkpatrick RJ (1976) Towards a kinetic model for the crystallization of magma bodies. J Geophys Res 81:2565–2571

Kittel C (1956) Introduction to solid state physics. Wiley, New York, 617 pp

Kohlstedt, DL, Goetze C, Durham WB (1976) Experimental deformation of single crystal olivine with application to flow in the mantle. In: Strens RGJ (ed) The physics and chemistry of minerals and rocks. Wiley, New York, pp 35–49

Komar PD (1976) Phenocryst interactions and the velocity profile of magma flowing through dikes and sills. Geol Soc Am Bull 87:1336–1342

Konowalow D (1881) Über die Dampfspannungen der Flüssigkeitsgemische. Wiedemann's Annales der Physik 14:34–52

Koster van Goos AF, Wyllie PJ (1966) Liquid immiscibility in the system Na_2O – Al_2O_3 – SiO_2 – CO_2 at pressures up to 1 kilobar. Am J Sci 264:234–255

Kozu S, Ueda J (1933) Thermal expansion of diopside. Proc Imp Acad (Tokyo) 9:317–319

Kreyszig E (1967) Advanced engineering mathematics. Wiley, New York, 898 pp

Kreyszig E (1970) Introductory mathematical statistics. Wiley, New York, 470 pp

Kudo AM, Weill DF (1970) An igneous plagioclase thermometer. Contrib Mineral Petrol 25:52–65

Kuno H, Yamasaki K, Iida C, Nagashima K (1957) Differentiation of hawaiian magmas. Japan J Geol Geography 28:179–218

Kurz MD, Jenkins WJ (1981) The distribution of helium in oceanic basalt glasses. Earth Planet Sci Lett 53:41–54

Kushiro I (1969) The system forsterite-diopside silica with and without water at high pressures. Am J Sci 267A:269–294

Kushiro I (1972) New method of determining liquidus boundaries with confirmation of incongruent melting of diopside and existence of iron free pigeonite at 1 atm. Carnegie Inst Wash Year book. 71:603–607

Kushiro I (1973) The system diopside – anorthite – albite: determination of composition of coexiding phases. Carnegie Inst Wash Year book 72:502–507

Kushiro I, Yoder HS (1974) Formation of eclogite from garnet lherzolite: liquidus relations in a portion of the system $MgSiO_3$ – $CaSiO_3$ – Al_2O_3 at high pressures. Carnegie Inst Wash Year book 73:266–269

Kushiro I (1976) Decrease in viscosity of some synthetic silicate melts at high pressures. Carnegie Inst Wash Yb 75, pp 611–614

Lambert IB, Wyllie PJ (1968) Stability of hornblende and a model for the low velocity sone. Nature 219:1240–1241

Larimer JW (1968) Experimental studies on the system Fe-MgO-SiO$_2$ and their bearing on petrology of chondritic meteorites. Geochim Cosmochim Acta 32:1187–1207

Lee T (1979) New isotopic clues to solar system formation. Rev Geophys Space Phys 17:1591–1611

Leeman WD, Lindstrom DJ (1978) Partitioning of Ni$_2^+$ between basaltic and synthetic melts and olivines – an experimental study. Geochim Cosmochim Acta 42:801–816

Lewis GN (1908) The osmotic pressure of concentrated solutions, and the laws of perfect solution. J Am Chem Soc 30:668–683

Lewis GN, Randall M (1923) Thermodynamics and the free energy of chemical substances. McGraw Hill, New York, 653 pp

Lewis GN, Randall M (1961) Thermodynamics. McGraw-Hill, New York, 723 pp

Loewinson-Lessing FY (1954) A historical survey of petrology. Oliver and Boyd, Edinburg, 112 p

Lofgren G (1974) An experimental study of plagioclase crystal morphology. Am J Sci 264:243–273

Long LE (1964) Rb-Sr chronology of the Carn Chainneag intrusion, Ross-shire, Scotland. J Geophys Res 69:1589–1597

Lugmair GW (1974) A new dating method (abstr.) Meteoritics 9:369

Lugmair GW, Scheinin NB, Marti K (1975) Search for extinct ^{146}Sm, 1. The isotopic abundance of ^{142}Nd in the Juvinas meteorite. Earth Planet Sci Lett 27:79–84

Luth WC, Jahns RH, Tuttle OF (1964) The granite system at pressures of 4 to 10 kilobars. J Geophys Res 69:759–773

Maaløe S (1974) The crystallization of simple pegmatites in the Moss area, Southern Norway. Norsk Geol Tidsskr 54, pp 149–167

Maaløe S (1975) The crystallization conditions of the Skaergaard intrusion, East Greenland. Thesis, Univ of Copenhagen, 190 pp

Maaløe S (1976) The origin of rhythmic layering, Mineral Mag 42:337–345

Maaløe S (1981) Magma accumulation in the ascending mantle. J Geol Soc London 138:223–236

Maaløe S (1982) Geochemical aspects of permeability controlled partial melting and fractional crystallization. Geochim Cosmochim Acta 46:43–57

Maaløe S (1984) Fractional crystallization and melting within binary systems with solid solution. Am J Sci 284:272–287

Maaløe S, Aoki K (1977) The major element composition of the upper mantle estimated from the composition of lherzolites. Contrib Mineral Petrol 63:161–173

Maaløe S, Jacobsson SP (1980) The PT phase relations of a primary oceanite from the Reykjanes peninsula, Iceland. Lithos 13:237–246

Maaløe S, Petersen TS (1976) Phase relations governing the derivation of alkaline basaltic magmas from primary magmas at high pressures. Lithos 9:243–252

Maaløe S, Petersen TS (1981) Petrogenesis of oceanic andesites. J Geophys Res 86:10273–10286

Maaløe S, Scheie Å (1982) The permeability controlled accumulation of primary magma. Contrib Mineral Petrol 81:350–357

Maaløe S, Wyllie PJ (1975) Water content of a granite magma deduced from the sequence of crystallization determined experimentally with water-undersaturated conditions. Contrib Mineral Petrol 52:175–191

Maaløe S, Wyllie PJ (1979) The join grossularite-pyrope at 30kbars and its petrological significance. Am J Sci 279:288–301

Maaløe S, Sørensen I, Hertogen J (1985) (to be published) The trachy basaltic suite of Jan Mayen. J Petrol

Malpas J (1978) Magma generation in the upper mantle, field evidence from ophiolite suites, and application to the generation of oceanic litosphere. Philos Trans R Soc Lond A288:527–546

Marsh BD (1976) Mechanics of Benioff zone magmatism. Geophys Monograph 19:337–352

Marsh BD (1978) On the cooling of andesitic magma. Phil Trans Roy Soc London A288:611–625

Marsh BD (1979) Island arc development: Some observations, experiments, and speculations. J Geol 87:687–713

Marsh BD (1982) On the mechanics of igneous diapirism, stoping, and zone melting. Am J Sci 282:808–855

Marsh BD (1984) Mechanics and energetics of magma formation and ascention. In: Studies in Geophysics (ed FR Boyd). National Acad Press, Washington: 67–83

Marsh BD, Kantha LH (1978) On the heat and mass transfer from an ascending magma. Earth Plan Sci Lett 39:435–443

Martin RF, Donnay G (1972) Hydroxyl in the mantle. Am Mineral 57:554–570

Mao HK, Bell PM (1978) Design and varieties of the megabar cell. Carnegie Inst Wash Year book 77:904–908

McBirney AR, Noyes RM (1979) Crystallization and layering of the Skaergaard intrusion. J Petrol 20:487–554

MacDonald GA (1949) Hawaiian petrographic province. Geol Soc Am Bull 60, pp 1541–1596

MacDonald GA (1968) Composition and origin of Hawaiian lavas. Geol Soc Am Mem 116:477–522

MacDonald GA, Abbott AT (1970) Volcanoes in the sea. Univ Press of Hawaii, Honolulu, 441 pp

MacDonald GA, Katsura T (1964) Chemical composition of Hawaiian lavas. J Petrol 5:82–133

McIntire WL (1963) Trace element partition coefficients: a review of theory and applications in geology. Geochim Cosmochim Acta 27:1209–1264

Margaritz M, Whitford DJ, James DE (1978) Oxygen isotopes and the origin of high $^{87}Sr/^{86}Sr$ andesites. Earth Planet Sci Lett 40:220–230

Mathez EA (1973) Refinement of the Kudo-Weill plagioclase thermometer and its application to basaltic rocks. Contrib Mineral Petrol 41:61–72

Matthews CS, Russel OG (1967) Pressure buildup and flow tests in wells. Soc Petrol Eng AIME Monograph, 171 pp

Mercier JCC, Nicolas A (1975) Textures and fabrics of upper mantle peridotites as illustrated by xenoliths from basalts. J Petrol 16:454–487

McKenzie DP, Roberts JM, Weiss NO (1974) Convection in the Earth's mantle: towards a numerical simulation. J Fluid Mech 62:465–538

McNown JS, Malaika J (1950) Effects of particle shape on settling velocity at low Reynolds numbers. Am Geophys Union Transactions 31:74–82

Menzies M, Murthy VR (1980) Enriched mantle: Nd and Sr isotopes in diopsides from kimberlite nodules. Nature 283:634–636

Mercier JC (1976) Single-pyroxene geothermometry and geobarometry. Am Mineral 61:603–615

Ming L, Bassett WA (1974) Laser heating in the diamond anvil press up to 2000°C sustained and 3000°C pulsed at pressures up to 260 kilobars. Rev Sci Instrum 45:1115–1118

Misener DJ (1976) Cationic diffusion in olivine to 1400°C and 35 kbar. In: Hofmann AW, Giletti BJ, Yoder HS, Young RA (eds) Geochemical transport and kinetics. Carnegie Inst Wash Pub 634:117–170

Mohr RE, Stout JH (1980) Multisystem nets for systems of n+3 phases. Am J Sci 280:143–172

Moorbath S, N'Nions RK, Pankhurst RJ, Gale NH, McGregor VR (1972) Further rubidium-strontium age determinations on the very early Precambrian rocks of the Godthåb region, West Greenland. Nature 240:78–82

Moore JG (1965) Petrology of deep-sea basalt near Hawaii. Am J Sci 263:40–52

Moore DR, Weiss NO (1973) Two-dimensional Rayleigh-Benard convection. Fluid Mech 58:289–312

Morey GW (1936) Heterogeneous equilibrium. Carnegie Inst Wash Year book 36:125

Morey GW, Bowen NL (1924) The binary system sodium meta-silicate-silica. J Phys Chem 28:1167–1179

Morey GW, Williamson EO (1918) Chemical equilibrium. Carnegie Inst Wash Year book 17:128

Morgan WJ (1972) Convection plumes and plate motions. Am Assoc Pet Geol Bull 56:203–213

Morse S (1980) Basalts and phase diagrams. Wiley, New York, 493 pp

Muenow DW, Graham DG, Liu NWK (1979) The abundance of volatiles in Hawaiian tholeiitic submarine basalts. Earth Planet Sci Lett 42:71–76

Murase T, McBirney AR (1973) Properties of some common igneous rocks and their melts at high temperatures. Geol Soc Am Bull 84:3563–3592

Mysen BO (1976) Partitioning of samarium and nickel between olivine, ortho-pyroxene, and liquid; prelaminary data at 20 kbar and 10125 °C. Earth Planet Sci Lett 31:1–7

Mysen BO, Boettcher AL (1975) Melting of a hydrous mantle: II. Geochemistry of crystals and liquids formed by anatexis of mantle peridotite at high pressures and high temperatures as a function of controlled activities of water, hydrogen, and carbon dioxide. J Petrol 16:549–593

Mysen BO, Kushiro I (1976) Compositional variation of coexisting phases with degree of melting of peridotite under upper mantle conditions. Carnegie Inst Wash Year book 75:546–555

Nehru CE, Wyllie PJ (1974) Electron microprobe measurement of pyroxenes coexisting with H_2O-undersaturated liquid in the join $CaMgSi_2O_6$-$Mg_2Si_2O_6$-H_2O at 30 kilobars, with applications to geothermometry. Contrib Mineral Petrol 48:221–228

Neuman H, Mead J, Vitaliano CJ (1954) Trace element variation during fractional crystallization as calculated from the distribution law. Geochim Cosmochim Acta 6:90–99

Newton RH (1935) Activity coefficients of gases. Ind Eng Chem 27:302–306

Newton RC, Sharp WE (1975) Stability of forsterite + CO_2 and its bearing on the role of CO_2 in the mantle. Earth Planet Sci Lett 26:239–244

Nielsen AE (1964) Kinetics of precipitation. Pergamon, Oxford, 151 pp

Niggli P (1937) Das Magma und seine Produkte. Akad. Verlagsgesellschaft, Leipzig 1937, 379 pp

Nocholds SR (1954) Average chemical compositions of some igneous rocks. Bull Geol Soc Am 65:1007–1032

O'Hara MC (1963) The join diopside-pyrope at 30 kilobars. Carnegie Insts Wash Year book 62:116–118

O'Hara MC (1968) The bearing of phase equilibria studies in synthetic and natural systems on the origin and evolution of basic and ultrabasic rocks. Earth Sci Rev 4:69–133

O'Hara MJ, Schairer JF (1963) The join diopside-pyrope at atmospheric pressure. Carnegie Inst Wash Year book 62:107–115

O'Hara MJ, Yoder MS (1967) Formation and fractionation of basic magmas at high pressures. Scott J Geol 3:67–117

Ohtani E, Kumazawa M, Kato T, Irifune T (1982) Melting of various silicates at elevated pressures. Adv Earth Planet Sci 12:259–270

O'Nions RK, Carter SR, Evensen NM (1979) Geochemical and cosmochemical applications of Nd isotope analysis. Am Rev Earth Planet Sci 7:11–38

O'Nions RK, Evensen NM, Hamilton PJ, Carter SR (1978) Melting of the mantle past and present: isotope and trace element evidence. Philos Trans R Soc Lond A 258:547–559

O'Nions RK, Hamilton PJ, Evensen NM (1977) Variations in [143]Nd/[144]Nd and [87]Sr/[86]Sr ratios in oceanic basalts. Earth Planet Sci Lett 34:13–22

Onsager L (1945) Theories and problems of liquid diffusion. Ann NY Acad Sci 46:241–265

Osborn EF (1959) Role of oxygen pressure in the crystallization and differentiation of basaltic magma. Am J Sci 257:609–647

Osborn EF (1979) The reaction principle. In: Yoder HS (ed) The evolution of the igneous rocks. Princeton Univ Press, Princeton, pp 133–169

Osborn EF, Schaierer JF (1941) The ternary system pseudowallastonite-akermanite-gehlenite. Am J Sci 139:715–763

Oxburgh ER, Turcotte DL (1968) Mid-ocean ridges and geotherm distribution during mantle convection. J Geophys Res 73:2643–2661

Oxburgh ER, Turcotte DL (1978) Mechanisms of continent drift Rep Prog Phys 41:1249–1312

Papanastasiou DA, Wasserburg GJ (1969) Initial strontium isotopic abundances and the resolution of small time differences in the formation of planetary objects. Earth Planet Sci Lett 5:361–376

Parmentier EM, Turcotte DL, Torrance KE (1975) Numerical experiments on the structure of mantle plumes. J Geophys Res 80:4417–4424

Pearce TH (1968) A contribution to the theory of variation diagrams. Contrib Mineral Petrol 10:142–157

Peirce BO, Foster RM (1963) A short table of integrals. Blaisdell, New York, 189 pp

Peltier WR (1980) Mantle connection and viscosity. In: Dziewonski A, Boschi E (eds) Physics of the Earth's interior. Elsevier, Amsterdam, pp 362–431

Peltier WR (1981) Surface plates and thermal plumes: separate scales of the mantle convective circulation. Am Geophys Union Geodyn Ser 5:229–248

Peltier WR, Jarvis GT (1982) Whole mantle convection and the thermal evolution of the Earth. Phys Earth Planet Int 29:281–304

Peterman ZE, Goldich SS, Hedge CE, Yardley DE (1975) Geochronology of the Rainy Lake Region, Minnesota-Ontario. Geol Soc Am Mem 135:193–215

Philpotts JA, Schnetzler CC (1970) Phenocryst-matrix partition coefficients for K, Rb, Sr and Ba, with applications to anorthosite and basalt genesis. Geochim Cosmochim Acta 34:307–322

Pressnall DC (1979) Fractional crystallization and partial fusion. In: Yoder HS (ed) The evolution of igneous rocks. Princeton Univ Press, Princeton, pp 59–75

Pressnal DC, Bateman PC (1973) Fusion relations in the system $NaAlSi_3O_8 - CaAl_2Si_2O_8 - KAlSi_3O_8 - SiO_2 - H_2O$ and generation of granitic magmas in the Sierra Nevada batholith. Geol Soc Am Bull 84:3181–3202

Pressnall DC, Dixon SA, Dixon JR, O'Donnell TH, Brenner NL, Schrock RL, Dycus DW (1978) Liquidus phase relations on the join diopside – forsterite – anorthite from 1 atm to 20 kbar: their bearing on the generation and crystallization of basaltic magma. Contrib Mineral Petrol 66:203–220

Ramberg H (1967) Gravity, deformation and the Earth's crust. Academic, London, 214 pp

Ramberg H (1972) Mantle diapirism and its tectonic and magmatic consequences. Phys Earth Planet Int 5:45–60

Raoult FM (1888) Über die Dampfdrucke ätherischer Lösungen. Z Phys Chem 2:353–373

Raleigh JWS (1896) Theoretical considerations respecting the separation of gases by diffusion and similar processes. Philos Mag 42:77–107

Rayleigh JWS (1902) On the destillation of binary mixtures. Philos Mag Ser. 6 vol 4, 521–537

Rayleigh JWS (1916) On convection currents in a horizontal layer of fluid when the higher temperature is on the under side. Phil Mag 32:529–546

Reid MJ, Gancarz AJ, Albee AL (1973) Constrained least square analysis of petrologic problems with an application to lunar sample 12046. Earth Planet Sci Lett 17:433–445

Reismann A (1970) Phase equilibria. Academic, New York, 541 pp

Reyer E (1888) Theoretische Geologie. E Schweizerbartsche Verlag, Stuttgart, 869 pp

Rhines FN (1956) Phase diagrams in metallurgy, McGraw-Hill, New York, 340 pp

Ricci JE (1951) The phase rule and heterogeneous equilibrium. D Van Nostrand, New York, 505 pp

Richardson FD (1974) Physical chemistry of melts in metallurgy I. Academic London, 289 pp

Richter DH, Moore JG (1966) Petrology of the Kilauea Iki Lava Lake, Hawaii. US Geol Surv Prof Paper 537-B, 26 pp

Richter DH, Murata KJ (1966) Petrography of the lavals of the 1959–1960 eruption of Kilauea volcano, Hawaii. US Geol Surv Prof Paper 537-D, 12 pp

Ringwood AE (1958) Constitution of the mantle. Geochim Cosmochim Acta 15:195–212

Ringwood AE (1966) Mineralogy o the mantle. In: Hurley PM (ed) Advances in Earth Sciences. MIT Press, Cambridge, Mass pp 287–356

Ringwood AE (1975) Composition and petrology of the Earth's mantle. McGraw-Hill, New York, 618 pp

Robertson JK, Wyllie PJ (1971) Rock-water systems, with special reference to the water-deficient region. Am J Sci 271:252–277

Robie RA, Waldbaum DR (1968) Thermodynamic properties of minerals and related substances at 298.15 °K (25.0 °C) and one atmosphere (1.013 bars) pressure and at higher temperatures. Geol Surv Bull 1259:256 pp

Robinson CS, Gilliland ER (1950) Elements of fractional destillation. McGraw-Hill New York, 492 pp

Roeder PL, Emslie RF (1970) Olivine-liquid equilibrium. Contrib Mineral Petrol 29:275–289

Roeder PL, Osborn EF (1966) Experimental data for the system $MgO-FeO-Fe_2O_3-CaAl_2Si_2O_8 - SiO_2$ and their petrological implications. Am J Sci 264:428–480

Roozeboom HWB (1893) Die Gleichgewichte von Lösungen zweier oder dreier Bestandteile mit festen Phasen: Komponenten, binäre und ternäre Verbindungen, in ihrem Zusammenhang dargestellt. Z Phys Chem 11:359–389

Roozeboom HWB (1901) Die heterogenen Gleichgewichte vom Standpunkte der Phasenlehre, Vol I. Friederich Vieweg, Braunschweig, 221 pp

Rutherford E, Soddy F (1902) The cause and nature of radioactivity Pt. 1. Phil Mag Ser 6 vol 4:370–396

Saxena SK (1973) Thermodynamics of rock-forming crystalline solutions. Springer, Berlin Heidelberg New York Tokyo, 188 pp

Scarfe CM, Mysen BO, Virgo D (1979) Changes in viscosity and density of melts of sodium disilicate, sodium metasilicate, and diopside composition with pressure. Carnegie Inst Wash Year book 78:547–551

Schairer JF (1957) Melting relations of the common rock-forming oxides. J Am Ceram Soc 40:215–235

Schairer JF, Yoder HS (1961) Crystallization in the system nepheline-forsterite-silica at 1 atm pressure. Carnegie Inst Wash Year book 60:141–144

Shilling JG, Winchester JW (1967) Rare-earth fractionation and magmatic processes. In: Runcorn SK (ed) Mantles of Earth and terrestial Planets. Interscience, New York, pp 267–283

Schreinemakers FAH (1919) In-, mono- and divariant equilibria 1. Proc Koninklijke Akad Wetensk, Amsterdam 18:116–126

Schreinemakers FAH (1912–1925) Papers by FAH Schreinemakers vol 1, 2 Pennsylvania State Univ Pennsylvania 1965, 579 pp

Schubert W, Froidevoux C, Yuen DA (1976) Oceanic lithosphere and asthenosphere: thermal and mechanical structure. J Geophys Res 81:3525–3540

Shaw DM (1970) Trace element fractionation during anatexis. Geochim Cosmochim Acta 34:237–243

Shaw HR (1965) Comments on viscosity, crystal settling, and convection in granitic magmas. Am J Sci 263:120–152

Shaw HR (1969) Rheology of basalt in the melting range. J Petrol 10:510–535

Shaw HR, Peck DL, Wright TL, Okamura R (1968) The viscosity of basaltic magma: an analysis of field measurements in Makaopuhi lava lake, Hawaii. Am J Sci 266:225–264

Shewmon PG (1963) Diffusion in solids. McGraw-Hill, New York, 203 pp

Snedecor GW and Cochran WC (1967) Statistical methods. Iowa State College Press, Ames, 593 pp

Stacey FO (1977) Physics of the Earth. Wiley New York, 414 pp

Stanton RL (1967) A numerical approach to the andesite problem. Proc Nederl Acad Wettensh Ser B, vol 70:176–191

Staudacher T, Allegre CJ (1982) Terrestial xenology. Earth Planet Sci Lett 60:389–406

Stewart OB (1967) Four-phase curve in the system $CaAl_2Si_2O_8 - SiO_2 - H_2O$ between 1 and 10 kilobars. Schweiz Mineral Petrog Mitt 47:35–59

Swanson CO (1924) A graphical solution of certain ratios in temperature concentration diagrams. Am J Sci 5 ser vol 39:233–238

Swanson SE (1977) Relation of nucleation and crystal-growth role to the development of granitic textures. Am Mineral 62:966–978

Tatsumoto M, Hedge CE, Engle AEJ (1965) Potassium, rubidium, strontium, thorium, uranium, and the ratio of strontium-87 to strontium-86 in oceanic tholeiitic basalts. Science 150:886

Tatsumoto M, Unruh DM, Desborough GA (1976) U-Th-Pb and Rb-Sr systematics of Allende and U-Th-Pb systematics of Orgueil. Geochim Cosmochim Acta 40:617–634

Temkin DE (1966) Molecular roughness of the crystal-melt boundary. In: Sirota NN, Gorskii FK, Varikash VM (eds) Crystallization processes. Consultants Bureau, New York, pp 15–23

Thorpe RS, Francis PW, Moorbath S (1979) Rare earth and strontium isotope evidence concerning the petrogenesis of north Chilean ignimbrites. Earth Planet Sci Lett 42: 359–367

Thorton CP, Tutle OF (1960) Chemistry of igneous rocks I. Differentiation index Am J Sci 258:664–684

Tilley CE, Yoder HS, Schairer JF (1965) Melting relations of volvanic tholeite and alkalic rock series. Carnegie Inst Wash Year book 64:69–82

Tilley CE, Yoder HS, Schairer JF (1965) Melting relations of volcanic tholeiite and alkalic rock series. Carnegie Inst Wash Year book 64:69–82

Torrance KE, Turcotte DL (1971) Structure of convection cells in the mantle. J Geophys Res 76:1154–1161

Treuil M (1973) Critéres pétrologiques, géochemiques, et structuraux de la genése et de la différentiation des magmas basaltiques: Example de l'Afar. Thése de Doctorrat d'Etat és Sciences, Université d'Orleans, 120 pp

Turcotte DL, Ahern JL (1978) A porous flow model for magma migration in the asthenosphere. J Geophys Res 83:767–772

Turnbull D, Fischer JC (1949) Rate of nucleation in condensed systems. J Chem Phys 17:71–73

Turner FJ (1968) Metamorphic Petrology. McGraw-Hill, New York, 403 pp

Turner JS (1973) Buoyancy effects in fluids. Cambridge University Press, 368 pp

Tuttle OF, Bowen NI (1958) Origin of granite in the light·of experimental studies in the system NaAlSi$_3$O$_8$ – KAlSi$_3$O$_8$ – SiO$_2$ – H$_2$O. Geol Soc Am Mem 74:153 pp

Tuttle OF, Gitting J (1966) Carbonatites. Interscience, New York, 591 pp

Twiss RJ (1976) Structural superplastic creep and linear viscosity in the Earth's mantle. Earth Planet Sci Lett 33:86–100

Uhlmann DR (1972) Crystal growth in glass forming systems – A review. In: Hench LL, Freiman SW (eds) Advances in nucleation and growth in glasses. Am Ceramic Soc Spec Pub 5:91–115

Ulmer GC (1971) Research techniques for high pressure and high temperature. Springer, Berlin Heidelberg New York, Tokyo, p 367

Van der Waals, J. O. (1873) Over de continuitet van den gas en vloeistoftoestand. Dissert Leiden

Vant Hoff JH (1887) Die Rolle des osmotischen Druckes in der Analogie zwischen Lösungen und Gasen. Z Phys Chem 1:481–508

Velarde MG, Normand C (1980) Convection. Sci Am 243:79–93

Vogt J (1904) Die Silikatschmelzlösungen mit besonderer Rücksicht auf die Mineralbildung und die Schmelzpunkterniedrigung. Vidensk Selsk Skr Kristiania, Math.-Phys. bl no. 8

Volmer H, Weber A (1926) Keimbildung in übersättigten Gebilden. Z Phys Chem 119: 277–301

Waff HS (1980) Effects of the gravitational field on liquid distribution in partial melts within the upper mantle. J Geophys Res 85:1815–1825

Waff HS, Bulau JR (1979) Equilibrium fluid distribution in an ultramafic partial melt under hydrostatic stress conditions. J Geophys Res 84:6109–6114

Wager LR (1961) A note on the origin of ophitic texture in the chilled olivine gabbro of the Skaergaard intrusion. Geol Mag 98:353–366

Wager LR and Brown GM (1960). Types of igneous cumulates. J Petrol 1, pp 73–85

Wager LR, Brown GM (1967) Layered igneous rocks. Oliver and Boyd, Edinburgh, 588 pp

Wager LR, Deer WA (1939) Geological investigations in East Greenland, Pt. III. The petrology of the Skaergard intrusion, Kangerdlugssuaq, East Greenland. East Greenland. Medd om Grønland 105. no 4:1–352

Wagstaff FE (1967) Crystallization kinetics of internally nucleated vitreous silica. GE Tech Rep 67-c-489, Schenectady

Wall FT (1974) Chemical thermodynamics. WH Freeman, San Francisco, 493 pp

Walton AG (1969) Nucleation in liquids and solutions. In: Zettlemayer $\bar{A}C$ (ed) Nucleation Marcel Dekker, New York, pp 225–307

Walton M (1960) Molecular diffusion rates in super critical water vapor estimated from viscosity data. Am J Sci 258:385–401

Wasserburg GJ, Papanastassiou DA, Nenow EV, Bauman CA (1969) A programmable magnetic field mass spectrometer with on-line data processing. Rev Sci Instr 40:288–295

Wasserburg GJ, Papanastassiou DA, Lee T (1980) Isotopic heterogenetics in the solar system. In: Early solar system processes and the present solar system. Soc Ital Fisica 73:144–191

Watts AB (1976) Gravity and bathymetry in the central pacific ocean. J Geophys Res 81:1533–1553

Weast RC (1968) Handbook of chemistry and physics. Cleveland Chemical Rubber, A 245

Weertman J (1970) The creep strength of the earth's mantle. Rev Geophys Space Phys 8:145–168

Weertman J (1971) Theory of water-filled crevasses in galciers applied to vertical magma transport beneath oceanic ridges. J Geophys Res 76:1171–1183

Weertman J (1972) Coalescence of magma pockets into large pods in the upper mantle. Geol Soc Am Bull 83:3531–3532

Weertman J (1974) Velocity of which liquid filled cracks move in the Earth's crust or glaciers. J Geophys Res 76:8544–8553

Weertman J, Weertman JR (1970) Mechanical properties, strongly temperature dependent. In: Calin RW (ed) Physical Metallurgy, Elsevier, New York, pp 983–1010

Wegener A (1915) Die Entstehung der Kontinente und Oceane. Sammlung Vieweg, Braunschweig, 94 pp

Wendtlandt RF, Mysen BO (1978) Melting phase relations of natural peridotite + CO_2 as a function of degree of partial melting at 15 and 30 kbar. Carnegie Inst Wash Year book 77:756–767

Wetherill GW (1975) Radiometric chronology of the early solar system. Annu Rev Nucl Sci 25:283–328

Wetherill GW (1981) The formation of the Earth from planetisimals. Scient American, June, pp 163–174

White WM, Hofmann AW (1982) Sr and Nd isotope geochemistry of oceanic basalts and mantle evolution. Nature 296:821–825

Whitehead JA, Luther DS (1975) Dynamics of laboratory diapir and plume models. J Geophys Res 80:705–717

Whitney JA (1975) The effects of pressure, temperature, and X_{H_2O} on phase assemblage in four synthetic rock compositions. J Geol 83:1–31

Wilkins RWT, Sabine W (1973) Water content of some nominally anhydrous silicates. Am Mineral 58:508–516

Wilkinson WL (1960) Non-Newtonian fluids. Pergamon, New York, 138 p

Williams DW, Kennedy GC (1969) Melting curve of diopside to 50 kilobars. J geophys Res 74:4359–4366

Williams RE (1968) Space-filling polyhedron: Its relation to aggregates of soap bubbles, plant cells, and metal crystallites. Science 161:276–277

Wright TL (1971) Chemistry of Kilauea and Mauna Loa Llava in space and time. US Geol Surv Prof Pap 735:40 pp

Wright TL, Fisher RS (1971) Origin of the differentiated and hybrid lavas of Kilauea volcano, Hawaii. J Petrol 12:1–65

Wright TL, Doherty PC (1970) A linear programming and last squares computer method for solving petrologic mixing problems. Geol Soc Am Bull 81:1995–2008

Woodruff DP (1973) The solid-liquid interface. Cambridge Univ Press, 182 p

Wulff G (1901) Zur Frage der Geschwindigkeit des Wachstums und der Auflösung der Krystallflächen. Z Krist 34:449–530

Wyllie PJ (1963) Effects of the change in slope occuring or liquids and solidus paths in the system diopside-anorthite-albite. Mineral Soc Am Spec Pap 1:204–212

Wyllie PJ (1966) Experimental studies of carbonate problems: the origin and differentiation of carbonate magmas. In: Tuttle OF, Gittins J (eds). Carbonatites Wiley, New York, pp 311–352

Wyllie PJ (1976) From crucibles through subduction to batholiths. In: Saxena SK, Bhattacharji S (eds) Energetics of geological processes. pp 389–433

Wyllie PJ (1977) Crustal anatexis: An experimental review. Tectonophysics 43:41–77

Wyllie PJ (1978) Mantle fluid compositions buffered in peridotite – CO_2 – H_2O by carbonates amphibole, and phlogopite. J Geol 86:687–713

Wyllie PJ (1979) Magmas and volatile components. Am Mineral 64:469–500

Wyllie PJ, Boettcher AL (1969) Liquidus phase relationships in the system CaO-CO_2 – H_2O to 10 kilobars pressure with petrological applications. Am J Sci 267A:489–508

Wyllie PJ, Huang WL (1976) Carbonation and melting reactions in the system CaO-MgO-SiO_2-CO_2 at mantle pressures with geophysical and petrological applications. Contrib Mineral Petrol 54:79–107

Wyllie PJ, Tuttle OF (1960) Experimental investigations of silicate systems containing two volatile components, Part 1. Geometrical considerations. Am J Sci 258:498–517

Wyllie PJ, Tuttle OF (1960) The system CaO-CO_2-H_2O and the origin of carbonatites. J Petrol 1:1–46

Wyllie PJ, Huang WL, Stern CL, Maaløe S (1976) Granitic magmas: Possible and impossible sources, water contents, and crystallization sequences. Can J Earth Sci 13:1007–1019

Yoder HS (1965) Diopside-anorthite-water at five and ten kilobars and its bearing on explosive volcanism. Carnegie Inst Wash Year book 64:82–89

Yoder HS (1967) Albite-anorthite-quartz-water at 5 kbars. Carnegie Inst Wash Year book 66:477–478

Yoder HS (1976): Generation of basaltic magma. Nat Acad Sci, Wash DC 265 pp

Yoder HS (ed) (1979) The evolution of the igneous rocks. Princeton Univ Press, Princeton, 588 pp

Yoder HS, Kushiro I (1969) Melting of a hydrous phase: Phlogopite. Am J Sci 267A:558–582

Yoder HS, Tilley CE (1962) Origin of basaltic magmas: An experimental study of natural and synthetic rock systems. J Petrol 3:342–532

Yoder HS, Stewart DB, Smith JR (1957) Ternary feldspars. Carnegie Inst Wash Year book 56:206–214

Zen E-an (1966) Construction of pressure-temperature diagrams for multicomponent systems after the method of Schreinemakers – a geometric approach. US Geol Surv Bull 1225:1–56

Zindler A, Jagoutz E, Goldstein S (1982) Nd, Sr and Pb isotopic systematics in a three-component mantle: A new perspective. Nature 298:519–523

Zindler A (1982) Nd and Sr isotopic studies of komatiites and related rocks. In: Komatiites Arndt NT, Nisbet EG (eds) George Allen and Unwin, London, pp 399–420

Subject Index

F. Liebau

Structural Chemistry of Silicates

Structure, Bonding, and Classification

1985. 136 figures. Approx. 400 pages.
ISBN 3-540-13747-5

Contents: Introduction. – Methods to describe the Atomic Structure of Silicates. – Chemical Bonds in Silicates. – Crystal Chemical Classification of Silicate Anions. – Nomenclature and Structural Formulae of Silicate Anions and Silicates. – Crystal Chemical Classification of Silicates: General Part. – Crystal Chemical Classification of Silicates: Special Part. – Other Classifications of Silicates. – General Rules for Silicate Anion Topology. – Conclusion.

Among the inorganic compounds found in the Earth's crust, the silicates are the most common making up more than 90% of the total. At the same time they are an important source of raw materials for the manufacture of such products as cement, glass and ceramics.

Structural Chemistry of Silicates is a comprehensive, richly illustrated survey of the structures of crystalline silicates designed to instill an appreciation of the significance of these compounds. It is the result of the author's own crystal-chemical classification of silicates, based on Bragg's classic system, and is an outstanding method of representing the connections between chemical compounds and silicate structures.

The book is recommended for third year students of chemistry, mineralogy geology, and materials science. It will also prove useful as a guide for researchers and industrial chemists in their persuit of deeper, more specialized knowledge.

Springer-Verlag
Berlin
Heidelberg
New York
Tokyo

Archaean Geochemistry

The Origin and Evolution of the Archaean Continental Crust

Editors: **A. Kröner, G. N. Hanson, A. M. Goodwin**

1984. 86 figures. X, 285 pages.
ISBN 3-540-13746-7
(Final Report of the IGCP-Project No. 92 (Archaean Geochemistry))

Contents: Mantle Chemistry and Accretion History of the Earth. – Geochemical Characteristics of Archaean Ultramafic and Mafic Volcanic Rocks: Implications for Mantle Composition and Evolution. – Archaean Sedimentary Rocks and Their Relation to the Composition of the Archaean Continental Crust. – Spatial and Temporal Variations of Archaean Metallogenic Associations in Terms of Evolution of Granitoid-Greenstone Terrains with Particular Emphasis on the Western Australian Shield. – Magma Mixing in Komatiitic Lavas from Munro Township, Ontario. – Oxygen Isotope Compositions of Minerals and Rocks and Chemical Alteration Patterns in Pillow Lavas from the Barberton Greenstone Belt, South Africa. – Petrology and Geochemistry of Layered Ultramafic to Mafic Complexes from the Archaean Craton of Karnataka, Southern India. – Pressures, Temperatures and Metamorphic Fluids Across an Unbroken Amphibolite Facies to Granulite Facies Transition in Southern Karnataka, India. – Origin of Archaean Charnockites from Southern India. – Radiometric Ages (Rb-Sr, Sm-Nd, U-Pb) and REE Geochemistry of Archaean Granulite Gneisses from Eastern Hebei Province, China. – The Most Ancient Rocks in the USSR Territory by U-Pb Data on Accessory Zircons. – Age and Evolution of the Early Precambrian Continental Crust of the Ukrainian Shield. – Significance of Early Archaean Mafic-Ultramafic Xenolith Patterns. – Subject Index.

Springer-Verlag
Berlin
Heidelberg
New York
Tokyo